Geographic Information Systems

A GUIDE TO THE TECHNOLOGY

Geographic Information Systems

A GUIDE TO THE TECHNOLOGY

John C. Antenucci

Kay Brown

Peter L. Croswell

Michael J. Kevany

with

Hugh Archer

VNR VAN NOSTRAND REINHOLD
New York

Copyright © 1991 by Van Nostrand Reinhold

Library of Congress Catalog Card Number 90-22557
ISBN 0-442-00756-6

Printed in the United States of America.

Van Nostrand Reinhold
115 Fifth Avenue
New York, New York 10003

Chapman and Hall
2–6 Boundary Row
London, SE1 8HN, England

Thomas Nelson Australia
102 Dodds Street
South Melbourne 3205
Victoria, Australia

Nelson Canada
1120 Birchmount Road
Scarborough, Ontario MIK 5G4, Canada

16 15 14 13 12 11 10 9 8 7 6 5 4 3 2 1

Library of Congress Cataloging-in-Publication Data

Geographic information systems : a guide to the technology / by John C.
  Antenucci . . . [et al.].
      p.   cm.
   Includes bibliographical references and index.
   ISBN 0-442-00756-6
   1. Geographic information systems.   I. Antenucci, John C.
G70.2.G44   1991
910'.2855369—dc20   90-22557
                                                                          CIP

In memory of James R. Anderson, a former Director in the Alaska Department of Natural Resources and GIS consultant to the City of Seattle, who died March 26, 1990. His farsighted and aggressive vision of GIS and allied technologies lead Alaska to a pioneering role in managing geographic information. His confidence in us as individuals and as a young company contributed to our own confidence and vision when it was needed most.

Contents

Preface

Geographic Information Systems: A Guide to the Technology was born of need—the need for a comprehensive work on the emerging field of geographic information management technology.

We encountered the need often in our work at PlanGraphics, Inc., a leading computer consulting firm that specializes in the design and implementation of geographic information systems (GIS) and assists organizations in using the technology to solve problems and perform work more efficiently. We needed it, our clients needed it, and it didn't exist.

Geographic information management technology—using computers to map, draw, store, and manipulate spatial data—evolved independently in many places with many variations. The technology is maturing and gaining broad acceptance, but it still lacks common structure and definition.

For example, the technology's many names confuse and probably deter broader recognition. In this book, we use the terms "geographic information management technology" and "geographic information system" or GIS to cover the array of computer systems designed to process maps and spatial data.

GIS is described by the computer industry press as an emerging standard application, soon to be on par with word pro-cessing, spreadsheets, and data base management. Recently, geographic information management technology came to the desktop, signaling a new era of increasing use and popularity. Enthusiastic users have long been the main source of grassroots support and growth.

We perceive at least three broad audiences for this book: 1) executives, managers, professionals, and other lay people; 2) practitioners and technicians; and 3) students and academicians. We recognize that *Geographic Information Systems* cannot completely satisfy the disparate needs of any of these groups, but we have tried to give each of these audiences a foundation upon which to build. Our purpose is two-fold. We hope to aid those considering, implementing, or using geographic information technology, and to advance understanding of the technology's potential and challenges.

Geographic information management technology encompasses many fields—including computer science, cartography, information management, telecommunications, geodesy, photogrammetry, and remote sensing—and is flavored with its applications of engineering, environmental analysis, land use planning, natural resource development, infrastructure management, and many, many others.

Geographic information technology

borrows concepts and techniques from many separate disciplines. This book is not a serious investigation or rigorous examination of any of the related fields, which are well documented elsewhere. One of the first books to comprehensively describe geographic information management technology, *Geographic Information Systems* touches only the corners of related fields that are significant to GIS. Readers may wish to seek out additional information to amplify the material presented here.

For pioneers, pathways often are full of obstacles. Frontiers pose hazards, as well as incredible opportunity. The frontier of geographic information management technology is an open land of opportunity that holds promise of far-reaching benefits for individuals, governments, industry, and society as a whole.

How to Read This Book. This book is a user's guide to geographic information technology. Some knowledge of computer and data processing concepts is helpful but not essential in understanding it. Explanations of basic concepts and terms make comprehensible to the layman the technical issues that affect GIS development, such as accuracy, scales, and projections. The book provides information and perspectives on the major requirements and issues surrounding geographic information management technology—building and maintaining a data base, selecting and upgrading hardware and software, using the technology to solve problems, funding, networking, providing access, and others.

Part I, Technology for the Information Age, gives an overview of geographic information management technology. Chapter 1 introduces the subject, defines key terms, discusses relevant technologies, and introduces some of the political, organizational, and technical issues affecting GIS development. Chapter 2 summarizes the history of the technology's evolution. Chapter 3 examines a broad range of applications and uses of GIS. Chapter 4 describes the benefits and costs of a GIS.

Part II, Systems Components, describes the elements of a geographic information system, including data bases, software, hardware, and data communication vehicles. Chapter 5 reviews data base concepts and development, and Chapter 6 discusses the various types of data commonly used in a GIS. Chapters 7 and 8 cover computer hardware and software, respectively. Chapter 9 reviews system configuration and data communication concepts important in geographic information systems.

Part III, Managing Change, examines how to make geographic information management technology work effectively. Chapter 10 lays out a step-by-step process for implementation. Chapter 11 covers emerging legal issues. Chapter 12, the final chapter, assesses the future of geographic information management technology.

Although geographic information management technology is being used throughout the world, in industrial as well as developing nations, this book focuses on the technology's use in North America. The text is illustrated with examples of products from operational systems. Boxed information both summarizes and expands upon key concepts.

For an overview of the field, executives, managers, professionals, and other lay people probably will find Chapters 1 and 3 of most help. Those interested in exploring the implementation of a GIS also should read Chapters 4, 10, and 11.

Practitioners of the technology probably will find Chapters 4, 5, 6, 7, 8, and 9 of most interest.

For a complete picture of the technology and its use, students and others will want to read all chapters, as well as the Glossary that defines the GIS terms used in this book.

References from the text grouped in the References section at the end of each chapter. Suggested Readings at the end of selected chapters will guide readers seeking additional information.

Acknowledgments

Geographic Information Systems: A Guide to the Technology emerges from several years of thought and as many years of research, writing, and editing. Over that extended period, many individuals contributed to a work that was thought useful in 1985, needed in 1988, and brought to press today.

Important contributions were made by many current and previous staff of Plan-Graphics, Inc. All deserve thanks. Of special note were the efforts of Leann Rodgers, Jani Sivills, and Amy Petit, who at various times struggled with text and edits; Jodi Kilcup, Vicke Bunker, and April DeLuca, who served as bridges between authors and publishers and handled tedious details; and Kaye Brothers and Holly Tincher who converted sketches to graphics.

Consulting staff provided needed research, advice, and criticism. Dan Thomas and Neil MacGaffey, who now follow their own pursuits, Annie Metcalf, Susan Keiley, Maury Foley, and Amy Ahner, who continue to contribute to our consulting activities, each helped in important and varying ways.

Of course, the publishing team is due their credit. Nola Hague's unflagging enthusiasm and confidence that the work should succeed led to a perfect working relationship with Van Nostrand Reinhold. Marjorie Spencer of VNR needs to be recognized for her contributions, coordination, and responsiveness, and McLaurine & Co. deserves praise for a measured edit.

Users of the technology and members of the GIS and AM/FM industry who are cited and uncited in this text provided the core of our experience; we thank them and wish them the best of luck in their respective endeavors in realizing the benefits of a most promising technology.

Part I

TECHNOLOGY FOR THE INFORMATION AGE

Chapter 1

Introduction

At the frontier of the information age, geographic information management technology is emerging as a powerful means to manage voluminous geographic data, to help cope with the information explosion, and to provide a foundation for solving the problems that beset planet Earth and its inhabitants.

Societies in general are becoming keenly aware of the need to manage information from a geographic perspective. This awareness has been brought about by the twentieth-century trends toward a global community and economy. At the same time, the often negative impact of advancing technology has shown the need for wise management of the earth's resources. Geographic Information Systems (GIS) provide the tools to help meet these challenges.

Throughout the world, governments, utilities, and businesses are investing billions of dollars in computer systems that store, manage, and analyze maps and geographic information. Geographic information management technology has grown in tandem with the widespread development and use of increasingly inexpensive, powerful computers.

Geographic information management technology has vast potential, recognized but unrealized, for solving environmental and human management problems—at least some of the world's complex social, economic, and environmental ills may yet prove susceptible to analysis and solution. Farsighted people throughout the world are beginning to apply geographic information management technology in attacking the afflictions of crime, poverty, disease, pollution, and urban congestion.

In the United States, interest in geographic information management technology may also be fueled by an increasing appreciation of the importance of geography as an academic discipline, and by the recognition that generations of Americans fail to grasp geography's fundamental concepts (Fig. 1-1).

A United Nations study of 30,000 students, age 10 to 14, in nine countries ranked American students next to last in their comprehension of foreign cultures. A decade after the Vietnam War, 95 percent of freshmen tested at a Midwestern college could not find Vietnam on a world map (Grosvenor 1988).

Failure to manage geographic information successfully also threatens the nation's economic competitiveness in an interdependent, international marketplace. Efforts to create order out of the chaos have progressed slowly, as geographers struggle to reconcile a myriad of separate and incompatible systems, data sets, and standards.

"Belatedly, the federal government realized that the successful formula for in-

Technology is the science of means. . . Geographic information technology is a means to make many types of work more efficient and workers more effective. It enables better decisions, based on better information.

3

BOX 1-1

We have got to think of how we are going to describe this earth if we are to address the increasingly complex problems that governments face, or indeed the world community faces. It is not *only* a land survey problem. It is not *only* a remote sensing problem. It is not *only* a forest inventory problem. We are all in the business of earth description; of "geography," because earth description is what the word geography means. We are really involved with building a composite description of the earth that will be used for decision-making purposes (Churchill 1986). . . . Mapped and related statistical data . . . form the greatest storehouse of knowledge about the condition of the living space of mankind (Tomlinson 1988).

> Roger F. Tomlinson, President
> Tomlinson Associates Ltd.
> Consulting Geographers

GIS is bringing about one of the greatest technological revolutions in the country, if not the world. . . . This business is going to approach the growth of the number of bits per chip, which is quadrupling every three years.

> John Bossler,
> Director of Center for Mapping
> Ohio State University

. . . GIS-related technologies provide us with a two-edged sword. In the near future, GIS technology, because of its low cost, high reliability, user friendliness and wide usefulness, will at last be adopted by the many potential users outside of the highly developed technological societies. This offers great promise for improving the future for billions of people (Dangermond and Smith, 1988).

> Jack Dangermond, President
> Environmental Systems Research
> Institute (ESRI)

Geographic information management technology has grown in tandem with the widespread development and use of increasingly inexpensive, powerful computers.

creasing national competitiveness and overall government effectiveness includes the successful management of government information resources," said Charles R. McClure (1988, p. 15), a professor at Syracuse University's School of Information Studies. "During the 1987 congressional hearings on scientific and technical information within the federal government, efforts to manage such information were characterized as 'in disarray' . . . important technical information frequently lies languishing in some unknown government office. As a result, policy-makers are unable to make informed decisions on issues that require technical information and empirical evidence."

Deterioration of infrastructure—water and sewer systems, bridges, roads, airports, and hazardous and solid-waste facilities—is a growing problem that could spur the use of geographic information

systems throughout the industrialized world. In the United States, Congress established the National Council on Public Works Improvement in 1984 to assess the condition of the public infrastructure, which was found inadequate to sustain a stable and growing economy. To keep America competitive, the Council said, capital spending on public infrastructure needs to increase by 100 percent (Giglio 1988). Geographic information technology is helping agencies at all levels of government meet the challenge of revitalizing aging facilities.

Technology is the science of means (Scruton 1982). Geographic information technology is a means to make many types of work more efficient and workers more effective. It enables better decisions, based on better information.

Currently, geographic information management technology is employed to

Doonesbury

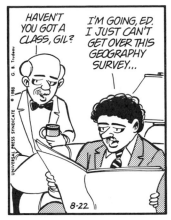

Doonesbury

Doonesbury

FIGURE 1-1 Doonesbury © G.B. Trudeau. Reprinted with permission of Universal Press Syndicate. All rights reserved.

conduct land use studies and environmental assessments, and to evaluate and develop timber, minerals, oil, gas, coal, and other resources. It is used to determine the best location for roads, hydroelectric dams, hazardous-waste facilities, residential settlements, and retail market outlets. Geographic information management technology helps governments and industry manage infrastructure such as sewer, water, electric, gas, telephone, rail, and road systems.

Governments use this technology to make maps, chart the most efficient routes for school buses and emergency vehicles, repair roads, collect taxes, analyze occurrence patterns of disease, and define election districts. Private companies rely on spatial and demographic analysis to help them locate industrial and commercial sites and to develop marketing plans for commodities and services. Private parcel services apply the technology to manage the collection and delivery of mail and packages.

Geographic information systems are used to fulfill special requests from citizens' groups for information affecting current issues and events. Citizens' demands for access to information are increasing at the state and local government levels as public records and other useful information become available. More usable information can be obtained rapidly by combining public data bases using GIS technology.

Interest in geographic information technology has increased significantly in the past five years and continues to grow. Some of this growth can be attributed to technological improvements in computers in general, and part to advances in geographic information technology and related fields.

KEY TERMS

Geographic information management technology has almost as many names and acronyms as uses. One common name is Geographic Information System (GIS).

Another is automated mapping/facilities management (AM/FM).

Although GIS has recently become more widely accepted as a generic term for the technology, the term *geographic information system* was first published in a 1965 Northwestern University discussion paper by Michael Dacey and Duane Marble (1965). Throughout this book, the term *geographic information management technology* is used broadly and generically to include many specific types of computer systems used for the mapping and processing of spatial information. This includes systems devoted to mapping and civil engineering drawing, the query and management of geographic databases, or more complex geographic analysis or modeling; the term *geographic information system* (or *GIS*) is used also in a broad sense to refer to all automated systems used primarily for the management of maps and geographic data.

Key terms associated with geographic information management technology include these:

- Automated Mapping (AM)
- Computer-Assisted or Computer-Aided Mapping (CAM)
- Computer-Aided Drafting (CAD)
- Computer-Aided Drafting and Design (CADD)
- Geographic Information System (GIS)
- Automated Mapping/Facilities Management (AM/FM)
- Geoprocessing and Network Analysis
- Land Information System (LIS)
- Multipurpose Cadastre

No universally accepted definition exists for these terms, which often are used interchangeably even though they denote different capabilities and concepts (Fig. 1-2).

Automated Mapping (AM) and Computer-Assisted Mapping (CAM) are technologies for drawing and producing maps (Figs. C-1, C-2*). They are designed for effi-

---

* Figures that appear in color are designated by a "C" preceding the figure number. Color figures are grouped in a section that follows page 82.

BOX 1-2   What is GIS?

Federal Interagency Coordinating Committee (1988): "A system of computer hardware, software, and procedures designed to support the capture, management, manipulation, analysis, modullary and display of spatially referenced data for solving complex planning and management problems."

Phil Parent (1988): GIS is a "system that contains spatially referenced data that can be analyzed and converted to information for a specific set of purposes, or application . . . The key feature of a GIS is the analysis of data to produce new information."

Francis Hanigan (1988): A GIS is "any information management system which can:

- Collect, store, and retrieve information based on its spatial location
- Identify locations within a targeted environment which meet specific criteria
- Explore relationships among data sets within that environment
- Analyze the related data spatially as an aid to making decisions about that environment
- Facilitate selecting and passing data to application-specific analytical models capable of assessing the impact of alternatives on the chosen environment
- Display the selected environment both graphically and numerically either before or after analysis."

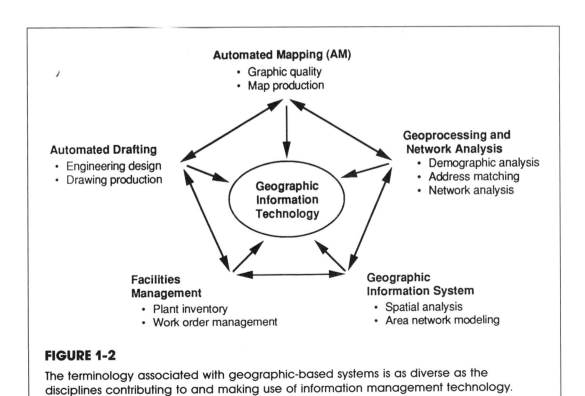

FIGURE 1-2

The terminology associated with geographic-based systems is as diverse as the disciplines contributing to and making use of information management technology.

cient graphic data processing and display. These systems generally are limited in their ability to store and manipulate any data other than graphic images, and they have little or no geographic analysis capability. While "CAM" occasionally is used to mean computer-assisted (or -aided) mapping, the term more frequently is used as an acronym for *computer-assisted manufacturing*.

Computer-Aided Drafting (CAD) and Computer-Aided Drafting and Design (CADD) systems are used in civil, mechanical, and other engineering design activities. These systems have interactive graphics capabilities similar to those of the mapping/GIS systems but, in addition, they include a variety of engineering calculation and analysis functions. CAD systems have graphic display and limited attribute-processing capabilities. CADD systems, in addition, incorporate some modeling of the graphic relationships and allow analysis of logical relationships. CAD systems are used for drafting graphic representations and structural plans (Fig. C-3); CADD systems are used, in part, to design infrastructure and facilities such as buildings, roads, bridges, and water and sewer systems (Fig. C-4).

Geographic Information Systems (GIS) and Automated Mapping/Facilities Management (AM/FM) systems process both graphic and nongraphic data. GIS and AM/FM systems handle graphic data in a manner similar to that of Automated Mapping systems, but they also have a sophisticated ability to store nongraphic attributes or geographically referenced data and link them with the graphic map features so that a wide range of information processing and display operations, including map production, can be accomplished.

In these systems, attributes are linked to a map or engineering drawing element through some common identifier. For example, a map depicting parcels (each identified by a parcel number) may be linked to an attribute data file containing ownership data, land use, and appraised value information for each parcel.

The focus is thus geographic analysis rather than mere display. GIS and AM/FM systems are used to perform a variety of spatial analyses, including overlaying combinations of features and recording resultant conditions, analyzing flows or other characteristics of networks, and defining districts in terms of specified criteria.

Although GIS and AM/FM systems have similar capabilities, GIS traditionally has referred to systems that emphasize spatial analysis and modeling (Fig. C-5), while AM/FM systems emphasize the management of geographically distributed facilities (Fig. C-6). Until recently, the term GIS was commonly used to describe systems found in government agencies, whereas the term AM/FM was used predominantly by utility companies.

Geoprocessing and network analysis systems are computer systems used primarily for geographic analysis; they depend on geographic identifiers, or *geocodes*. A geocode is a spatial index (e.g., coordinates) identifying unique points, lines, or areas. The most common application is the assignment of planning, census, or statistical area identifiers to data records that contain addresses, using a geographic reference file such as a Dual Independent Map Encoding (DIME) or topological (TIGER) file.

DIME files were introduced by the United States Bureau of the Census to define the geographies for 1970 and 1980 censuses; TIGER files were developed for the 1990 census.

While typically not oriented to map production, geoprocessing systems may include mapping capabilities (Fig. C-7). They also may include the capability to perform network or district aggregation or analysis using geocode data.

Systems that rely on the connectivity of various linear elements, such as utility circuit networks and routing street networks (for routing emergency vehicles), evolved independently. These systems have characteristics similar to the early geoprocessing systems.

Land Information Systems (LIS) include a wide-ranging set of data bases, both graphic and text, describing physical or legal characteristics of land areas. Sometimes referred to as land records systems (LRS), the most common forms of land information systems are those that describe property ownership, land value, tax assessment, and property boundaries. They are typically managed by local governments in the United States, and are generally maintained by provincial or central government bodies elsewhere. LIS also are employed by the real estate and property divisions of companies with large land holdings for tracking the natural resources and environmental features of land.

Traditionally those types of data have existed in map and tabular form. The emergence of geographic information systems has had a significant influence on the automation of LIS (Fig. C-8).

Multipurpose Cadastre is a term popularized by a panel sponsored by the National Research Council's Committee on Geodesy in the report titled *Need for a Multipurpose Cadastre* (1980). The term refers to the general utility of land ownership information and embodies the logical integration of legal (e.g., property ownership or cadastre), physical (e.g., topography, man-made features), and cultural (e.g., land use, demographics) characteristics of the landscape into a common and accurate reference framework. This reference framework is typically established through rigorous geodetic and survey control standards. Figure 1-3 shows an example of a multipurpose cadastre.

CAPABILITIES

Uses of geographic information technology vary widely. This section discusses some of the technology's major capabilities. Specific applications of geographic information technology are discussed in Chapter 3.

Cartographic capability allows accurate maps and engineering drawings to be produced efficiently. This capability includes digitizing (converting analog products to digital form), graphic display generation, interactive graphic manipulation (e.g., add, modify, delete, create window), and plotting (Fig. C-9).

Data management capability enables the efficient storage and manipulation of geographic data, both graphic and nongraphic. Storage and retrieval of nongraphic data that are linked to graphic images is sometimes called *attribute processing*. Attribute processing can select data and produce graphics and reports on the basis of attribute values (Fig. C-10).

Analytical capability permits sophisticated processing and interpretation of spatial data. Examples of analytical capability include the following:

- Engineering functions needed for the planning, design, construction, and management of roads, buildings, bridges, and waterways (Fig. C-11)

- Computation operations like measuring the distance between points or the length of an element, and determining area size or volume

- Functions that rely on complex algorithms for engineering and demographic analysis and modeling and are used for such tasks as evaluating storm water flow, balancing electricity loads, and assessing property values with Computer-Aided Mass Appraisal (Fig. C-12)

- The definition and evaluation of many types of districts, such as school attendance, election, and sales and service areas. Some systems support the creation of districts merely by changing the graphic images of potential districts through the computer's interactive capability. Others generate districts automatically from attribute data (e.g., student registration) and control factors (e.g., maximum travel time)

- Area tabulation and plotting, which allow production of density or thematic maps showing expenditures, maintenance activities, or incidents by election or administrative district. Area tabulation and plotting draws on a system's cartographic, data management, and an-

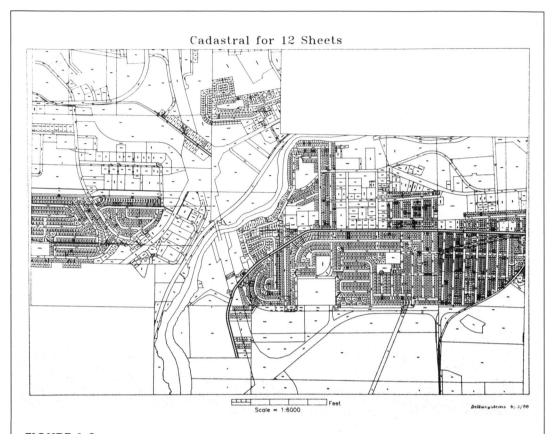

Cadastral for 12 Sheets

Scale = 1:6000   Feet

Deltasystems 6/3/88

FIGURE 1-3

Parcel maps of an area provide more than a representation of ownership. The maps serve a wide range of purposes if they are constructed with a known level of positional accuracy and integrated with other cultural and physical information. Such databases are frequently used as a multipurpose cadastre (courtesy of Genasys).

alytical capabilities (Figs. C-13, C-14, C-15)

- Polygon processing, in which sets of polygons representing one data theme (e.g., soil types) are overlaid on another (e.g., parcels). This capability goes farther than merely overlaying polygons from different data sets; it actually constructs new polygons from the union of the sets and generates new data (e.g., area of composite attributes) that can be further processed.

Collectively, these capabilities give managers an enhanced ability to manipulate and use data more effectively. Graphic presentations are especially powerful for conveying information. The capability to produce maps or engineering drawings showing particular combinations of features for selected geographic areas at appropriate scales is one of the technology's most valuable features.

RELEVANT TECHNOLOGIES

A number of fields and related technologies—including computer science, information management, cartography, geodesy, photogrammetry, and data communications—influence and shape geographic information technology. As those fields and their technologies advance, the advances are incorporated in geographic information technology. As data base management software becomes more advanced, its benefits accrue to GIS. As computer processors become even faster, less expensive, and more powerful, GIS technology becomes more useful, and more accessible, to more people. Significant technological improvements in many related fields cumulatively have made geographic information technology much more practical.

The relationships of geographic information technology to other fields and technologies are discussed briefly below.

Computer Science

Computer science provides the technologies for data capture, manipulation, storage, and output. As such, computer science encompasses hardware, machine processing capabilities, software development, and computer programming languages. In the last several years, the technology has become better suited to the specific needs of GIS. Processors are faster, the cost of processing is lower, and the capabilities of the graphic devices are substantially improved.

Greater performance is resulting from improvements such as faster and higher capacity storage media, higher level query and programming languages, and the incorporation of standards in operating systems and data communications. Advances in hardware and software will continue to have a direct impact on the use and development of geographic information systems.

Information Management

As information science progresses in its ability to code the logical and mathematical relationships that govern the processes used to associate and manipulate data, GIS software developers are incorporating

those improvements. Organizations now can maintain large data bases in part because designs for both graphic and nongraphic data bases are more sophisticated, making access to large volumes of data easier, faster, and less expensive.

GIS developers have adopted specialized structures for managing nongraphic data. Confronted with data bases of enormous size, computer scientists have developed approaches to data organization such as quad tree structures that index and store data more efficiently. Other structures, such as network and relational data bases, handle large arrays of data, building and retaining connectivity and relationships between data sets.

In addition, data storage costs have fallen sharply. With the maturation of optical disk technology, storage will become less costly by orders of magnitude.

Nonalgorithmic digital records systems, which store scanned images, now record and manage some GIS data. These will likely soon contain audio records as well as video and digital data.

Cartography

Cartography is the science or art of making maps (*Webster's*, Third Edition). An ancient skill, cartography has contributed to GIS the general conventions used to produce map products. These general standards include accuracy and precision; map projection and coordinate referencing schemes; and the graphic appearance of the map through symbols, linework, and annotation. Such cartographic conventions, as they have developed over time, continue to guide the construction of graphic data elements within a GIS database.*

Knowledgeable cartographers with access to GIS technology enjoy a greatly expanded ability to produce maps that effec-

---

* We distinguish in this book between *data base* (two words) and *database* (one word). Data base refers to a large collection of data, of which a subset or specific grouping of data is a database. See Chapter 5 for a fuller discussion.

Knowledgeable cartographers with access to GIS technology enjoy a greatly explanded ability to produce maps that effectively communicate geographic issues and ideas.

tively communicate geographic issues and ideas. The role of the cartographer's conventions cannot be ignored in developing GIS map products. Certainly every map does not have to be a work of art, but to be useful and communicative, each map must adhere to the general principles of cartography.

Geodesy, Photogrammetry, and Remote Sensing

Geodesy and photogrammetry frequently are used in geographic information management technology to establish a foundation (i.e., base maps) for other spatial data. Geodesy is a branch of applied mathematics that observes and measures the shape and size of the earth or a large part of its surface, and determines the exact location of points on its surface (*Webster's*). Photogrammetry is the science of making reliable measurements by the use of photographs (usually aerial) in surveying and mapmaking (*Webster's*).

Geodesy provides a framework for large-area data bases. Surveying, a related field, is used to define the extent of land features in small- and large-area data bases. Surveying is a branch of applied mathematics that provides techniques for determining the area of any portion of the earth's surface, the lengths and directions of the bounding lines, the contour of the surface, and accurate delineation of the whole on paper (*Webster's*).

Over the past twenty years, knowledge of the earth's shape has improved as a result of space programs and the development of earth satellites. One such program, NAVSTAR, a military satellite system to support naval and aerial navigation, has contributed substantially to civilian geodetic surveying. The civilian application, known as the Global Positioning System (GPS), uses a series or "constellation" of satellites of known position and earth-based satellite sensors or tracking devices. By simultaneously recording the positions of multiple satellites with multiple sensors, computations can yield very accurate (within 5 centimeters) locational information for each sensor. By locating or

moving the sensors throughout an area, a network of geographic coordinates can be established to support geodesy, surveying, and photogrammetric mapping.

Photogrammetric mapping techniques, which can produce a high level of accuracy based on survey control, are commonly used to gather the raw data needed to create base maps. Aerial photography is a mainstay in the photogrammetric sciences, and improvements in camera and film types, lenses, and emulsions allow higher flights and produce photographs with better resolution and more precision at a lower cost.

Remote sensing, a discipline that evolved from photogrammetry, is the analysis and interpretation of images gathered through techniques that do not require direct contact with the subject. Remote sensing of the earth's resources uses aerial or space photographs, electronic scanners, and other devices to collect data about the earth's surface or subsurface. In the last two decades, remote sensing has increasingly made use of digital techniques for collecting and processing spatial data. Remote sensing products have provided important input to geographic information systems.

Great strides have been made in the use of remotely sensed data acquired from satellites (Fig. C-16). The Landsat program, initiated by the National Aeronautics and Space Administration (NASA) in the early 1970s, provided images useful for evaluating the earth's resources. The early Landsat platforms, then called ERTS, or Earth Resources Technology Satellites, produced images that aided regional analysis, although the sensors lacked the resolution required for the most detailed mapping or analysis.

The increasing spatial and spectral (light band) resolution of satellite sensors made in the late 1970s and 1980s has extended the usefulness of their images. In particular, the Thematic Mapper sensor of Landsat 4 and 5 increased ground resolution from the 79 meters of previous Landsat sensors to 28 meters. Since 1986, Landsat has been owned and operated by Earth Observation Satellite Company (EOSAT), a private firm. Landsat 6 will be launched in 1992. ·

The French government, in cooperation with the European Space Agency, has developed the SPOT satellite (Satellite Probotoire de l'Observation de la Terre) and is currently marketing data collected by scanners with resolution in excess of 20 meters (Fig. C-17). Other countries, including the Soviet Union, Japan, and India, have also initiated ambitious domestic remote-sensing satellite programs.

Data Communications

Tremendous advances in data communications and networking have expanded the flexibility of computer technology by moving away from centralized, mainframe configurations toward more distribution of processing power and data. Hardware advances have increased computer performance while reducing the size and dependency on strict environmental controls. Software developments have brought about complex networks that use sophisticated communication protocols to transfer large volumes of data and, at the same time, provide the necessary error recovery and security demanded of computer networks.

Continuing advances in data communication media have allowed greater flexibility in network configurations. High-speed local-area networks and advanced remote communications techniques, such as digital phone systems, microwave, and satellite communications, offer many reasonably priced network options.

Perhaps the most important trend, which began in the early 1980s, is the growth of communications standards that facilitate networking of devices from many different vendors.

ISSUES

Several factors—institutional as well as technical—continue to constrain the usefulness of geographic information technology.

Technical Issues

Technical issues dominated the emergence of GIS in the 1970s and its maturation in the early 1980s. In some instances, expectations of the technology were more advanced than the technology itself, and in other cases, the user's lack of experience in adapting the technology to a particular application was the challenge. Within the context of pioneering systems—and the simultaneous educating of an industry, both vendors and users—progress was steady despite periodic setbacks.

Technological issues paled in importance compared with the impact of the organizational and procedural concerns associated with GIS implementation. Although there is much room for technical enhancements and developments, even today most system implementations are not exploiting the available technology fully. Still, GIS is rooted in technology, and the following discussion highlights key considerations.

What System to Buy? Among other important activities, organizations beginning to implement geographic information technology must make choices from an overwhelming array of software and hardware on the market.

The trend since the late 1970s has been away from user-developed systems and toward integrated sets of hardware and software installed by a vendor. The term *turnkey* as used in the GIS world has several meanings.

In the early days of GIS, turnkey systems were successfully promoted on the basis of their ability to make maps immediately. Turnkey systems were attractive to many organizations because they localized the control of activities formerly held by the Data Processing (DP) Department. In theory, users no longer had to go to a DP department and wait to be put on the priority list, and then wait again for programmers to write applications or for operators to run a data set against a program. However, none of the systems was truly turnkey in the sense that a user could make maps effortlessly, because building the

Institutional as well as technical issues continue to constrain the usefulness of geographic information technology.

In many organizations, too much emphasis is placed on acquisition of the hardware and software, and too little emphasis on the design, creation, and maintenance of the data base.

data base and writing application programs were required.

Another driving force behind the interest in turnkey systems was the user's desire to minimize risk. Early computer systems (not just in GIS) posed numerous integration and support problems. Users who lacked expertise in sophisticated computer systems welcomed the security of a turnkey approach.

From the perspective of the seller, turnkey represents a marketing opportunity whereby a system with components from various vendors is assembled by an integrator who guarantees it will work regardless of the component sources. In some limited instances, vendors offer integrated packages of their own hardware and software.

The framework approach offers yet another possibility for system acquisition. In this instance, a vendor sells an operating system of complementary software with basic functional capabilities for geo-based analysis. However, the user organization must rely on either internal sources or third-party software packages to enhance the application potential of that system.

Advantages and disadvantages accompany all these approaches. With a turnkey system that uses hardware and software from a single vendor, the hardware and software are known to operate together. A single vendor can deliver both the training and the maintenance on the software and hardware. When something goes wrong, it is obvious whom to call for help.

When a system is put together by a system integrator, a user organization will probably sign multiple contracts with multiple vendors for the various components of the system. One of the vendors assumes the responsibility for the system's operation, sometimes with significant cost implications associated with performance penalties.

A disadvantage of the single-vendor approach is that it may limit the ability of user organizations to select the most appropriate mix of technology for their needs. With system-integrator turnkeys, the user chooses from a much wider range of hardware and software components, but must rely on the capability of the integrator and its working arrangements with vendors of the system's components.

Some companies have taken the opposite approach—they sell only software or hardware components. The general advantage of this approach is that software and hardware packages can be acquired at lower cost. User organizations must choose the components and provide support themselves for design and implementation of the system, user training, and other functions sometimes performed by system integrators.

Cost-Effective Construction of Large Data Bases. Analog data (e.g., paper and mylar maps, aerial photographs) must be converted to a digital format that can be manipulated by the computer—a process integral to GIS. In many organizations, too much emphasis is placed on acquisition of the hardware and software, and too little emphasis on the design, creation, and maintenance of the data base. Data conversion is often an enormous task. The hardware and software may represent as little as 5 percent of a project's total cost—seldom more than 20 percent of the total cost. The data base is by far the costliest facet of a typical GIS implementation.

The cost of building the data base can vary considerably, depending on the number of features included, the accuracy to be achieved, the condition of source materials, the availability of existing digital files, and whether new data must be acquired.

The preoccupation with computer technology and lack of focus on the task of data base development hindered development for some of the early systems. The city of Milwaukee, for example, did a pilot project for six months in which it estimated the time and cost to build the data base internally. On the basis of the pilot, they projected that it would take twice as long and cost twice as much as initially expected; later in the process those estimates of time and cost were doubled again. It eventually took seven years to build the first data base.

"Part of the reason it took longer than expected was [that] the vendors told us we could pull people off the street to do the digitizing," said William Huxhold, who managed Milwaukee's GIS. "We spent a long time using federal CETA [Comprehensive Employment Training Act] employees, but we found they didn't understand map conventions. We were misled by overzealous selling and oversimplification on the part of the vendors."

Many of the early systems had only one or two workstations, and entering the data in-house was slow and tedious. Another problem was that once a large volume of maps had been digitized, maintaining those maps consumed existing memory capacity and curtailed progress on digitizing new maps. Milwaukee managed the problem by working the equipment three shifts a day with staff added for the conversion.

The data conversion industry, which transforms data from analog to digital, has become larger, as well as more experienced, reliable, and affordable since its evolution from the keypunching shops of the 1960s. However, problems remain with the conversion process, including its requirements of extensive manual labor and capital-intensive equipment, which result in high costs.

Encouraging developments on the horizon include

- Use of scanning technology, which still has not fulfilled its promise but is steadily improving
- Greater use of *raster* databases as background to vector databases

Compatibility with Other Systems and Data.

For decades, competition between vendors has stymied efforts to establish uniform standards for computer hardware and software that would allow flexible networks and data sharing. This problem inhibits GIS as well as the computer industry in general.

The industry press reports that barriers are about to be dismantled, that effortless networking is on the horizon, that computers and operating system software are, at long last, increasingly compatible (Gannes, 1988; Mitchell, 1990). As of the early 1990s, however, many obstacles to the adoption of uniform standards throughout the industry remain. The ability to use existing geographically related data may be critical to the cost-effectiveness of a GIS project. Since the inception of GIS, its developers have not viewed standardized technology as a priority, and as a result, GIS systems do not cross-communicate well. Compatibility problems are encountered at all levels—in the hardware, software, data structures, and data communications. Those problems are beginning to be resolved, but compatibility is still uncommon.

For GIS, the necessity of data compatibility also has not been recognized. Until recently it was universally assumed that data would reside in the dedicated GIS computer. In the early days of GIS, data was obtained from wherever possible—either by extracting it from paper records or transferring it from another digital source file—and loaded into the GIS. Those processes often resulted in the maintenance of two or more file systems containing some of the same information, and thereby created the potential for inconsistent information due to lags in updating the data and other factors. Now the trend is toward transferring data from other systems to the GIS on an as-needed basis, with the data, at times, residing permanently (for maintenance and updating) on a system other than the GIS.

Resolution of Discrepancies Between Similar Sets of Data.

An issue in both the technical and institutional spheres, this problem reflects the fact that data bases have long been maintained both manually and digitally with substantial redundancy and inconsistency. Similar digital data from various sources frequently appear inconsistent. The user organization then confronts the challenge of judging which information is correct.

As systems gain the participation of more and more agencies, the probability of

> The trend is toward transferring data from other systems to the GIS on an as-needed basis, with the data residing permanently (for maintenance and updating) on a system other than the GIS.

data discrepancies increases and the following questions must be resolved: Whose data will be accepted as the baseline data? Whose data is best? Although determining which data is "best" is usually a technical issue, the resolution of which data to use is frequently a political concern.

Quality Control. Explicit quality control procedures must be exercised throughout the process of building and maintaining a GIS data base. Without rigorous quality control, the integrity of the data will not be assured, and users will not be able to depend on its reliability, significantly diminishing its value to the system.

Institutional Issues

Expectations. Expectations of a system's capabilities influence its success or failure. For various reasons, users and organizations sometimes have unrealistic expectations of the benefits a GIS can provide. At times, GIS has been hyped by some more eager to make sales than to address an organization's needs. In other instances, advocates within an organization have oversold the benefits and underestimated the resource requirements necessary for successful implementation. It is important for an organization to identify clearly what it wants to accomplish and how it functions before making the commitment to develop a GIS. An objective appraisal of needs is essential to determine whether a GIS suits the circumstances.

> It is important for an organization to identify clearly what it wants to accomplish and how it functions before making the commitment to develop a GIS.

If management or users are misled by exaggerated expectations, the probability of success of the system will be considerably lessened. Even if the system performs as it is optimally capable, management and user support may decrease significantly if that performance falls short of (exaggerated) expectations.

Staffing and Staff Training. Qualified people are essential to the implementation of a GIS. In-house staff may be trained to operate the system or, if necessary, additional support can be acquired through new staff

or consultants for the implementation phases. System implementation includes initial training for key staff members and provision for subsequent refresher courses. As key staff members become more competent with the selected system, the user organization should develop in-house training programs for other staff.

Systems continue to fail because of qualitative and quantitative staffing inadequacies. Emphasis should be placed on quality staffing at both the managerial and technical level. A dedicated, motivated staff working in an environment that provides continuity is essential.

Organizations often have difficulty finding and keeping individuals who are competent in the management, implementation, and operation of geographic information systems. Employees' propensity for turnover underscores the need, regardless of the overall financial environment, to provide the staff with incentives to stay and contribute to the system's success.

Turf Wars. The introduction of geographic information technology requires cooperation and perhaps redefined responsibilities of individuals and groups within an organization. Unfortunately, organizations and individuals find change threatening and may resist what they perceive as loss of control over information within their purview. Users may resent the access that geographic information technology provides to outsiders (Why are they looking over my shoulder at my information?). Others may prefer to develop a GIS within their own subunit rather than participate in an organization-wide project. Agencies whose budgets are supplemented by selling maps may be reluctant to surrender the activity for a common effort. Jealousy, mistrust, and other emotions may prevent cooperation and impede progress.

The organizational impact of the system often is not foreseen and frequently is given too little weight in the implementation plan. Introduction of so powerful and impersonal a tool as a large GIS demands a great deal of organizational and personal change. In failing to appreciate that, management also may fail to provide sufficient resources within the system to change attitudes and overcome inertia, as well as to incorporate needed changes as they arise.

Multiagency Systems. When multiple organizations collaborate to create a GIS, it is possible to eliminate redundant activities among separate units. This is a major benefit of a multiorganization system.

Although systems developed by multiple agencies offer the unequaled opportunity for cost-effective GIS development, potential problems caused by the threat of change are magnified in multiagency systems. Multiagency systems usually require a new form of management that is responsive to all participants, such as a user committee or direct top management control.

In a multiorganization system, an equitable method is needed to allocate initial and operating costs among users. The allocation method may assign percentages of cost based on the amount of data an organization contributes, the extent of an organization's actual or anticipated use, or some other measure. It may be necessary to make provision for recovery of initial costs, accumulation of replacement and upgrade funds, and estimated operating and maintenance costs.

Financing. How a system is financed is an important issue for many organizations. The source of funds varies, depending on whether the organization is public or private, among other factors. The availability of funds may determine whether or not a system is successful. System acquisition and data base development are major initial costs. When organizations attempt to develop systems with limited operating budgets, problems are likely to result from shortcuts taken in the design and implementation phases.

Access to Information. As a public agency's GIS reaches maturity, it will probably have to develop procedures for public access to the data base. These procedures must fit within the legal framework of governmen-

tal open-records laws and still allow for reasonable cost recovery from joint users of the system.

Public policies on access to GIS information are evolving. Some policies provide for open and free access to all data, while others permit fees to recover the costs involved in establishing and maintaining the system. An associated concern of public agencies is their ability to maintain reasonable control over the resources required to respond to public information requests.

Marketing and selling GIS data relate to the issue of access. Private companies are building data bases, often using data from public sources, and marketing the data in association with a value-added service or product; some retain their data in a proprietary way, either not disclosing it at all or under strict licensing arrangements. Complicated public-policy issues arise when private companies enter in joint ventures with public agencies to build a network of data bases, as in a consortium project that includes more than one local government agency or private utility, since the system's data is a blend of public and private records. Can a public agency's GIS data and products be sold, and what is their value? This issue is discussed in more detail in Chapter 12.

Waste Created by Failure to Integrate. While the benefits of multiagency, integrated systems seem apparent, relatively few such systems have been successfully developed to date. Some of the issues are summarized in this excerpt from the newsletter of the Urban and Regional Information Systems Association (URISA):

> Most transportation planners and analysts share another unspoken URISA paradigm that integrated, comprehensive information systems are always more cost-effective than disjointed data collection activities and independent data bases. The savings from more efficient data collection and the benefits of greater understanding from crosstabulating everything with everything else have been stated in every URISA Proceedings

When multiple organizations collaborate to create a GIS, it is possible to eliminate redundant activities among separate units. This is a major benefit of a multiorganization system.

for the last two decades, yet precious few examples exist of comprehensive, integrated information systems that really achieve the benefits after years of development and mega-bucks of investment. In fact, integration costs money and time, and raises difficult institutional questions such as: Who pays for the information collection, management, and access; who updates the files; who sets definitions of variables; who determines minimally acceptable data reliability; who has access to the results; and how can you meet a diverse clientele's needs without reducing the information to a uselessly low common denominator or a hopelessly complex cross-referenced system?

(*URISA News* 1986)

Whether integration of systems is realistic in any particular circumstance, the growing perception is that failure to integrate wastes public dollars and resources.

Public and Private Political Constraints. Geographic information systems developed by public agencies almost always must gain the backing of political appointees and elected officials.

Public officials may be reluctant to embrace a GIS project, considering the substantial resources and long-range commitment. Appointed agency officials as well tend to think in terms of what they can accomplish within a relatively brief electoral term.

Typically, it is difficult to fund a project for more than one year at a time, resulting in the ongoing threat of discontinuity. In addition, since government employees frequently are paid less than their counterparts in industry, there is a fair amount of attrition. As people become better educated in the use and development of high technology, they are able to demand a higher wage.

The proliferation of GIS technology has coincided with a time of shrinking government budgets, a coincidence that has created problems for many systems. Planned system upgrades have been delayed or canceled. Adequate staffing often is not available. Perhaps the most serious long-term problem, and one that newcomers may be able to avoid, has been the tendency to resort to expedient, short-sighted decisions. Data bases established from readily available but inaccurate source material, lack of adequate geodetic control, and other shortcuts continue to haunt organizations that fail to consider the long-term consequences.

Politics also influence GIS development in the private sector, which has its own peculiarities. Unlike most public officials, private-sector executives quantify the value of investments based on the rate of return. The private sector is affected by swings in profitability that may be completely unrelated to GIS or its benefits. Private-sector GIS projects have to compete with other priorities, and they are subjected to external market forces that may require a change of direction or a shift in emphasis that drains away resources. Mandates can shift overnight.

Both public and private systems typically need a champion at a high level in the organization to be successful. In some organizations an advocate emerges; in others, the selection of an effective project leader is left to happenstance.

SUMMARY

Full realization of the potential of geographic information technology requires progress on all fronts—technical, organizational, and political. While continuing technological breakthroughs offer hope for overcoming impediments such as the high cost of converting geographic information to digital form, the human and organizational barriers to using the technology have received inadequate attention. People make GIS work. The commitment and vision of individuals transform organizations when they coalesce the resources needed to open the GIS toolbox. People make the technology relevant by applying it to a particular job's needs.

REFERENCES

Churchill G. (1986, October). Roger Tomlinson, GIS pioneer. *Photogrammetric Engineering & Remote Sensing 52:* 1695.

Committee on Geodesy. Panel on a Multipurpose Cadastre, National Academy of Sciences (1980). *Need for a Multipurpose Cadastre,* Washington, DC: National Academy Press.

Dacey M., Marble D. (1965). Some comments on certain aspects of geographic information systems. *Technical Report No. 2,* Evanston, IL: Department of Geography, Northwestern University.

Dangermond J., Smith L. K. (1988, July). Geographic information systems and the revolution in cartography: The nature of the role played by a commercial organization. *The American Cartographer 15:* 309.

FICCDC Technology Working Group. A process for evaluating Geographic Information Systems. *Technical Report 1.* U.S. Geological Survey Open-File Report 88–105 (1200).

Ganneo S. (1988, August). Tremors from the computer quake. *Fortune 118,* no 3: 43.

Giglio J. M. (1988, July). Infrastructure in trouble. *State Government News 31:* 8.

Grosvenor G. M. (1988, August). Mapping a course for success. *State Government News 31:* 14–16.

Hanigan F. (1988). GIS by any other name is still . . . *The GIS Forum 1:* 6.

McClure C. R. (1988, October). Managing public information. *State Government News 31:* 15.

Mitchell W. (1990, February). Enterprise networks: The multivendor networks of the 1990's. No. 2, pp. 69–72.

Parent P. J. (1988). Geographic Information Systems: Evolution, academic involvement and issues arising from the proliferation of information. Master's thesis, University of California, Santa Barbara: 1.

Scruton R. (1982). *A Dictionary of Political Thought.* London: Macmillan Press, s.v. "technology."

Tomlinson R. F. (1988, July). The impact of the transition from digital cartographic representation. *The American Cartographer 15:* 249.

URISA News (1986, December). Transportation SIG is on the move. 5, 6.

Webster's New World Dictionary of the American Language, 2nd college ed. s.v. "geodesy."

Webster's New International Dictionary, 3rd ed. s.v. "cartography," "geodesy," "photogrammetry," "surveying."

Wellar Graf T. Bibliography on Urban and Regional Information Systems: Focus on Geographic Perspectives. Exchange Bibliography 316, 317, Montecello, IL: Council of Planning Libraries, Sept. 1972, 22; as cited in Parent (1988, October): 29.

Chapter **2**

Evolution of the Technology

Geographic information management technology has evolved over the last four decades through the significant contributions of hundreds of people and organizations. The history of GIS development is partially documented in the published conference papers of the Urban and Regional Information Systems Association (URISA), AM/FM International, the American Congress on Surveying and Mapping (ACSM), and the computer industry press. However, little has been written specifically on the evolution of the field as a whole (Parent 1988).

Major developments in the evolution of modern GIS, primarily from a North American perspective, are summarized in this chapter.

ORIGIN OF GEOGRAPHIC INFORMATION SYSTEMS

The roots of geographic information management technology date to the mid-eighteenth century, when cartography developed and the first accurate base maps were produced. Thematic mapping came about soon after (Robinson 1982 in Parent 1988). The eighteenth century also saw the refinement of lithographic techniques and

the early development of statistical techniques, number theory, and advanced mathematics (Parent 1988, p. 4).

Various developments over the next two hundred years did affect GIS, but the emergence of the first electronic computers in the late 1940s marked the beginning of the computer era and a time of rapid evolution for the technology (Fig. 2-1). Although geographic information systems are not necessarily computer-based, most systems today rely increasingly on automation.

Geographic information technology developed through numerous parallel but independent efforts spanning multiple disciplines. The major types of geographic systems—drafting systems (graphics capabilities), analysis-oriented systems (spatial analysis tools), and statistical systems (database management) (Fig. 2-2)—evolved separately, beginning in the 1950s and 1960s.

Computer-oriented design (CAD) grew out of systems developed to automate engineering drafting and the design of integrated electric circuits. Early GIS developers sought solutions to the problems of handling spatially referenced data, for the most part directed to land use, natural resources, and environmental analysis. Geoprocessing systems developed around

The roots of geographic information management technology date to the mid-eighteenth century, when cartography developed and the first accurate base maps were produced.

20

the United States Bureau of the Census DIME file technology discussed in Chapter 1.

Very Early Users

In the 1950s, the first isolated attempts to automate *thematic mapping* began in the United States, Britain, and other parts of the world. In the mid-1950s, British botanists preparing an atlas of British flora used punched cards and a modified tabulator to produce more than 2,000 maps, thereby realizing more than a thirtyfold savings in time over manual production. Some of the first thematic maps from line printers were produced by meteorologists for forecasting and portraying crude land contours (Coppock 1988). By the late 1950s, meteorologists, geophysicists, and geologists had incorporated computer-generated maps into their work (Rhind 1977, Vol. 2 in Parent 1988).[6] The United States military developed graphic display capabilities in the 1950s as part of the Semi-Automatic Ground Environment (SAGE) air-defense system, which converted radar data into computer-generated pictures (Machover 1978 in Parent 1988).

A transportation study commissioned in the late 1950s by the city of Chicago led to development of the Cartographatron by the Armour Research Foundation in order to portray graphically the volume of traffic over certain routes. Phillip Parent identifies the Chicago transportation study as perhaps the first automated graphic output of spatial data for planning purposes (1988, pp. 11, 12).

By the late 1960s, the Canadian government had developed a major system, the Canada Geographic Information System, that contained information on agriculture, forestry, wildlife, recreation capabilities, census divisions, and land use. Canada was the first nation with the ability to carry out continent-wide geographical analysis from high-resolution map data concerning many aspects of its natural resources and socioeconomic conditions (Tomlinson 1988).

The United States oil industry at-

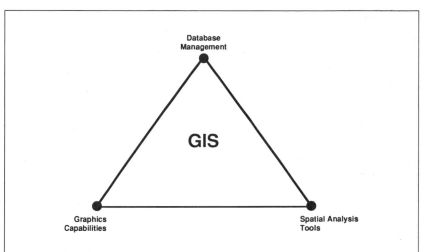

FIGURE 2-1

Geographic information systems rely on the integration of three distinct aspects of computer technology: data base management (of graphic and nongraphic data); routines for manipulating, displaying, and plotting graphic representations of the data; and algorithms and techniques that facilitate spatial analysis.

tempted throughout the 1960s to develop computer systems to map geologic and geophysical data used in exploration, as well as facilities and other data.

Large gas and electric utilities began building rudimentary automated mapping and facility management systems in the late 1960s. Public Service Company of Colorado (PSCo), a gas and electric utility, was one of the first in the United States to actively explore geographic information technology. A pilot project to convert the gas and electric distribution information for an 800-square-mile area of PSCo's subsidiary, Cheyenne Light Fuel and Power Company, lasted from 1968 through 1972. From 1972 through 1977, PSCo and IBM undertook a joint project to refine the PSCo pilot system. "After the initial study work the project almost disappeared, but was kept alive on a limited basis at PSCo and by 'Skunk Works' at IBM and resurfaced in the late seventies at San Diego Gas and Electric Company," said consultant Hank Emery (1988), who helped develop the PSCo system. "Refinements were made. Wisconsin Public Service Development and Integrated Information Systems

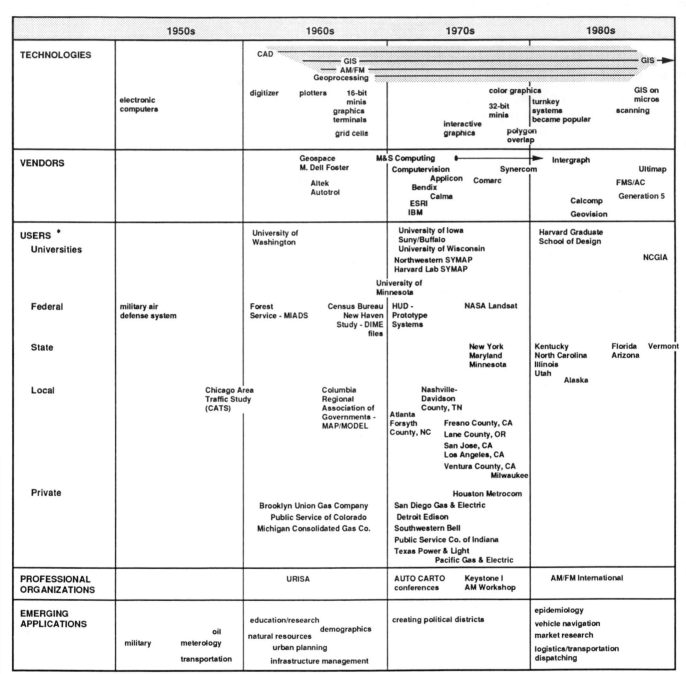

	1950s	1960s	1970s	1980s
TECHNOLOGIES	electronic computers	CAD GIS AM/FM Geoprocessing digitizer plotters 16-bit minis graphics terminals grid cells		color graphics 32-bit minis interactive graphics polygon overlap	GIS GIS on micros turnkey systems became popular scanning
VENDORS		Geospace M. Dell Foster Altek Autotrol	M&S Computing Computervision Applicon Bendix Calma ESRI IBM	Synercom Comarc	Intergraph Ultimap FMS/AC Calcomp Generation 5 Geovision
USERS * Universities		University of Washington	University of Iowa Suny/Buffalo University of Wisconsin Northwestern SYMAP Harvard Lab SYMAP University of Minnesota	Harvard Graduate School of Design NCGIA
Federal	military air defense system	Forest Service - MIADS Census Bureau New Haven Study - DIME files	HUD - Prototype Systems NASA Landsat	
State			New York Maryland Minnesota	Kentucky Florida Vermont North Carolina Arizona Illinois Utah Alaska
Local		Chicago Area Traffic Study (CATS) Columbia Regional Association of Governments - MAP/MODEL	Nashville-Davidson County, TN Atlanta Forsyth County, NC Fresno County, CA Lane County, OR San Jose, CA Los Angeles, CA Ventura County, CA Milwaukee	
Private		Brooklyn Union Gas Company Public Service of Colorado Michigan Consolidated Gas Co.	Houston Metrocom San Diego Gas & Electric Detroit Edison Southwestern Bell Public Service Co. of Indiana Texas Power & Light Pacific Gas & Electric	
PROFESSIONAL ORGANIZATIONS		URISA	AUTO CARTO conferences Keystone I AM Workshop	AM/FM International
EMERGING APPLICATIONS	military oil meterology transportation	education/research demographics natural resources urban planning infrastructure management	creating political districts	epidemiology vehicle navigation market research logistics/transportation dispatching

* among many others

FIGURE 2-2

A sampling of events and participants in the evolution of geographic information system technology.

became involved both as users and vendors of the system, which is now called Geographic Facilities Information System (GFIS) by IBM."

CAD systems were becoming useful for circuit and architectural design and drafting applications. URISA, which was incorporated as a nonprofit organization in 1968 and has become one of the major professional organizations for users in the field, held its first conference in Los Angeles in 1963.

Universities

A great deal of theoretical work on spatial relationships and geography was done during the 1950s and 1960s, and much of the activity in the United States occurred at universities. Computer-aided drafting technology that was developed at the Massachusetts Institute of Technology (MIT) in the late 1950s and early 1960s eventually became a standard for operations in the civil engineering industry.

Eventually, though not immediately, the development of quantitative geographical analysis—the so-called quantitative revolution—would have a major impact on GIS. In Sweden, geographer Torsten Hagerstrand was one of the first scientists to use the computer in spatial simulation research (Kao 1962 in Parent 1988). Harold McCarty at the University of Iowa and William Garrison at the University of Washington were also among the first to develop quantitative methods of geographical analysis (Tomlinson 1988, p. 252).

The University of Washington, in particular, became a center for early GIS research and development. A number of students who studied there under William Garrison and Edgar Horwood helped spread GIS concepts throughout the country; several became prominent in the field. Many of the "next-generation" GIS centers—such as Northwestern, Harvard's Laboratory for Computer Graphics and Spatial Analysis, and the GIS Laboratory at SUNY/Buffalo—were directed and staffed by professors who were graduates

of the University of Washington program. Dr. Horwood and his students developed the "tape mapping program" in 1962 as part of the Spokane Community Renewal Project; it was a prototypical GIS that provided tools to address urban planning issues (Parent 1988, pp. 27, 28).

F. Stuart Chapin, Jr., professor of planning at the University of North Carolina, influenced GIS development with his work on urban land use planning. Chapin developed a numerical cellular model for spatial data and began to use it for planning analysis in the mid-1950s.

Northwestern University became a center of theoretical work as Drs. Garrison, Richard Morrill, Duane Marble, and others joined the Northwestern faculty (Parent 1988, p. 27). In the early 1960s, Dr. Howard Fisher of Northwestern's Technology Institute developed the Synagraphic Mapping System, or SYMAP, which could compose diverse spatial data into a map, a graph, or other visual display. Fisher refined SYMAP during his tenure as the first director of the Harvard Laboratory of Computer Graphics (Parent 1988, p. 15).

Much of the pioneering work in the development of software and data structures was done at the Laboratory for Computer Graphics and Spatial Analysis at the Harvard Graduate School of Design, founded in 1965. Work at the Harvard Lab advanced the technology significantly. The Harvard systems—SYMAP and others developed subsequently—were perhaps the most widely used of the grid-based packages and served as models for others (Tomlinson 1984 in Parent 1988).

On the West Coast, Sam Arms at the University of Oregon, an early researcher in digitizing and overlaying polygons, created a parcel data base with graphic representation of parcels in Lane County, Oregon. The resulting system was an analytical tool that produced rudimentary maps for planning. MAP/MODEL, developed in 1963 by the Columbia Regional Association of Governments in Portland and transferred to the University of Oregon in 1967, was a general-purpose, user-oriented GIS that could store and retrieve spatial data, as well as perform limited overlay functions (Arms 1970 in Parent 1988).

The University of Minnesota's Center for Urban and Regional Affairs began developing a statewide GIS data base in 1967 called the Land Management Information Center (LMIC), which was transferred to the state planning agency ten years later and still serves a broad range of users.

Role of Federal Government

In the early 1960s, the United States Forest Service's Berkeley office became a leader in GIS with the development of MIADS, a system that could store and retrieve attributes of a given grid cell, perform simple overlay functions and mathematical calculations, and generate output in a graphic format. MIADS may have been the first "full-service" GIS for the natural resources environment in the United States (Parent 1988, p. 10).

The United States military expanded its use of geographic information technology in the 1960s as it continued to develop applications for the air defense system and other purposes. A major contribution by the military was its work in the field of *interactive graphics*, the display of maps or graphic images on a monitor or screen, coupled with the ability of the operator to carry on a direct, immediate dialogue with the system to control and select information to be displayed, among other functions. A model designed in the early 1960s by the Rand Corporation allowed display of airplanes tracked by radar. Pointing an electronic "pen" at an airplane image on a display screen retrieved and displayed attribute data about that aircraft. As part of the North American Air Defense Command, the systems were installed across North America. The computers on which these systems operated used vacuum tubes, filled a large room, and required a tremendous quantity of electricity for power and cooling. Even so, the computer provided less processing power than a small PC today.

Development of interactive graphics technology in the defense world led to many other uses. By the mid-1960s, the Central Intelligence Agency, the Army, Navy, and the Defense Mapping Agency were developing systems to support their activities (Parent 1988, p. 14).

Within the same time period, federal domestic programs increased their support of GIS. In the civilian sector, lower level technology was developed that would influence GIS for the next two decades. The Geographic Base File Dual Independent Map Encoding (GBF DIME) file concept was developed by the United States Bureau of the Census in its 1967 New Haven, Connecticut, Census Use Study. This concept incorporated the first popular use of topology in geographic information management. Use of DIME file technology eventually spread around the world. The Department of Housing and Urban Development became involved with geographic information management technology in the late 1960s, when it sponsored a study of urban and regional information systems by System Development Corporation (U.S. Department of Housing and Urban Development 1969) that led to the formation of the federal Urban Information Systems Inter-Agency Committee (USAC). The USAC, which was composed of nine federal agencies, funded the prototype development of integrated municipal information systems in cities around the country. Some of those projects addressed the management of geographic information with limited or no graphic capabilities.

Technologies

The digitizer, which became the standard instrument for digital capture of locational information (i.e., x and y coordinates), was invented in Great Britain in the late 1950s by Dr. Ray Boyle in collaboration with Dr. David Bickmore (Rhinol 1988). Bickmore later founded an Experimental Cartography Unit (ECU) that made far-reaching advances in many aspects of digital cartography, including the use of plotters for cartographic representation.

By the early 1960s, digitizers and plotters became available commercially, with graphics terminals and minicomputers following soon after (Boyle 1980 in Parent 1988). At about the same time Dr. Ivan Sutherland of MIT developed the "Sketchpad" system. It used interactive graphics on the Whirlwind computer, which had an early cathode-ray tube (CRT) that eventually evolved into the Digital Equipment Corporation's DEC 338, perhaps the first intelligent graphics terminal commercially available (Machover 1978 in Parent 1988). Also around the same time, General Motors was working on Design Augmented by Computer (DAC1), which eventually became a key feature in designing GM vehicles and a model for the IBM 2250 graphics console introduced in 1964 (Parent 1988, p. 19).

The 1967 New Haven Census Use Study published plots made on a monochrome photographic plotter from the Geospace Corporation. The Geospace Plotter, one of the first operational graphic plotters, produced images transmitted from a CRT screen and imprinted on a rotating drum covered with photosensitive paper.

"One of the most striking features of these early [GIS] developments, especially in the 1960s, was the way in which initiatives were occurring independently in many places, often without reference to, and even in ignorance of, related work," wrote Professor J. Terry Coppock (1988), a British geographer. The phenomenon of parallel invention in digital cartography was also noted by Nicholas Chrisman (1988), assistant professor of geography at the University of Washington and one of the original members of the Harvard Laboratory for Computer Graphics and Spatial Analysis.

GROWTH OF THE TECHNOLOGY

GIS technology made major progress and began to gain acceptance in some fields in the 1970s. Some of the advances resulted from responses to issues and problems of the times, such as the rising concern about the environment.

Environmental Concerns

Interest in protecting the environment continues to be a dominant issue propelling GIS applications. Global pollution of the air and oceans, depletion of underground aquifers and the ozone, and disposal of growing quantities of hazardous and nuclear wastes as well as common garbage, are some of the problems that practitioners hope GIS applications will solve. From early in its history, GIS technology has been seen as the means of conducing more rational analysis of factors that influence human health, as well as the ecological and economic impacts of land use.

Awareness of Systems

Ian McHarg's 1969 book *Design with Nature* raised public awareness of the interaction among natural systems and laid the groundwork for analyzing complex spatial relationships. A landscape architect, McHarg popularized the idea of capability and suitability analysis using mylar map overlays. He dismissed computer mapping technology in the 1970s but now embraces it because of improvements in the accurate representation of data and aesthetics of computer graphics. Speaking at a 1988 URISA conference in Atlanta, McHarg asserted, "We've got to see the computer as a device by which we can enlarge our sensibilities, our sensitivity, our understanding, and therefore affect our modifications of the environment to enhance human health and well being."

Technologies and Vendors

One pioneer GIS developer was Jim Meadlock, now CEO of Intergraph Corporation. In the late 1960s, Meadlock worked as an IBM manager overseeing computer programming for federal missile systems.

BOX 2-1 Reflections on the History of GIS Development: An Interview With Jack Dangermond

Jack Dangermond, president of Environmental Systems Research Institute (ESRI) and one of the people whose efforts have changed the world's capabilities for handling geographical data, shares his perspective on GIS history in this exchange:

Q. Who were the people who made a significant impact on early GIS?

A. Geographers, planners, regional scientists, and landscape architects. Their initial contributions came largely in the form of theoretical notions and concepts about spatial models and relationships. When computer technologies began to arrive, many continued to expand their research by using concepts of digital data models and software tools. These early leaders initially had the vision of modeling geographic phenomena and patterns associated with them.

Dr. Stuart Chapin at North Carolina was one of the first urban and regional planners in this field to develop a number of spatial data models for land use forecasting and planning analysis in the middle fifties and early sixties. His GIS-related work involved spatial analysis of residential and other land use activities and why they evolved in certain patterns.

Another early leader from the field of geography was Dr. Waldo Tobler, [whose contribution] focused on mathematical data models and spatial analysis algorithms for geographic and digital cartography.

There was also a great deal of initial work done in Great Britain using cellular and vector tools for spatial analysis and display. Perhaps the most well known of the researchers in the display field is Dr. David Bickmore, whose work with the Experimental Digital Cartography Lab in the very early sixties resulted in some of the first high-quality vector maps with computers and focused on the use of plotters for cartographic representation. This group also integrated some of the first coordinate digitizer technology for spatial data capture. Dr. Ray Boyle, a British engineer who was very interested in the subject of map data capture, actually invented one of the first map digitizers in the late fifties. His patents were major footprints in the evolution of these types of instruments.

We saw the early 1900s emerge with many debates involving parametric versus integrated models for landscape. The parametric or thematic approach had as its basis the concept that landscapes could be broken up into measurable components and mapped as layers (e.g., soil, geology, vegetation). This contrasted with the concepts that landscapes should be mapped and described in terms of morphological units which could be observed and managed as an integrated whole. This parametric approach was greatly aided with the breakthrough technology of aerial photography, which allowed people to observe, measure, and record separate parameters on maps. In the United States, one of the first institutions to make large-scale use of parametric mapping was the U.S. Soil Conservation Service, [which] thematically mapped soils and related natural factors as a way to spatially predict agricultural changes.

Q. What about activities in North America?

A. Decades later, more popularized versions of land analysis were reintroduced by several landscape architects in the United States (i.e., Ian McHarg and Philip Lewis). This popularization of landscape analysis by the design professional has laid the groundwork for many environmental planning applications today; both manual and automated approaches.

The Canadians developed one of the most comprehensive natural resource inventory systems of its time [and] some of the first computer technology for its automation and management. These efforts were led by Dr. Roger Tomlinson as part of the Canadian Geographic Information System (CGIS). Many believe Dr. Tomlinson was the first to develop a fully operational GIS as we know it today.

Another individual who made sustaining contribution is Dr. Charlie Miller. Dr. Miller developed coordinate geometry at MIT as a set of mathematical tools during the middle 1960s. These tools, initially designed for civil engineers, have spread throughout the world and have become the underpinning for mathematically describing geometrics, which we use in virtually all of our most advanced COGO and graphics systems.

Then there was Howard Fisher, an architect who, in 1963, started making maps with a computer by over-printing alphanumeric characters. He was the person that inspired SYMAP. He developed it first at Northwestern in 1963 and 1964, and then received a Ford Foundation Grant to develop a research facility at Harvard, which later became known as the Laboratory for Computer Graphics and Spatial Analysis. For nearly fifteen years, this laboratory was a major contributor to various aspects of this technology.

Howard Fisher founded [the lab] and participated in its management until about 1970. Other participants included Dr. William Warntz, Alan Schmidt, Dr. Brian Berry, Eric Tiecholtz, and Denis White.

In the lab, lots of experimental work was done contributing to the development of SYMAP. SYMAP had a dramatic effect on the way people made maps or thought about automated mapping. The technology opened up a vision that automated mapping, and the data structures that they were built on could provide a framework for spatial analysis. This stimulated many to work further on concepts of spatial relationships and spatial models. One individual who contributed greatly to this field was Dr. Carl Steinitz. He was the chief motivation for grid cell GIS technology development by the Harvard Landscape Architecture Department. He worked with David Sinton and many graduate students to organize SYMAP into an efficient cellular GIS package, later called IMGRID. They ultimately developed an efficient grid-cell package that used primitive graphics and was able to quickly process and display spatial information. Technical improvements to SYMAP and related mapping and spatial modeling programs have evolved over a period of approximately twenty years.

He left IBM in 1969 to start a company that was initially called M&S Computing and later Intergraph.

M&S was formed to enable government agencies to use digital computers for the real-time guidance of missiles, said Keith Schonrock, a technical consultant who was one of M&S's first employees, now on Intergraph's board of directors (InterVue 1988). In 1973, M&S obtained a contract from NASA's Marshall Space Flight Center to develop interactive graphics capability. M&S Computing produced its first stand-alone graphics system, which cost about $100,000, under a contract with the Army Missile Command.

"The delivery of the system to the Army was a major turning point," said Schonrock. "It was at that time that M&S management woke up to the fact that maybe there was a market for packaged turnkey systems consisting of storage, memory, computational horsepower, and a display interface with application-oriented software."

M&S Computing was one of the first to develop an interactive automated mapping system for a local government—

TOPOGRAPHIC ASPECT

☐ PEAK
⊡ PIT
▨ SOUTH ASPECT
☰ SOUTHWEST ASPECT
◪ SOUTHEAST ASPECT
⊟ WEST ASPECT
⊠ EAST ASPECT
▦ NORTHWEST ASPECT
◼ NORTHEAST ASPECT
■ NORTH ASPECT

FIGURE 2-3

The MAGI database was composed of 88,000 91.8-acre grid cells (2,000' × 2,000'), each cross-referenced to a database containing factual and interpretive information on physical, environmental, cultural, and administrative features.

Nashville, Tennessee. The Nashville system, begun in 1973, initially used batch processing of punch cards on the city's mainframe, but the company entering the data went bankrupt and was unable to fulfill its contract. M&S persuaded the Nashville government to switch to an interactive system running on a Digital Equipment Corporation PDP computer. M&S provided the system and also contracted to digitize a significant portion of the base maps for the city.

M&S Computing marketed its systems heavily and grew, selling many M&S and then (after renaming itself) Intergraph systems; it was a prime mover behind the revolution in automation of government mapping operations.

During the same era, other companies, including Autotrol, ComputerVision, and Calma, also developed interactive graphic systems. Calma developed a system at Brooklyn Union Gas and Computervision sold systems in Los Angeles. Both were principally used for graphic update and production. The Calma system was one of the first that had attribute processing capability—using "buckets" that stored attribute data.

Most of the early attempts at GIS and thematic mapping used a *raster*, as opposed to a *vector*-based, format. In raster format, graphic images are stored as values for uniform grid cells or pixels. In vector systems, graphic images are described by x and y coordinates or as rules for computation of coordinates or display images.

Once the potential for using computers to overlay and compare data was recognized, the first attempts used grid-cell systems, where a code for the data type (e.g., slope and vegetation) was assigned to each grid cell. Cells corresponding to the same location could be compared or "overlaid" and specific combinations of data could be identified (Fig. 2-3).

Grid-cell systems embodied generalizations of spatial data. Capturing and processing data representing actual boundaries became the challenge. Environmental Systems Research Institute (ESRI) developed one of the first effective polygon overlay systems, where polygons incorporated homogeneous data content and actual boundaries could be overlaid. The resultant combinations also retained their homogeneity.

These early geographic information systems focused on processing attribute data and geographic analysis, and had only rudimentary graphic and mapping capabilities. Graphics were used to display the results of spatial analysis and were not of very high quality, nor did they have well-developed interactive capabilities for data entry and manipulation.

In the meantime, CAD technology was developing on another track, in part because automated mapping (in its then-current state of development) was not very profitable. The CAD/CAM industry served the more lucrative industries, much as electronics, car manufacturing, and aerospace development, that sought drafting and engineering drawing efficiencies. Calma, ComputerVision, and Autotrol drifted out of automated mapping to a large extent, leaving M & S virtually alone in the market for quite some time.

Phenomenal increases in computer processor speed and power in the 1970s and 1980s had a major influence on GIS development. While early systems used punch cards and mainframe computers, it was not until 32-bit minicomputers became available and affordable in the mid-1970s that GIS technology began to spread.

In the 1970s, a number of systems matured and others materialized, including the automated design systems of McDonnell Douglas and Boeing, IBM's CADAM, the commercial GIS software of Comarc and Synercom, and the public domain Map Overlay Statistical System (MOSS). Each had particular strengths relative to the others. Many of the systems at this juncture incorporated data base management capabilities. Limited polygon overlay capability was supported by only a few systems. The automated mapping and drafting systems gradually were linked with other analytical and engineering capabilities, and began functioning more like design systems.

Early Users

The use of GIS technology was initiated in many government and private organizations during the 1970s. Then systems began to be adopted by a broader range of users, typically in a pilot or demonstration mode.

The New York Department of Natural Resources was the first state government to build a GIS around a statewide inventory of land use and land cover. New York was followed by Minnesota, Maryland, and Texas. Minnesota's system evolved from efforts to monitor and control lakeshore development. Maryland's system was developed to support land use planning and policy initiatives (Fig. C-18). The primary goal of Texas's Natural Resources Information System was to share geographic data between agencies. A number of states, including Kentucky, North Carolina, Illinois, Utah, and Alaska, began acquiring and implementing systems in the mid and late 1970s. Most of the state systems resulted from increasing concern for comprehensive land use planning and environmental protection. In some instances, development efforts were funded in part by federal programs, such as the Department of Housing and Urban Development's "701" Comprehensive Planning Assistance Grants, Interior's Land and Water Conservation, Surface Coal Mining Reclamation Programs, and the Environmental Protection Agency's Section "208" Non-Point Pollution Control Grants. By this time there was also a clear shift from internal development to acquisition of commercially available systems.

Many local government and utility systems also were developed. Efforts to implement systems began in the early 1970s in several local governments, including Atlanta, Georgia; Forsythe County, North Carolina; and Nashville, Tennessee. The Forsythe County system, initiated in

Early geographic information systems focused on processing attribute data and geographic analysis, and had only rudimentary graphic and mapping capabilities.

Phenomenal increases in computer processor speed and power in the 1970s and 1980s had a major influence on GIS development.

1973, was designed to automate and maintain tax assessment maps and to synchronize them with records kept by the register of deeds. The early Nashville system was abandoned before completion in the late 1970s; a second effort to develop a system also faltered in the early 1980s. A third attempt launched in 1985 resulted in five Nashville-area organizations forming the Metropolitan Area Geographic Information Consortium (MAGIC), a $10-million system expected to be complete in 1991.

Milwaukee, Wisconsin, and Virginia Beach, Virginia, followed in the next wave of local system developments. In 1977, Milwaukee and Virginia Beach purchased M & S Computing Interactive Graphics Systems that operated on Digital Equipment Corporation PDP 11/34 computers. The Milwaukee data conversion effort was completed in 1983. Numerous applications were developed that use the nongraphic data from other systems. Virginia Beach acquired a major system upgrade in the mid-1980s and extended data conversion to include compilation of a new, more accurate planimetric and topographic base.

Federal Users in the 1970s

In the United States, investments by the federal government in technology for military and space applications produced useful innovations for GIS such as satellite land data.

Launch of the Landsat satellites, beginning with Landsat 1 (originally known as ERTS-1) in 1972, opened up a major new source and method of collecting data useful for GIS. In the mid 1970s, the National Aeronautics and Space Administration funded substantial work related to the Landsat satellite program, including the development of the Maryland Automated Geographic Information (MAGI) system.

A series of national labs was operated by the Department of Energy and the Atomic Energy Commission. In the late 1970s, both Argon and Oak Ridge National Laboratories were involved in applied research, using GIS for siting of electric power plants, including nuclear facilities, in the coastal zone of Maryland. Oak Ridge has continued its research and application of geographic information systems with particular emphasis on their use in federal programs.

The Census Bureau's DIME file technology continued to develop and stimulate ideas for using computers to make maps from census data. The Interior Department agencies have long produced a wealth of spatial information, and in the 1970s they began to experiment with and develop techniques for disseminating digital data, most of which were nongraphic. Interior Department agencies, including the United States Geological Survey (USGS), United States Fish and Wildlife Service, the National Park Service, and the Bureau of Land Management, played various roles throughout the 1970s in developing GIS for natural resource management and environmental assessment. In particular, the USGS played a preeminent role in providing data for use by other federal, state, and local agencies.

The National Cartographic Information Center (NCIC) in Reston, Virginia, is an arm of the USGS that distributed a full range of maps, map byproducts, and other cartographic information produced by USGS, long before digital products were available. Initially, the digital products dispensed by the Interior Department were basically nongraphic databases, such as stream flow information and water quality information. NCIC now distributes graphic products such as digital line graph (DLG) and digital elevation model (DEM) files of USGS map products as well.

A number of other federal agencies actively developed geographic data and GIS applications in the 1970s, including the Department of Agriculture's Forest Service and the Soil Conservation Service; the United States Army Corps of Engineers and the Defense Mapping Agency in the Department of Defense; and the Environmental Protection Agency. The Tennessee Valley Authority developed systems for mapping its service area and managing its extensive land holdings.

Role of Professional Organizations

Much of the knowledge about geographic information technology was transmitted to organizations and users through professional organizations and conferences. URISA, formed in 1963, held annual conferences that provided recurrent opportunities to exchange information on GIS. The American Congress on Surveying and Mapping (ACSM) held a series of conferences called AUTO-CARTO on the automation of cartography beginning in 1974. AM/FM International grew out of a series of "Keystone" conferences on automated mapping sponsored by the Kellogg Corporation in the late 1970s and early 1980s.

THE TECHNOLOGY MATURES

Geographic information technology matured in the 1980s, and it is now applied much more successfully, reliably, and productively by many more organizations.

Technologies

During the early 1980s, several hardware advances affected the evolution of the technology:

- *Display devices* evolved from storage tubes to *refresh graphics* and from the vector mode to the lower-resolution but higher-speed raster mode of operation
- *Color graphics displays* contributed the important ability to differentiate among types of features or attribute values, and they also provide a more appealing display in general
- *Electrostatic* and *ink jet plotters* have become popular printing devices, the former especially useful when fast, inexpensive plots are required or production volume calls for high speeds

The linkage of data bases to graphics was another milestone for GIS. Comarc Design Systems, ESRI, and Intergraph were the first firms to integrate data base management technology with graphics.

Comarc's in-house system, called Geographic Data Management System (GDMS), was oriented primarily toward the needs of the timber industry. ESRI developed a commercial *relational data base*. Intergraph's research produced a *hierarchical data base*.

The ability of systems to integrate data from different sources has greatly improved. The early 1980s saw linkages developed between dedicated GIS systems and mainframe computers, which store parcel files and other sets of geographically related data. Often that linkage was difficult to accomplish because of the incompatibilities between the GIS minicomputer and the mainframe.

System architecture, interface devices, software, and other solutions were developed in the mid-1980s to allow on-line linkage between commonly used systems. A very appealing concept is a *distributed network*, in which the data are stored in one system but directly accessible to others. In practical application, however, distributed networks are only now becoming widely used.

By the late 1980s, the technologies of the three main types of geo-based systems had merged to various degrees. Most systems today offer powerful capabilities for graphic production, processing of attributes, and analysis.

Users

Many new systems have been initiated at all levels of government and within private industry.

Multiuser systems, such as those in Knoxville and Nashville, Tennessee, and Indianapolis, Indiana, have been developed to minimize cost and maximize access to common data bases.

Within the federal government, Interior Department agencies have expanded their use of digital mapping, and USGS is making more databases available in digital form. The operating agencies such as Fish and Wildlife, BLM, and the National Park Service have used GIS increasingly as an analytical tool. The Forest Service and the

> "You are hard-pressed to find a place that isn't getting into GIS in some way, that isn't at some stage in the process."
>
> John McLaughlin, GIS consultant to the UN

> By the late 1980s, the technologies of the three main types of geo-based systems had merged to various degrees. Most systems today offer powerful capabilities for graphic production, processing of attributes, and analysis.

Environmental Protection Agency made major commitments in the mid-1980s to build GIS networks throughout the country. Use of GIS in other federal agencies also grew in the 1980s.

International Developments

GIS has taken root throughout the world. "You are hard-pressed to find a place in the world that isn't getting into GIS in some way, that isn't at some stage in the process," said University of New Brunswick professor John McLaughlin, a consultant to the United Nations on GIS technology.

The United Nations Development Program promotes the use of geographic information technology around the world by hosting regional cartographic conferences. It also provides funds, technical assistance, and training to the governments of about one dozen developing countries for development of digital land information systems; other countries receive GIS development funds through bilateral aid agreements.

"We're the spark at the beginning that gets programs started," said cartographer Max Dehenseler of the Department of Technical Cooperation for Development. "We respond to requests from developing countries for training and expertise."

Interest has grown in the creation of global data bases. A conference on global data bases sponsored by the International Geographic Union was held in London in 1987. In response to growing concerns about population growth and stress on the environment, the United Nations Environmental Programme (UNEP) began development of a Global Resource Information Database (GRID).

Universities

A number of universities have been influential in GIS development and have developed GIS curricula over the years. In addition to those previously mentioned are Ohio State University, Penn State University, Portland State University, State University of New York at Buffalo, the University of California at Santa Barbara, the University of Florida, the University of Maine, the University of Maryland, the University of New Brunswick, the University of Washington, the University of Wisconsin, and Yale University.

REFERENCES

Arms S. (1970). *Map/Model System: System Description and User's Guide*. Eugene, OR: University of Oregon, Bureau of Governmental Research and Service.

Boyle A. R. (1980). Development of equipment and techniques. In D. R. F. Taylor (Ed.), *The Computer in Contemporary Cartography*, vol. 1, Progress in Contemporary Cartography (series). New York: John Wiley & Sons, pp. 39–58; as cited in Parent (1988).

Chrisman N. (1988). The risks of software innovation: A case study of the Harvard Lab. *The American Cartographer 15:* 291.

Coppock J. T. (1988). The analogue to digital revolution: A view from an unreconstructed geographer. *The American Cartographer 15:* 265.

Emery H. (1988). AM/FM: A retrospective look. *Emery DataGraphic Quarterly:* 2.

Intergraph, Inc. (1988). In the beginning: An interview with Keith Schonrock. *InterVue 7:* 4–7.

Kao R. C. (1962). The use of computers in the processing and analysis of geographic information. *The Geographical Review 53:* 530–547; as cited in Parent (1988).

Machover C. (1978). A brief, personal history of computer graphics. *Computer 11:* 38–45; as cited in Parent (1988).

Parent P. J. (1988). Geographic Information Systems: Evolution, academic involvement and issues arising from the proliferation of information. Master's thesis, University of California, Santa Barbara.

Robinson A. H. (1982). *Early Thematic Mapping in the History of Cartography*. Chicago: University of Chicago Press, as cited in Parent (1988).

Rhind D. (1977) Computer aided cartography.

Transactions of the Institute of British Geographers 2: 71–97, as cited in Parent (1988).

Rhind D. (1988) Personality as a factor in the development of a discipline: The example of computer-assisted cartography. *The American Cartographer 15:* 279–280.

Robinson A. H. (1982). *Early Thematic Mapping in the History of Cartography.* Chicago: University of Chicago Press, as cited in Parent (1988).

Tomlinson R. F. (1984). "Geographical Information Systems—A New Frontier." Keynote address, *International Symposium of Spatial Data Handling.* Zurich, pp. 1–14; as cited in Parent (1988).

Tomlinson R. F. (1988). The impact of the transition from digital cartographic representation. *The American Cartographer 15:* 252; 257.

U.S. Department of Housing and Urban Development. (1969) *Urban and Regional Information Systems: Support for Planning in Metropolitan Areas.* U.S. Government Printing Office, p. 450.

(1988). In the beginning: An interview with Keith Schonrock. *InterVue,* 7, 4–7.

Chapter 3

Applications

People in many fields of endeavor have discovered ways to use geographic information management technology. A few applications have reached maturity and gained broad acceptance; many others are yet emerging. This chapter discusses the uses that government agencies, utilities, and private businesses are now making of geographic information technology and provides examples of operational systems.

Spatial data applications may be divided into a dozen broad categories, several of which are likely to be used simultaneously by many public agencies and businesses. Dr. Francis L. Hanigan (1988) identified the following categories of use:

- *Business applications.* The ability of GIS technology to help make better business decisions through the rapid portrayal of geo-demographic data and interactive business site analysis is quietly winning the technology a place in the business community.
- *Election administration and redistricting.* The decision-making processes associated with these applications are highly political, very visible, and quite sophisticated. GIS technology's ability to portray and analyze geo-demographic data used in these processes offers the opportunity for more consistent analysis of data and the ability to make more credible political decisions.

- *Infrastructure management.* Those responsible for the development, maintenance, and management of community water, sewer, gas, electric, and telecommunication utilities were among the first to appreciate the potential of GIS technology.
- *Map and database publishing.* Federal mapping agencies have taken the lead in automated cartography. The Defense Mapping Agency, the Department of Interior, the United States Bureau of the Census, and other federal agencies have massive digital mapping programs underway. Civilian map publishers are just beginning to automate. The cottage industry devoted to the enhancement and application of GBF/DIME/TIGER files and other census data will continue to grow.
- *Oil, gas, and mineral exploration.* Geophysicists and geologists have begun seriously to discover the value of GIS technology as a tool for enhancing sophisticated exploration models.
- *Public health and safety.* Public health officials as well as crime and arson analysts have long appreciated the usefulness of maps for tracking the spread of communicable diseases and for displaying patterns of criminal activity. GIS technology offers the ability to perform these tasks more quickly, more frequently, with a higher level of credibility, and at significantly lower costs than now possible.

34

- *Real estate information management.* State, county, and municipal taxing authorities have been quick to recognize GIS technology's potential as a tool for developing and maintaining complete and accurate real property inventories.
- *Renewable resources management.* The growing demand for environmentally sensitive management of our water, air, soil, forests, and wildlife has caused state and federal environmental agencies to turn to GIS technology as a practical way to manage resource data required to carry out their mandates. Private-sector applications lag far behind those in the public sector.
- *Surveying and mapping.* Because many people think of the data-conversion and mapping companies as vendors rather than users of GIS technology, they fail to recognize that the efficient building of base maps and the conversion of existing facility records is a highly specialized application and operational tool, for which GIS is especially suited.
- *Transportation and logistics.* GIS technology offers over-the-road freight haulers, local delivery services, and railroads new tools for competing efficiently. Transportation planners, state highway departments, and metropolitan transit authorities are also benefiting from the technology.
- *Urban and regional planning.* The ability to provide planners with rapid access to masses of data has earned GIS technology wide acceptance within the conceptual planning community.
- *Research and education.* The National Center for Geographic Information and Analysis, a consortium of three universities funded by the National Science Foundation, is undertaking a broad range of GIS research projects, including one that will focus on the use and social, economic, and institutional value of information.

Other applications are illustrated in the accompanying plates and figures. Within those general areas outlined by Dr. Hanigan, government (federal, state, and local), utilities (mature and developing systems), and private industry (oil and gas, forestry, transportation, marketing and sales, and real estate) have developed specific applications in ways discussed below.

GOVERNMENT APPLICATIONS

GIS use is growing rapidly in agencies at all levels of government in the United States. As a group, government employees form the largest category of GIS users, resulting in more advanced and widespread GIS use in public agencies generally than in the private sector. This section discusses GIS uses within federal, state, and local governments.

Federal Government

Federal agencies use GIS extensively. At least sixty-two federal organizations use or intend to use GIS, and eighteen now use it operationally, according to a 1990 survey by the Federal Interagency Coordinating Committee on Digital Cartography (FICCDC 1988). The FICCDC was formed by the Office of Management and Budget in 1983 to coordinate and report on the digital cartographic activities of federal agencies, and to develop standards and specifications for the production of digital cartographic data. The FICCDC initially focused its attention on coordinating the production of digital data germane to USGS map products. Recently, it has expanded its sphere of involvement to deal with other federal cartographic and geographic data sets and is working to encourage coordination and exchange of digital spatial data throughout the federal agencies.

The FICCDC's 1987 annual report noted that although GIS technology has been available for more than a decade, the level of acceptance and use had remained relatively low within federal agencies. However, the last several years of technological development coupled with increasing availability of spatial data have produced an explosive growth in the use of GIS (FICCDC 1990). The FICCDC estimated that federal agencies spent between $26 and $46 million on GIS in fiscal year 1987, between $50 and $55 million in fiscal year 1988, and between $60 and $70 million in fiscal year 1989. For fiscal year 1992 four federal organizations estimated their GIS funding requirements to be in excess of $10 million, five organizations estimated needs between $5 million and $10 million, and an additional 20 organizations budgeted in excess of $1 million.

General Types of Applications. The FICCDC identified the following ten general uses of GIS within the federal government:

1. *Terrain maps*—Topographic maps, elevation models, slope maps, aspect maps, shaded relief maps, and perspective views (Figs. C-19, C-20)
2. *Data display and analysis*—Map data access and display, merging and integrating data bases, and image maps
3. *Navigation systems*—Air traffic control systems, in-flight navigational data, nautical cartographic data, and air route structure data
4. *Network simulation models*—National Rail Network Model and National Rail Defense Essential Network
5. *Mineral resource assessment*—Geologic maps, mineral and mineral fuel resource assessment, engineering geology, and geologic hazards
6. *Environmental assessment and monitoring*—Environmental impact assessment, irrigation suitability, irrigation development monitoring, contamination and pollution studies, soil conservation, and flood mapping (Fig. C-20)
7. *Thematic map data*—Socioeconomic, demographic, soil, and other characteristics at the state, county, congressional district, and Standard Metropolitan Statistical Area levels; National Airspace System; and feature analysis
8. *Land and water resource inventory, planning, and management*—Site design and screening, road design, land management, prime farm and forest land sur-

veys, cultural land mass information, wetland maps, habitat assessment, and water quality and quantity studies (Fig. C-22)
9. *Trainers and simulators*—Radar trainers and flight simulators
10. *Base map development*—Analysis, plotting, modification, and revision of quadrangle maps, aeronautical charts, marine information, cadastral maps, ocean surveys, base charting, and navigation

The many and varied uses of GIS in federal agencies involve federal planners, biologists, scientists, researchers, foresters, air-traffic controllers, engineers, geographers, geologists, ecologists, mappers, managers, and others. Some of the ways they use GIS to answer questions, solve problems, make decisions, and carry out agency functions are discussed below.

Mapping. Federal agencies produce tens of thousands of maps each year. The major mapping agencies are the Department of Interior's United States Geological Survey, National Mapping Division (USGS-NMD), which produces basic topographic maps for the nation; the Department of Commerce's National Oceanic and Atmospheric Administration (NOAA) and Office of Charting and Geodetic Services (C&GS), National Ocean Service (NOS), which produces domestic nautical and aeronautical charts; and the Defense Mapping Agency (DMA), which maps foreign lands and prepares aeronautical charts for the United States military branches.

The goal of the USGS National Mapping Program is to produce and maintain topographic maps for the entire United States; initial coverage is expected to be complete by 1990. Recognizing the demand for digital data creation, the Survey has adopted a modernization plan and expects to complete the transition from manual to predominantly digital operations by the year 2000.

For almost a decade, the USGS has been collecting information for the National Digital Cartographic Data Base (NDCDB), a collection of digital data organized by scale-specific, product-specific cartographic units. The NDCDB contains information on transportation, hydrography, boundaries, public land survey system (PLSS), topography, land use and cover, and geographic names. Two major types of digital data are produced: elevation data, distributed as digital elevation models (DEMs), and planimetric and contour data, distributed as digital line graphs (DLGs). An increasing portion of the nation is covered by digital data produced from existing USGS hard-copy maps (Jannace and Ogrosky 1987).

The base maps and digital data produced by the USGS are used by agencies at all levels of government, as well as by private industry, often as a land base or base map to which other information is added.

Within the Department of Commerce, the Office of Charting and Geodetic Services (C&GS) of the NOS charts the waters of the United States and its territories and conducts geodetic, hydrographic, and photogrammetric surveys for civil and military uses. NOAA uses C & GS data along with other data to ensure the availability and distribution of navigation charts and other information necessary for safe marine operations. Aeronautical support products, including visual and instrument area route and departure charts, and radio facility charts, also are produced.

Like other federal agencies, NOS is computerizing its mapping operations. The present automated system handles about 10 percent of the workload. The agency has plans to build a much larger digital national marine navigation data base that could be partially operational by 1990 and thus is now converting its voluminous storehouse of analog data to digital form (Hull 1987).

The Defense Mapping Agency (DMA) provides maps, precise positioning data, digital data, nautical and other charts, and marine navigational data to the United States Armed Services, the Agency for International Development, the Central Intelligence Agency, the National Aeronautics and Space Administration, and other federal agencies (Hull 1987, p. 214). The DMA continues to develop and implement a worldwide digital data base that will produce twenty-eight different products (hard copy and digital) in an all-digital environment.

Land and Water Resource Inventory, Planning, and Management: Environmental Assessment and Monitoring. Many federal agencies are using GIS for land management and environmental assessment. The major users include the Department of Interior's Bureau of Land Management (BLM), Fish and Wildlife Service (FWS), National Park Service (NPS), Bureau of Indian Affairs (BIA), Bureau of Reclamation (BOR), and USGS; the Department of Agriculture's Forest Service (FS) and Soil Conservation Service; NOAA, in the Department of Commerce; several organizations in the Department of Energy; the Tennessee Valley Authority; the Agency for International Development; and the Army Corps of Engineers, Army Engineer Topographic Laboratories, and Naval Ocean Research and Development Activity. Other federal agencies are planning or developing environmental and land-use planning applications.

The Bureau of Land Management's Oregon State Office is using GIS technology as the primary means for development of its Resource Management Plans, prepared once each decade for five management districts in western Oregon to determine land and resource use on federal lands. When completed, the Western Oregon Digital Data Base will contain about twenty types of natural resources data and a new, accurate, large-scale base map (Federal Digital Cartography Newsletter 1988).

The Fish and Wildlife Service (FWS) is making broad use of GIS technology for fish and wildlife species analysis and modeling, habitat suitability determinations, national wetlands inventory mapping, refuge vegetation/terrain mapping, refuge master planning, and land records information management (Fig. C-23). In doing so, the FWS is integrating many types of planimetric and thematic data to portray land-use conflicts, monitor changes in wetlands, and prepare impact statements.

The FWS is building a digital National Wetlands Inventory (NWI) data base that now covers about 5 percent of the continental United States. Statewide data bases have been completed for New Jersey, Delaware, and Maryland, and are in progress for Illinois and Indiana. NWI digital data also are available for portions of fourteen other states.

The FWS is beginning to use GIS to determine the effects of hazardous waste on federal endangered species, national wildlife refuges, and certain migratory birds as required by the Comprehensive Environmental Response, Compensation, and Liability Act (CERCLA), or "Superfund" legislation.

The USGS Water Resources Division, the largest GIS user within the Survey, has GIS facilities in more than forty locations across the United States linked with GEONET, its distributed information system.

One example of a cooperative, interagency project is the digital data base being built to study and assist clean-up efforts in the Elizabeth River drainage basin in Virginia. The Elizabeth River drains into Chesapeake Bay, the nation's largest estuary. The area is heavily industrialized and the river is so polluted that in most of it harvesting of shellfish for direct marketing has been banned for more than fifty years. The USGS and the Environmental Protection Agency are using a GIS to integrate more than sixty science and natural resource data bases; as many as fifteen federal, state, and local agencies are contributing data to the Elizabeth River data base (Federal Digital Cartography Newsletter, 1987/88).

NOAA's National Environmental Satellite, Data, and Information Service is using a GIS to produce atlases showing the distribution of climatic variables such as ice, drought, temperature, precipitation, sunshine, and length of growing season. NOAA's National Geophysical Data Center is constructing global data bases of thematic and geophysical data with a GIS.

Using GIS to help react to hazardous material spills is an application developed by NOAA's Hazardous Materials Response Branch. CAMEO II, the latest version of the software, is used by about 1,700 local firefighters, police departments, pollution-control organizations, hazardous materials response teams, county governments, and private companies. The Hazardous Materials Response Branch, which aids the Coast Guard, developed the program in 1986 to help identify chemical spills and meet "first-response" needs at chemical accidents. The system was subsequently enhanced with the passage of the Superfund Amendment and Reauthorization Act (SARA), Title III of which requires communities to plan for hazardous materials emergencies. The data base includes information on 2,600 chemicals, as well as maps and information on the locations and quantities of hazardous materials storage sites, which enables the system to calculate the likely path of a chemical plume and identify areas for evacuation.

GIS applications at the Department of Energy's Oak Ridge Laboratory include environmental monitoring and assessments, such as acid rain analysis and global climate studies. In a pilot test of GIS, the Federal Insurance Administration of the Federal Emergency Management Agency is combining digital cartographic base data and digital flood insurance rate maps to produce a directory of flood-prone properties.

Military applications of GIS include terrain modeling and analysis, base planning, and environmental studies. In support of the INF (Intermediate-range Nuclear Forces) treaty between the United States and the Soviet Union to eliminate intermediate-range and shorter-range missile systems, the Department of the Army used the results of a GIS analysis to assess the environmental impacts of destroying Pershing missile rocket motors (Geographic Resources Analysis Support System Newsletter 1988).

The Army Corps of Engineers uses GIS for environmental planning and land use management activities. For example, with a GIS the Portland, Oregon, District identified the most suitable locations for offshore disposal of dredged materials at the mouth of the Columbia River, one of the district's largest estuaries. The Portland District also has used GIS for wildlife habitat suitability and river bank erosion analyses (1988).

Demographic Data. In preparation for the 1990 Decennial Census of the United States, the Bureau of the Census, in cooperation with the USGS, developed a major new digital data base called the Topological Integrated Geographic Encoding and Referencing (TIGER) System.

TIGER is the first comprehensive digital street map of the United States, covering 100 percent of the country's land area. TIGER contains digital data at a scale of $1:100,000$ for every street and road in the nation, the range of address numbers located along each section of every street in the 345 largest urban areas, all railroads and the names of their operating companies, all significant hydrographic features and their associated names, American Indian reservations, military bases, the boundaries, names, and numeric codes for all geographic areas used by the Census Bureau to tabulate the results of the 1990 census, and other information. TIGER covers the fifty states, the District of Columbia, Puerto Rico, the Virgin Islands of the United States, Guam, American Samoa, and the Northern Marina Islands.

The Census Bureau's digital data are used in numerous planning and analysis programs in government agencies and the private sector. A major use for the 1990 census data will be election-district reapportionment throughout the country.

The Division of Preventive Medicine, Department of Epidemiology at Walter Reed Army Institute of Research, is using ZIP Code Boundary Files developed by Geographic Data Technology, Inc. (GDT), of Lyme, NH, to map the human immunodeficiency virus (HIV), which causes AIDS, by zip code in urban areas (*GDT News* 1988, vol. 4).

The Soil Conservation Service (SCS) also is creating digital thematic data for GIS applications. SCS has created three national geographic data bases at different scales to improve the storage, manipula-

---

BOX 3-1   TIGER Maps the Nation

While the 1980 Census DIME files have been used widely for demographic analysis, the TIGER files also will be used for operations. "The USA spends a half-trillion dollars a year on logistics: moving things and people from place to place. Applying computer algorithms and digital maps to logistic operations typically yields a 5 to 15 percent improvement in efficiency. Take 5 percent of a half-trillion dollars, and you'll see why a nationwide TIGER digital street map is important," said Donald F. Cooke, founder and president of Geographic Data Technology, Inc. (GDT). Cooke was a member of the New Haven Census Use Study Research Group that made the first DIME files in 1967. GDT develops and markets products based on DIME and TIGER data, and the company digitized files for TIGER under contract to the Census Bureau.

"This is why I contend that the value of TIGER products will far outweigh the value of data tabulations from the 1990 enumeration. TIGER represents a $182 million investment by the government with a potential payback measured in billions," Cooke added. "The full nationwide coverage [the TIGER files] provided will permit development of logistics management, districting, and referral systems. By the end of the 1990s most major logistics and audiotext systems will depend on digital street maps for their operation." (*GDT News* 1988, Summer).

TIGER is the first comprehensive digital street map of the United States, covering 100 percent of the country's land area.

---

tion, and retrieval of soil map information. The geographic data bases are linked to an attribute data base that catalogues over twenty-five soil properties for more than 15,300 soil series recognized in the United States. Attribute data include particle size

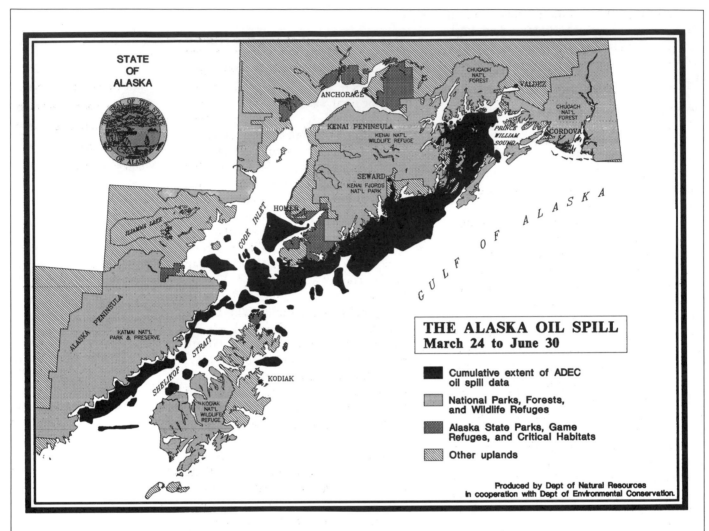

FIGURE 3-1

The State of Alaska tracked, from successive flight observations and beach surveys, the extent of the 1989 Exxon Valdez oil spill. GIS technology was used to plan cleanup and monitoring activities and will support the state's litigation efforts (courtesy of Alaska Department of Natural Resources).

distribution, bulk density, available water capability, soil reaction, salinity, organic matter, and subsidence characteristics.

Soils data can be used for a wide range of applications, including siting and management of sanitary facilities, building development, recreation development, watershed resource planning and management, and farm and ranch conservation planning.

Mineral Resource Assessment. Two Interior Department agencies, the USGS and the Bureau of Mines (BOM), each use a GIS to manipulate and analyze geologic data. A mineral assessment program in eleven western states is using a GIS to identify areas of mineral mining potential, along with restrictions imposed by law, regulations, and stipulations of management agencies (1987/88).

The Survey's Geologic Division is using a GIS to assess the mineral resource potential of selected areas in the Paducah quadrangle area of Illinois, Kentucky, Missouri, and Indiana. GIS techniques are being used to merge and analyze multiple digital geoscience and cartographic data sets, to model the occurrence of metallic and industrial minerals, oil and gas, coal, and water.

Navigation Systems. The National Ocean Service is developing electronic charts that can be used for real-time navigational analysis. The United States Postal Service is using digital census data to develop a program for interactive modeling of delivery and collection routes. The Federal Highway Administration is involved in cooperative research efforts with other government agencies and the private sector to develop electronic navigational systems that would ease traffic congestion.

Network Simulation Models: Trainers and Simulators. The military uses GIS radar trainers and flight simulators to replicate battlefield situations. The United States Army Engineer Topographic Laboratories are developing geographic-based tactical decision aids and terrain data displays. Substantial effort has been invested in three-dimensional terrain modeling of foreign lands to support United States missile programs.

Ongoing GIS Research. The National Center for Geographic Information and Analysis was established in 1988 with a $5.5-million National Science Foundation (NSF) grant to a consortium of three universities: the University of California, Santa Barbara, State University of New York at Buffalo, and the University of Maine at Orono. The center will be located at UC Santa Barbara, with research and education programs at the other universities.

The NSF has defined three roles for the center. First, the center will conduct basic research on ways to improve and use GIS. Second, the center will promote the spread of GIS-based analysis and help increase the number of GIS experts available to use the technology. Third, the center will act as a clearinghouse for information on GIS research, applications, and education, and will study the social, legal, and institutional impacts of the spreading use of GIS technology.

State Government

Management of geographic information can be viewed as a subset of the larger field of information resources management (IRM), which deals with the creation, production, collection, management, distribution, and retrieval of information. Syracuse University's School of Information Studies reports that 90 percent of state government employees engage in some form of information handling, while states spend an estimated $20 billion annually on the management and use of information technologies (Syracuse University 1989).

The use of geographic information technology is widespread within state governments. At least forty-five states are using a GIS or a related spatial system in at least one state agency, and at least thirty-four states have more than one geographic information system.

States' uses of GIS include natural resources, transportation, planning, public lands management, and property assessment applications.

Resource and Land Use Issues. States are using GIS to support public policy decisions regarding lands and water, natural resources, fish and wildlife inventories, planning and management; and environmental assessment and monitoring. At least thirty-eight states have developed some ability to use geographic information technology for natural resources and land-use analysis and management (Fig. 3-1).

GIS is used to identify and analyze the development of minerals, coal, oil and gas, and timber resources; to prepare land-use plans; to support environmental regulatory programs; and to manage historical and cultural resources. GIS is also used to determine the best location for hazardous-

At least thirty-eight states have developed some ability to use geographic information technology for natural resources and land-use management.

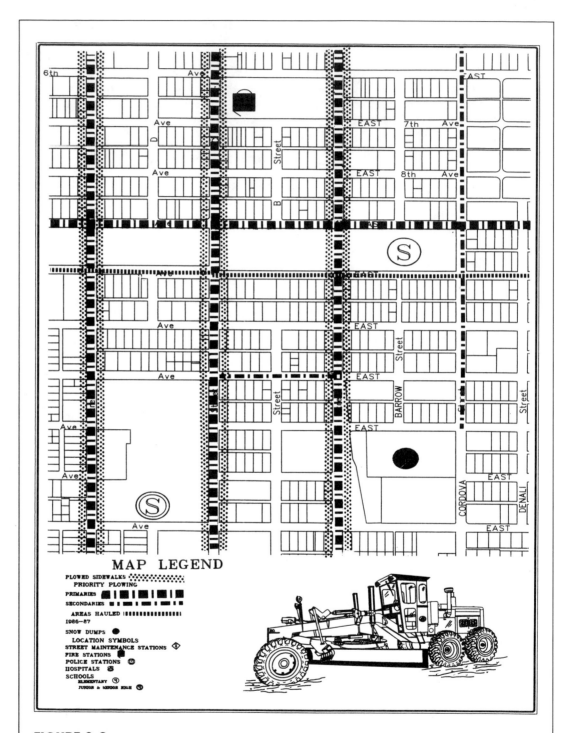

FIGURE 3-2

Municipality of Anchorage, Alaska, uses GIS to plan and optimize snow removal activity (courtesy of Municipality of Anchorage).

waste disposal sites. In other instances, it is used to manage fish and wildlife and to evaluate the effects of development on wildlife habitats.

Many agencies in Minnesota make use of the Land Management Information Center (LMIC) within the Minnesota State Planning Agency. It coordinates the activities of state departments and is also used by local and regional agencies. Several divisions of the Department of Natural Resources (DNR) routinely use the LMIC to support decisions on land acquisition and disposal, recreation expenditures, and water management, among other applications. LMIC is used by more than 100 environmental planning projects each year (Licht 1987).

In addition, several Minnesota agencies have in-house systems. The Department of Military Affairs uses a GIS at Camp Ripley to manage activities that include siting small-arms ranges, landfill location, assessing the impact of training activities on the land, and scheduling timber sales. The DNR Division of Forestry uses a GIS to inventory and manage forests on state lands (Metropolitan Council Data Center 1988).

Maryland's Automated Geographic Information (MAGI) System is used widely by state agencies for resource management and planning. MAGI was established in 1973 to serve as the primary data base in preparation of a state land-use plan. MAGI has been used by the Department of State Planning and other agencies for agriculture land mapping; coastal use studies; power plant, sanitary landfill, and coastal facility siting; habitat studies of the wild turkey, bald eagle, and other wildlife; analysis of the potential for commercial forest productivity; highway corridor studies; and oil spill contingency plans (Antenucci 1982).

The Alaska Department of Natural Resources began developing a GIS in 1977. It is used primarily to prepare land-use plans, analyze development alternatives and potential for land settlement and resource development, prepare resource sale offerings, make maps to support litigation,

analyze land trades, and manage university and mental health facility lands. DNR also is building an automated land records system to track property ownership and land status. In addition to these systems, the DNR Division of Oil and Gas uses a geoprocessing system to evaluate the oil and gas potential of state lands proposed for oil and gas leasing. The Division's geophysicists and geologists process seismic data and produce geologic-structure and contour maps showing the potential locations of oil and gas deposits. The analysis is used to project expected bids and to select bidding methods. The maps are used at confidential management briefings before and after lease sales.

The Illinois Department of Energy and Natural Resources uses its system for a variety of applications, including mining permit review and tracking, and site evaluation for hazardous-waste materials. Like a number of other states, Illinois used its GIS to help prepare a 1988 application for the United States Department of Energy's supercolliding/superconducting facility competition.

Transportation. Geographic information technology is used by at least twenty-nine state departments of transportation. Many use it for engineering drafting and design. Fewer use it extensively for mapping and geographic analysis—Michigan, Florida, and Ohio among the first.

Specific transportation applications of GIS include road design, highway mapping, pavement and maintenance management, capital budget planning, analysis of accident data and traffic volumes, and routing and dispatching vehicles. Many states are now attempting to link existing road log data with their mapping and drafting systems (Fig. 3-2).

Tax Appraisal and Land Records. Keeping track of boundaries, property ownership, taxation, and assessment information is a major use of GIS for state and local governments. In most states, property tax assessment and collection is regulated by the state and carried out by local govern-

Specific transportation applications of GIS include road design, highway mapping, pavement and maintenance management, capital budget planning, analysis of accident data and traffic volumes, and routing and dispatching vehicles.

ments. Several, including New York, North Carolina, Kansas, and Virginia, have developed programs to help local governments update and modernize their land records.

The state of New York was one of the first to develop a comprehensive local-property mapping program. Legislation passed in 1970 amended the Real Property Tax Law and required each county to prepare and maintain tax maps. The legislation directed the New York State Board of Equalization and Assessment to establish rules and provide guidance in preparation of the new tax maps, and also to consider other potential land uses of the maps. New York provided one dollar per parcel to help counties obtain new tax maps that conformed to the new regulations. Of New York's fifty-seven counties, fifty had completed tax mapping projects by late 1988, eighteen years after the program began.

North Carolina initiated a program in 1977 to modernize and establish greater uniformity in local land-records systems. The North Carolina Land Records Management Program provides, at the request of a county governing body, technical assistance, training, and advice on standards and specifications for preparation of county base maps and cadastral maps. Adoption of standards is voluntary. The program is authorized by statute to provide matching grants to counties of up to 50 percent of the costs of a land-records project. As of late 1988, 38 of 100 counties had completed their mapping projects, and 35 counties had work in progress. The state has put about $3 million into the program since 1977, and the counties have spent about $32 million.

In response to legislation passed in 1985, the state of Kansas undertook a major computer-assisted mass appraisal mapping project costing about $65 million and covering 80,000 square miles and 1.5 million parcels of property. The Kansas Department of Revenue, Division of Property Valuation, established guidelines for the property mapping effort of the 105 counties. The project, completed in 1989, has

resulted in uniform, computerized property ownership maps for the entire state (Donatello in Williams 1987). Many counties took the initiative and developed broader GIS capabilities during the three-year program.

Redistricting. Using GIS to redraw election districts based on new census data will become a common application in the early 1990s.

Minnesota was one of several states to use GIS for redrawing both Congressional and state legislative districts following the 1980 census. The majority and minority caucuses of the state legislature used the state Land Management Information Center's GIS for analyses of new election districts. When the legislators were unable to agree quickly enough, the federal District Court stepped in, purchased its own workstation, hired a consultant, and developed a plan using the center's GIS.

Other states using GIS for reapportionment following the 1980 census included Florida, Maryland, and New York.

Emergency Medical Services. The Texas Departments of Health and Highways and Public Transportation are implementing a statewide automated geographic information system that could serve more than 1,000 emergency medical service providers. The system was developed during 1987 and 1988 by the Lyndon B. Johnson School of Public Affairs, University of Texas at Austin, in cooperation with the state agencies. As reported by Yves Renaud, Pascale Rorde, and David Eaton (unpublished paper), the system will automate the collection, transfer, and analysis of hundreds of variables related to calls for emergency medical care and responses of paramedic teams.

Strategically situated as the middle layer of government, states are playing an increasing role in geographic information coordination. A number of state governments are assisting local governments with geographic information collection and automation. In several states, legislative and executive advisory committees have recommended establishment of an agency or office to coordinate data standards, access, and exchange, and to offer technical assistance. In many states, the proliferation of GIS at the local level, as well as recognition of the inherent inefficiency of developing multiple, incompatible systems, is driving the push for interstate coordination.

Wisconsin has demonstrated leadership by designing a model for coordination of geographic information. A Wisconsin law governing land information management practices was passed in 1989 and 1990 after years of concerted effort by professionals in government, industry, and academia. The Wisconsin Land Information Program will improve methods for managing and using land-related information, fund land-record development by local governments, and will enable sharing of geographic information among all interested organizations.

Grants to local governments for land-records modernization are to be funded with a staged increase in the real estate recording fee collected by county registers of deeds, effective July 1990. To qualify for funding, local governments must establish and maintain a land information office, develop a plan for land-records modernization, and get approval from the Wisconsin Land Information Board (WLIB), which oversees the program. The grants must be spent exclusively to implement board-approved plans, and counties must provide a 25 percent match. Most state departments must prepare biannual action plans for geographic data sharing.

"We're still working out what 'land-records modernization' means, what the plan content should be, and what they have to do to qualify," said William Holland, executive director of the WLIB. "Our goal is to ensure that information collected under the program can be shared. The key is systems integration. We won't specify hardware and software, but we'll address issues such as quality and data interchange formats. The standards will be tied to the information counties have to de-

velop and map to meet statutory requirements, such as wetlands and soils."

Holland said he has received inquiries about the new law from all over the country. "We're trying to develop a comprehensive system in a decentralized fashion, from the bottom up. I think that's the only way to make it work, since the majority of land records are collected and maintained at the local level. Besides, centralized anything just doesn't fly in Wisconsin. Wisconsin is an independent state."

The state of Vermont legislated statewide land-use planning goals and is using GIS technology to coordinate the planning effort. The 1988 Growth Management Act (Act 200) appropriates nearly $5 million and requires that a statewide GIS be established in support of the effort. State and local data bases will be developed to meet standards for accuracy and content established by the governor. A network of twelve regional planning agencies will coordinate local and state information into regional land-use plans promoting "appropriate development." GIS will also serve more common applications as a decision-support tool for several state agencies. Vermont is one of several states leading a new trend in statewide land-use planning programs that will manage such information-intensive efforts with a GIS.

Local Government

GIS technology helps make sense of the municipality—an interdependent network of people, buildings, systems, and services linked by geography. It offers the potential for a quantum leap in the quality, accuracy, accessibility, and timeliness of information used by municipal employees from the mayor to the desk clerk. When local government has better, more accessible information, the private sector and general public benefit as well.

Local government has numerous, diverse uses for GIS, including tax assessment and collection, maintenance of property records, dispatch of emergency service vehicles, planning, zoning, permitting, code enforcement, traffic engineer-

GIS technology helps make sense of the municipality—an interdependent network of people, buildings, systems, and services linked by geography. It offers a quantum leap in the quality, accuracy, accessibility, and timeliness of information used by employees from the mayor to the desk clerk.

ing, operation of public works and utilities, road maintenance, school-bus routing and district design, voter registration and election district apportionment, delivery of public health programs, economic development, disaster relief planning and response, map production, and engineering design and drafting. In fact, GIS offers benefits to virtually all functions of local government (Figs. C-24–28).

Local governments already are one of the largest GIS user groups, and an increasing number of counties, cities, municipalities, and boroughs will use GIS in the future as the technology becomes less costly. Though hindered by database development costs, the relatively new micro-based systems have made the technology accessible even to very small jurisdictions. Although GIS has been used in some local governments since the early 1970s, its use

spread dramatically in the mid and late 1980s.

Local government's geographic information systems are used not only by agency managers and employees, but also by elected officials, attorneys, developers, financial institutions, realtors, and the general public. The systems are sometimes operated by a consortium of functional departments and agencies, by an individual department as a service to others, or by an individual department for its own purpose. Coordination of all the potential users of a system rarely occurs, although multiagency systems are increasing (Fig. 3-3).

One of the pioneer municipal users of GIS, the city of Milwaukee, Wisconsin, has a mature system with many applications in service delivery, management, and policy analysis functions. The primary uses in-

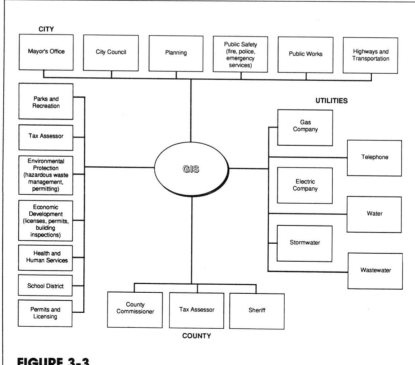

FIGURE 3-3

Cooperative programs between cities, counties, and public and private utilities are becoming increasingly popular as a technique for sharing the cost and effect of developing and maintaining databases, and at times, GIS processors.

clude city base map maintenance, land planning, and building permit processing.

Milwaukee's GIS helps policy makers address current issues. For example, a health policy committee noticed an increase in lead poisoning among children and suggested the possible need to remove lead-based paint in residences. To help officials see the extent of the problem, the GIS was used to identify and map the concentrations of houses built before 1950 where children under seven years old resided.

Milwaukee's system is also used routinely to redraw special service areas and districts. After the 1980 census, the GIS helped the election commission redraw aldermanic district and voting ward boundaries more than twice as fast and for less than half the cost of the manual method used in 1970. Figure C-29 illustrates a third application of the Milwaukee system, describing the distribution of liquor licenses.

Other examples of local systems in use include Virginia Beach, Virginia, a community of 350,000 that uses a system for zoning, transportation routing, and utilities planning, among other applications. The city tax assessor virtually paid for the system by using it to find land that had not been on the tax rolls. (*Louisville Courier Journal* 1986).

Wyandotte County (Kansas) Surveyor Murray Rhodes said the cost of the system has been amortized many times over. "The effort has been both technically successful and cost effective. In 1984, for example, the County collected $500,000 in delinquent taxes that would not have been collected under the manual system" (*IBM Update* 1987).

The Police Department of the city of Tacoma, Washington, uses geographic information technology to select, display, and analyze criminal activity. The Police Department's crime analysis and research and development sections use microcomputers to gain access to crime, demographic, street, and other data sets maintained by the Law Enforcement Support Agency and three other departments. More recently, the city has initiated a design for a multiagency GIS (Anderson et al., 1987).

The Los Angeles City Planning Department developed a prototypical system for emergency preparedness planning and operations. While the Los Angeles system emphasizes earthquake preparedness, it can be expanded to cover other potential hazards. The project was funded by the Federal Emergency Management Agency and the State Office of Emergency Services through the Southern California Earthquake Preparedness Project. The project developed several applications, including dam flood risk assessment and response planning, allocation of shelter facilities, provision of services to handicapped persons, and hazardous building mitigation (Johnson 1987).

The United Way of Minneapolis is using GIS to assess needs for its services, such as determining areas lacking services for battered children, and for demographic analysis, such as identifying the locations of low-income single parents (Regional Mapping Consortium 1988).

The Yakima Indian Nation is building a GIS that will help manage its 1.3-million-acre reservation in central Washington. The data base will contain information about farm, forest, and range lands; wildlife; water resources; natural foods, herbs, and medicines; and agricultural, timber, and grazing leases. Carroll Palmer, Yakima Nation's director of natural resources, expects it to take about five years to build a reservation-wide data base covering their major resources. The Yakima Nation's Department of Natural Resources decided to acquire its own system when a similar but more limited effort by the Bureau of Indian Affairs (BIA) failed to produce useful data after four years.

Theresa Valentine, BIA's GIS coordinator for the Yakima Nation, said that forestry, soils, land status, and planimetric data have been digitized for about three-fourths of the reservation, but the data have not been available due to changes in hardware and software by BIA at the national level. A migration from one software system to another necessitated a cleanup and verification of the data. The BIA's data base will be available for the Yakima Nation to use and develop further for land planning and specific management needs, Valentine said. The system will give tribal decision makers useful information in graphic form that can be comprehended and used more easily than technical statistics.

Multiagency Systems. In a number of areas around the country, local land bases and systems are being developed by consortia of agencies, and sometimes private utilities, with overlapping jurisdictions.

During 1984 and 1985, ambitious projects were initiated in Knoxville–Knox County, Tennessee, and soon were followed by similar efforts in Nashville–Davidson County, Tennessee; Sacramento, California; Louisville–Jefferson County, Kentucky; Little Rock, Arkansas; Indianapolis, Indiana; and suburban Washington, D.C., among others. Though each differed in composition, the common element of the group was their ability to mesh a diversity of responsibilities and interests to construct areawide geographic databases using compatible—if not identical—computer systems. Many of these systems are now moving into operational phases with cities, counties, and public and private utilities sharing both the costs and fruits of a common design. The concept of consortium projects is now an accepted management approach to GIS development efforts.

One of the earliest efforts at a citywide system was initiated by the city of Long Beach, California, which began planning such a GIS in the early 1980s. Today the system is used for a broad range of applications.

The New York City Planning Department operates a GIS that links eighteen agencies to a common land base and produces an array of thematic maps used widely by agencies, businesses, and the public for planning and management. A portion of the system, Geosupport, is based on the Census Bureau's DIME file. The Bureau of Electrical Control, which maintains the city's 320,000 street and park lights, developed a management system using Geosupport that dramatically improved service and reduced complaints (Urweider 1986). The Board of Education now uses the system to determine the eligibility of public and private schoolchildren for subsidized busing. Other uses include water, sewer, and road management, and analysis of air and noise pollution and crime incidence (Juhl and Wallick 1986).

Microcomputer Systems and Intelligent Workstations. Though not restricted to local governments, the advent of microcomputers and intelligent workstations is speeding the use of geographic information management technology in local government agencies beyond that in any other sector.

As reflected in these examples, most GIS activity to date has been in metropolitan rather than nonmetropolitan areas. The American Farmland Trust, a national nonprofit organization committed to protecting agricultural lands, found in a 1985 survey that only 25 of about 63,000 nonmetropolitan local governments in the United States had established their own geographic information systems (Maizel and White 1986).

West Columbia, a small city of 20,000 people in central South Carolina, exemplifies a low-cost, microcomputer system being used by the local government for municipal utility management.

West Columbia's population has doubled since 1980, making it one of the fastest-growing communities in the state. The municipal water service provides 25,000 connections in a 50-square-mile area; the sewer service area covers 16,000 connections in a 30-square-mile area. During one six-month period, high growth forced the city to turn down 800 applications for new water taps.

In 1986, West Columbia commissioned a consulting firm to develop "a computerized, interactive facilities system mapping plan." During the pilot project, the city staff verified the preliminary water and sewer maps with field inspections, and discovered valves and water lines that the maintenance staff was unaware of, as well as several potential problems in the sewer system. As the base map, West Columbia's system uses 1:200 planimetric maps available from the regional planning council. The system's land and data bases cover the entire city service area. The system was developed to run on IBM® AT or compatible microcomputers so that it could use existing in-house computer resources. In one instance, the system was used to detect and analyze a water line break. It tackled a problem that would normally take six hours to solve and solved in only two. Since becoming operational, West Columbia's system won the South Carolina Municipal Association's Municipal Achievement Award in Recognition of Superior Performance and Innovative Programs Leading to Improved Quality of Life.

League City, Texas, is in a rice-growing rural area located twenty miles south of Houston. Its population numbers some 25,000 residents. League City is building a geographic information system using PC-based software developed specifically for small municipalities. Their Engineering and Planning Department began the system, which also will be used by other local departments, for water, sanitary, and street maintenance management. The hardware and software cost about $40,000. The 57-square-mile land base, 70 percent complete in mid-1988, is being digitized by department employees on one digitizing station that runs sixteen to twenty hours a day.

Over the last months of 1987, the department did a pilot study of one subdivision that included data to support every application of potential interest (e.g., water, sanitary, storm, street lights, spot elevations, driveways, sidewalks). The pilot helped the department set priorities, fore-

cast expected benefits, and reduce the project to a manageable size.

The South Florida Water Management District has developed a microcomputer-based network that allows five Regional Planning Councils and 137 local governments to access data in the district's extensive cartographic data base, which it began developing in 1978 on a CADD system. The district's responsibilities include granting permits for and regulating existing and new uses of surface and ground water supplies and the development of a regional water-use plan. The system's automated data is used for hydrologic modeling, data analysis, weather analysis, accounting and civil engineering CAD applications, mapping, and planning. In 1983, the district began installing microcomputers and PC-based software throughout the organization; a communications link between the CADD system and the PC system provides broad access to the agency's automated data files (Brown and Reel 1987).

Nearby, the East Central Florida Regional Planning Council is using a microcomputer-based GIS to help counties in the region meet the requirements of a state-mandated growth management law. Florida's 1985 Local Government Comprehensive Planning and Land Development Regulation Act increased the scope and detail of local comprehensive plans and supporting maps. The Council's system also uses software that runs on a PC microcomputer (Gilbrook and Sheldon 1987).

UTILITY APPLICATIONS

The electric, gas, water, wastewater, and telephone utilities that have embraced geographic information technology cite productivity gains and improved customer service as benefits of AM/FM systems.

In the utility world, the term *automated mapping/facilities management* describes a geographic information system that integrates nongraphic facilities management

TABLE 3-1 Example Organizations and Their Systems

Organizations	Software	Hardware
NOAA	CAMEO II	Macintosh with HyperCard
Minnesota Land Management Information Center	ESRI	Prime
Alaska Department of Natural Resources	Synercom ESRI	Digital
Illinois Department of Energy and Natural Resources	ESRI	Prime
City of Milwaukee, Wisconsin	Intergraph	Intergraph
Virginia Beach, Virginia	Intergraph	Intergraph
Police Department, City of Tacoma	ESRI	Prime
Planning Department, City of Los Angeles, California	ESRI	Prime
Department of Public Works, Anchorage, Alaska	Synercom, ESRI	Digital
Department of Planning, Anchorage, Alaska	ESRI	Prime
Yakima Indian Nation	ESRI	IBM-PC
Bureau of Indian Affairs, Yakima Nation	MOSS, ARC/INFO	
North Slope Borough, Alaska	ARC/INFO	Prime
City of Long Beach, California	Synercom	Digital
Department of Planning, New York City, New York	Synercom	Digital
West Columbia, South Carolina	AutoCAD, FMS/AC	IBM-PC Compatibles
League City, Texas	TerraCAD	IBM-PC
South Florida Water Management District	ComputerVision AutoCAD	SUN IBM-PC
East Central Florida Regional Planning Council	ERDAS	IBM-PC
Wisconsin Public Services	GFIS, GSC	IBM
Portland General Electric	Intergraph	Intergraph
Texas Utility Services	Intergraph	Intergraph
San Diego Gas and Electric	GFIS, GSC	IBM
United Telephone Company of Florida	Intergraph	Intergraph
Consolidated Gas Transmission of West Virginia	Synercom	Digital
Wisconsin Gas Company	GFIS	IBM
Denver Water Department	Synercom	Digital
Salt River Project	GFIS Intergraph	IBM Intergraph
Nashville MAGIC, Tennessee	Synercom	Digital
Knoxville, Tennessee	Intergraph	Intergraph
TransAtlantic Utilities Corporation	Intergraph	Intergraph
Public Services of Indiana	Synercom	Digital
Potomac Edison Company	AutoCAD	IBM-PC
Houston Lighting and Power	Synercom	Digital
Pacific Bell	Synercom McDonnell Douglas	Digital Digital
Bell Canada	Synercom	Digital
Amtrak	Calcomp	
Scott Paper Company	GeoBased	IBM-PC

information into a database that is tied to facilities maps. In some instances, the AM/FM system also stores the connectivity relationships between facilities, simplifying the creation of utility network models.

Map production and facility inventory are among the most common uses of geographic information management technology by utilities. Utilities also use the technology for engineering, material control and inventory, troubleshooting, planning, construction, order dispatching, maintenance management, emergency response, and financial and accounting functions (Fig. 3-4).

Utilities have been active in the GIS technology's development since the mid-1970s, when at least several dozen major United States utilities began exploring options for automating map production and updating aging maps. Some of the early leaders were Wisconsin Public Service, Public Service of Colorado, San Diego Gas and Electric Company, Brooklyn Union Gas, Detroit Edison, United Telephone Company of Florida, and Washington Gas Light.

Although a number of systems are under development, conversion of land and facility databases has been completed in relatively few cases. Some systems have been under development for more than ten years, but few have yet developed integrated networks of digital information throughout the company.

Mature Utility Systems

Those GIS systems that have reached maturity in utility applications contribute substantial benefits to their organizations (Figs. C-30–32).

Gas and Electric Utilities. Wisconsin Public Service (WPS), a gas and electric utility with more than 12,500 square miles of service territory, has developed a large number of applications packages for its mainframe system that a subsidiary development company, Geographic Systems Corporation (GSC), markets to other organizations. The system was initiated in

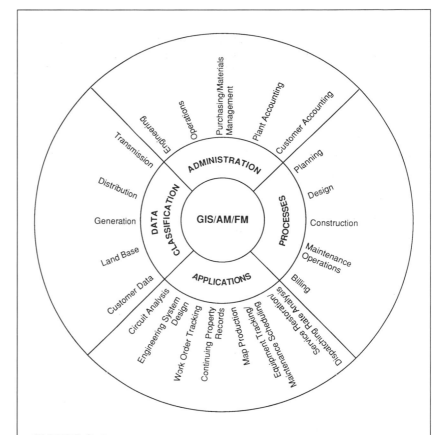

FIGURE 3-4

Automated mapping/facilities management and GIS capabilities can integrate information assets for key planning, operational, and management functions.

1981, and conversion of the land base and gas facilities was completed in the spring of 1988. WPS expects to complete conversion of the distribution network by 1992 and facilities attribute data by 1995.

Jack Bernard, vice-president and general manager of GSC, said a postimplementation audit found that mapping alone has not and would not pay for the system. Major benefits were found, however, in information facilities management areas where they were never expected.

The WPS system cost $5.4 million; its implementation has produced net savings annually of about $2 million. As soon as the system was operational, WPS found it reached a break-even point long before the data conversion was complete.

Utilities have been active in the GIS technology's development since the mid-1970s, when at least several dozen major United States utilities began exploring options for automating map production and updating aging maps.

BOX 3-2   North Slope Borough Maps Subsistence Use

The Alaska North Slope Borough, geographically the largest local government in the country, designed a geographic information system to respond to pressure placed on the subsistence lifestyle of Inupiat Eskimos by oil development. The borough purchased a GIS in 1983 and used it to prepare comprehensive land-use and coastal management plans. The data base contains planimetric oil and gas facility, fish and wildlife habitat and ranges, subsistence use, and other data.

Subsistence data was most important to the local government in this situation, because of the competing pressures between development and preservation advocates. The Inupiats found that the GIS helped them express their concerns and geographic interests in concrete terms. Using the GIS, the Inupiats captured information about live nature and extent of subsistence activity. They were then able to present their case using the same technology and media used by the oil industry and other developers.

Portland General Electric (PGE) Company, which serves about 500,000 customers in a 3,500-square-mile area of northwest Oregon, began instituting an AM/FM system more than ten years ago when faced with the need to redraw its fifty-year-old facility maps. The old maps were at a scale too small to show underground facilities, and a large backlog of underground work orders had accumulated. One consequence was that power outages were prolonged while crews waited for job sketches showing the location of underground utilities, which then had to be pieced together and related to points on a map. With each passing year, the maps were more inadequate and out-of-date.

After a feasibility study and pilot project, PGE purchased an AM/FM system in 1979. Data conversion was completed in 1985. Today PGE operates an enhanced system that integrates the land-based information with databases containing nongraphic information about its facilities.

According to Edward Sipp, a consultant to electric utilities, the system has helped PGE increase its productivity and efficiency. Map maintenance costs are much lower than with the old manual system, and maps are more accurate and legible. Data is available to all departments throughout the company, and data integrity has improved due to the elimination of redundant recordkeeping. With accurate maps, crews are more efficient in the field. In addition, applications are now possible that could not easily be performed manually, such as outage and voltage analysis. Applications implemented at PGE include these:

- Facilities maps
- Street light maps and billing
- Feeder maps, loading, and expansion
- Subdivision maps
- Wire mileage reporting
- Circuit tracing
- Distribution equipment inventory
- Transformer load monitor
- Pole rental billing
- Capacitor and switch location and status
- Outage tracing and support
- Voltage profile
- Load growth impact
- Fault studies
- Phase balancing
- Optimal voltage and power factor correction
- Marketing analysis (using Census Bureau DIME files)
- Ad-hoc maps (e.g., maps of poles needing inspection)
- Ad-hoc reporting (e.g., list of devices by size and age)

Texas Utilities Services, Inc. (TUS) of Dallas transformed its operations with an

AM/FM system that supports the daily activities of estimating, accounting, mapping, and tracking work authorizations through multiple departments. Custom software was developed by TUS and implemented at three sites of Texas Electric Service. Automation began in 1982, and the conversion of 18,000 square miles of land base and 12,000 distribution facility maps was completed in 1987. The major thrust of the system was to develop an estimating system for the engineers.

San Diego Gas and Electric (SDG&E) has been a leader in applying automated mapping technology in the utility industry. In 1977, SDG&E began to develop an integrated system that would 1) reduce the cost of maintaining facility records; 2) store the utility's standardized records in a corporate database; 3) improve the availability of facilities data; and 4) generate up-to-date maps rapidly and accurately.

Following a pilot program, an economic analysis determined that the system could save SDG&E $1 million a year in labor costs alone, and that system payback would occur about three years after conversion of the utility's land and electric data (*AM/FM Scribe* 1987, February). The system was installed in 1983, and by 1988 it was operational in three of four districts. Conversion of the land and facility data was completed in May 1989. (*1990 Referral Directory AM/FM International*). The system provides information for numerous applications, including work order processing support, circuit analysis, and tax reporting. Like Wisconsin Public Service, SDG&E founded a subsidiary, Integrated Information Systems (IIS), to market its extensive mainframe application programs.

Telephone. United Telephone Company of Florida was one of the first telephone companies in the country to complete automation of its facility records and land base. The project was initiated in 1982, and data conversion of 65,000 facility maps was completed in 1986. The major goal was to provide a more responsive and accurate engineering and investment records system by consolidating and providing for single-entry updating of outside plant engineering, accounting, and investment records. The system is used throughout the production process, beginning with work order design. The system allows generation of special reports and provides audit tracking of the investment records for purposes of rate establishment (*AM/FM Scribe* 1987, March).

Developing Utility Systems

AM/FM systems are under development or study in many utilities. As of 1988, about 100 United States electric utilities were actively pursuing AM/FM technology, according to Dr. John Bossler, director of Ohio State University's Center for Mapping.

Gas, Electric, and Water Utilities. Consolidated Gas Transmission Corporation of West Virginia, whose service area also covers parts of New York, Pennsylvania, and Ohio, began implementing an AM/FM system in 1979 and completed data conversion in early 1986. The system integrates information from three different map sets at different scales, one of which dates back to the 1890s (*Design Graphics World* 1986, January). The system's major uses include map production, engineering, and management applications.

Wisconsin Gas Company, which serves 430,000 customers in a 900-square-mile area, began data conversion for an AM/FM data base in 1988 after eleven years of study. The system will use a third-party software package for load and network analysis, among other uses.

Among other gas and electric utilities with AM/FM systems under development are Carolina Power and Light Company of Raleigh, North Carolina; Commonwealth Gas Company of Southborough, Massachusetts; Duke Power Company of Charlotte, North Carolina; Michigan Consolidated Gas of Detroit; and Northeast Utilities of Hartford, Connecticut.

The Denver Water Department, which serves more than 1 million customers in a 305-square-mile area, purchased a GIS in

BOX 3-3 Anchorage Builds Citywide GIS Network

Anchorage's Departments of Public Works and Economic Development and Planning are employing geographic information technology in some unusual ways that provide productivity benefits to municipal agencies and give the public new access to geographic information.

Initiated in 1983 using Synercom software, the Public Works System includes 900 quarter-section base maps at a scale of 1" = 100' covering 200 square miles. Custom maps and digital data disks are produced on request and sold for a fee as a public counter service in the Department of Public Works. Portions of the data base can be obtained in digital or hard-copy form. The Public Works system has won several national awards, including the URISA Exemplary Systems in Government award.

The Economic Development and Planning Departments system, which runs ARC/INFO software is used for land-use analysis and comprehensive planning. The data base was digitized primarily from 1" = 500' maps and includes parcel lines and descriptive information about each parcel such as land use, ownership, number of residential units, and tax parcel identification numbers. Other information in the data base includes housing stock inventory, demographic and economic data, zoning boundaries, census tracts, flood plains, avalanche zones, wetlands, and other environmental data.

The Planning Department sells some data to the public, mostly "ready-made" base maps of varying detail and scale. Due to staff shortages, customized mapping is limited to municipal-related planning and analysis, but most private-sector map requests are filled from an archive of more than 800 maps.

The Anchorage School District is using the Public Works base maps and ARC/INFO software, for school boundary and enrollment analysis, bus routing, and to evaluate school size in relation to enrollment to determine whether to build or close schools.

In 1989 Anchorage began merging its system into a municipal-wide GIS network with ARC/INFO the primary software running on DEC VAX platforms. The Public Works system is undergoing major conversion from Synercom to ARC/INFO that is expected to be complete by the fall of 1991. Base maps and parcel maps still will not be compatible, but their eventual merger is planned.

The Public Works Department has promoted data exchange among public agencies and utilities within the department's service area for a number of years and it has provided base maps and data base access to two city-owned utilities—Municipal Light and Power, and the Anchorage Telephone Utility—as well as to the Economic Development and Planning Department and School District. In addition, the state-owned Alaska Railroad purchases data from Public Works.

The Anchorage systems have been used for many successful applications. Below are examples:

- The Anchorage Assembly used GIS to analyze the impact of various versions of an adult entertainment ordinance adopted in November 1988 (Fig. C-35).
- The Economic Development and Planning Department developed a GIS model to identify residential, school, and church parcels expected to be affected by undesirable levels of traffic noise. The model is also used to review building permits and other development decisions.
- The Public Works Department created a comprehensive atlas to help locate and clear snow-covered storm drains.

- Traffic hazards at school crossings are analyzed using the GIS.
- Public Works used its GIS to create snowplow route maps and street-sweeping subdistricts by calculating mileage and determining the most efficient uses of equipment.
- Mapping capital budget proposals and project information for the Municipal Assembly is a routine use of GIS by the Public Works Department.
- The GIS was used to analyze site locations for the main Olympic stadium proposed in the city's 1994 Olympics bid.

1985. By 1988, it had converted about 40 percent of the land base and facility maps and had implemented several applications, supporting reservoir volume calculations, network analysis schematics, and plan and profile drawings.

The Salt River Project, which provides electric and water services to about 500,000 customers in central Arizona, began development of an AM/FM system in 1980. Conversion of electric facility and land data is to be completed by 1994.

Consortia. Because shared systems are more economical, some utilities are pursuing joint mapping and records projects with other utilities and local governments. In many instances, the utilities have found significant cost savings by having access to reliable digital base maps that are then routinely updated by the public-sector organizations.

As an example, Nashville's Metropolitan Area Geographic Information System (MAGIC) is being developed by five organizations within the metropolitan government of Nashville and Davidson County. Each member of the consortium has been allocated a percentage of the total cost. Nashville Electric Service contributes 35 percent and Nashville Gas Corporation provides an additional 10 percent. Departments of Nashville-Davidson County provide the balance of the funding as follows: Water and Waste Water, 20%; Planning, 17.5%; and Public Works, 17.5%.

While full-scale conversion of planimetric, topographic, and parcel maps is underway, MAGIC is developing a pilot project covering 6 square miles in which thirty-four layers of information for the five organizations will be digitized. The GIS will "mature," with most of the basic information having been entered by 1992, at a cost of about $10 million. Nashville Electric Service (NES) will use MAGIC as a facility information tool for load management, identifying outages and maintaining an inventory of fixed assets, among other applications. Tom Green, former assistant general manager of NES, recognized the system as an important productivity tool— one that will help the utility process orders more efficiently and provide better service to the customers. With the manual system, the same data were handled and mapped five times. With the automated system, the information will pass from mapping, to design, to operations, directly on to construction, then to accounting and closing.

Across the state of Tennessee, the city of Knoxville, Knox County, and the Knoxville Utilities Board, which manages electric, water, gas and wastewater utilities, began a joint mapping project in 1984 that became functional in 1988. With the joint system, Knoxville Utilities has access to an accurate base map in a form that is maintained for the most part transactionally by city and county agencies.

Incremental Approach. Many utilities are pursuing less than full AM/FM systems, or they are pursuing technological upgrades incrementally.

TransAlta Utilities Corporation of Calgary, Alberta, Canada, upgraded its system as technology improved, moving from an automated drafting system that it installed in 1975 to an automated mapping system that began data conversion in 1979. TransAlta initiated an AM/FM development project in 1985 (Reid 1986). By 1988, it had converted all of its rural facilities maps.

Public Service Indiana (PSI), which started using its system in 1978, is reaping the benefits of increased efficiency and new analytical capabilities, said Patricia Mitchell, who manages a field support group for the electric utility. The PSI mapping system was instituted to make the company's information more accessible and accurate, and to support mapping, planning, and engineering functions. PSI has recently upgraded its hardware and software and plans to integrate its automated plant accounting and facility records with the mapping system.

Wisconsin Electric Power Company (WEP) and Wisconsin Natural Gas are building a joint automated mapping system. Data conversion is to be completed by 1990, sixteen years after WEP began the initial investigation of an automated mapping system. While map conversion continues, the companies are working to tie existing mainframe databases to the automated mapping system. WEP is starting to see benefits from the system—improved access to information, quicker response to user requests, and a reduction in personnel required to keep facility maps up to date—said John Zaganczyk, WEP map records supervisor.

The Potomac Edison Company, an electric utility that serves 300,000 customers in Maryland, Virginia, and West Virginia, is using personal computers to map land base and facilities and perform data extraction. Richard A. Cushwa, supervisor of engineering services, notes that its FM system is not sophisticated, but it has been workable, useful, and inexpensive.

In the early 1980s, the Allegheny Power System, of which Potomac Edison is an operating company, formed a facilities working group to study the need for AM/FM. Detailed studies determined that AM/FM was not feasible at all three of the member companies, and so AM/FM development was put on hold. Meanwhile, manual facilities mapping proceeded at Potomac Edison.

In 1986, Potomac Edison purchased its first PC-based CAD workstation for drafting functions. Later that year, the utility was convinced of its potential for electric facilities mapping. Late in 1986, the detailed facilities mapping of 80 percent of Potomac Edison's service territory was completed.

Houston Lighting and Power Company (HL&P) has completed data conversion of substations, transmission, and overhead distribution lines in its automated mapping system, and is working to complete conversion of the underground distribution system. HL&P already has realized several benefits from the system, and is considering expansion to accommodate additional applications, said John T. Chambers, a supervisor in the automated mapping section (Chambers 1987).

Telephone Companies. Some of the largest AM/FM projects are being developed by the telephone industry, where use of geographic information technology is fairly widespread. Most telephone companies have developed independent systems for market analysis as well as for mapping and facilities management. Deregulation of the telephone industry and breakup of the Bell system has spurred the use of geographic information technology for "bypass" analysis—determining where a company could be vulnerable to attempts by competitors offering reduced cost of services, to attract customers within a concentrated area.

Early leaders in implementing AM/FM within the telephone industry were Southern Bell, Southwestern Bell, British Columbia Tel, and United Telephone of Florida. A regional Bell company is likely to spend in the range of $150 million for a system, including data conversion.

BellSouth, one of seven regional companies formed from the divestiture of

Some of the largest AM/FM projects are being developed by the telephone industry, where use of geographic information technology is fairly widespread.

AT&T, provides telephone service in parts of nine southeastern states through its operating companies, Southern Bell and South Central Bell. BellSouth is developing geographic information systems to map and manage outside plant facilities throughout its service areas. Another system is being developed for market and rate setting research and analysis. As of 1990, 100 percent of the land base and about 75 percent of the facilities maps had been automated. (*1990 Referral Directory AM/FM International*).

Pacific Bell began investigation of AM/FM in 1984 and developed an interdepartmental system linking automated data on inside-plant, outside-plant, and real estate functions.

"We saw the technology as allowing Pacific Bell engineers to engineer the capital assets of the company, rather than just the outside-plant facilities," said John Porter, a consultant and former outside-plant project manager who helped design and implement a pilot AM/FM system for Pacific Bell. Its strategy was to get data into the hands of the engineers who needed to manage the company's data assets, whether in graphic or nongraphic form. Only 20 percent of this asset management solution is graphic; 80 percent is marshalling nongraphic data, most of it in a mainframe environment.

The primary justification for the system was that it can significantly reduce the cost of engineering, primarily in labor savings. Second, the system improves the speed and quality of telephone network engineering; and third, it supports an image of the corporation as a technological leader in the telecommunications environment.

"This technology enables a telephone company to reduce its records-posting and drafting forces—my estimation is by 90 percent," Porter said. Manual engineering for a central office takes about 18 months, but "our estimate was that in a mechanized environment, using the tools we were building, it could be designed in less than a month. Well, that's an enormous improvement of both competitive advan-

tage and the ability to meet an immediate customer need. . . . Being able to engineer faster, that's a quality item. To the extent that the data has fewer errors . . . fewer mistakes on the engineering print, fewer construction work orders will be done incorrectly," Porter said.

By 1990, the Pacific Bell pilot project had converted about four percent of the facilities maps and land base of 50,000 square miles. Full implementation of the system, if pursued, could take twelve years.

Bell Canada completed a pilot project in 1988 that converted about 2 percent of the firm's service area, which covers most of the provinces of Quebec and Ontario, as a first step toward creation of the corporate Outside Plant Data Base. After an evaluation of costs and benefits, the company will decide whether to proceed with full-scale conversion.

Among other telecommunications companies implementing geographic information systems are Contel Service Corporation, which serves 2.3 million customers across the country; GTE Data Services, Inc., which serves the GTE Telephone Companies; and Southern New England Telephone Company, which serves 1.5 million customers in Connecticut.

Railroads. Amtrak began implementing a comprehensive GIS in 1983. Amtrak has built a digital data base containing accurate maps and more than fifty fields of information on the corporation's land holdings. The system enabled consolidation of four sets of original manual maps covering the same areas that contained information about property conveyance and leasing. Editing of the digital map and attribute data base, the last step in the data conversion process, is expected to be completed in 1991.

The Amtrak system is used primarily by the real estate management and engineering groups, and is evolving toward a full facilities management system. Amtrak's Real Estate Department uses the system to manage leasing of its land and air rights. The Engineering Department uses

---

BOX 3-4   City of Long Beach Maps Plan of Action

Bordering Los Angeles on the north, Long Beach is named for its fifteen miles of beaches and waterways. Covering almost fifty square miles and home to more than 370,000 people, the city is probably best known for its commercial port (the second largest in the nation) and the largest municipally operated marina in the world [and as the home of the ocean liner *Queen Mary*, the flying boat *Spruce Goose*, and a Grand Prix automobile race]. Less well-known, perhaps, is the fact that the city has substantial oil properties with a maze of underground pipelines—a mapping nightmare.

Before 1983, most of the spatial data management was carried out by the Engineering Bureau of the city's Public Works Department. With only a five-person staff, the bureau was responsible for maintaining more than 4,000 maps relating to storm drains, sanitary sewers, oil pipelines, street lighting, subdivision records, house numbering, zoning, and drainage, among others.

A serious pipeline rupture and subsequent explosion required the city to quickly identify the owner and location of the pipeline [to turn it off]. The difficulties Long Beach experienced in identifying the ruptured pipeline and its owner contributed to a mandate for new safety requirements that called for an entirely new map inventory. Long Beach took the then-extraordinary step of pursuing a digital mapping and database management system.

The city conducted a planning and design study during which each city department identified and ranked in order of priority the data types and sources it needed to do its job. After a vigorous vendor selection process, the city selected a GIS and the full system was installed in July 1983. A comprehensive land base was constructed using photogrammetric techniques. Digital maps of the planimetric features, including roadway lines, driveways, sidewalks, railroads, building outlines, contours, spot heights, and trees were included. In response to the earlier pipeline rupture and explosion, a detailed database was developed. It contained twenty-five attributes for each pipeline section, including ownership, emergency contacts, material, thickness, outside diameter, flow direction, surrounding soil condition, and depth. At present, seventy-eight layers exist in the primary database for the city, and the list continues to grow. In 1990 the city undertook a comprehensive evaluation of its needs and current technology. As a result, it initiated programs to expand the systems' capabilities.

---

it for engineering design of facilities and electrical systems, mechanical design of rolling stock, track layout, and master planning. In addition to adding facilities information, the department plans to link the GIS data base to an optical disk system storing original deeds.

"Historically, any time the railroads had a problem, they threw people at it. That's cost-prohibitive now," said Daniel Brunner, Amtrak's former director of corporate real estate. "This system makes it possible to do more with less."

Union Pacific Railroad Company, a freight line serving the Midwest and Western states, is converting eighty-year-old maps to a GIS data base containing its tracks, bridges, and other facilities. About two years into data conversion, the system was also being used in construction and field work, as well as signal, mechanical, and electrical applications. In 1990 a large number of railroads—ranging from short lines to large carriers such as Kansas City Southern Railroad and CSX—initiated requirements analysis for AM/FM Systems.

PRIVATE INDUSTRY APPLICATIONS

Geographic information technology has been under development in private companies, particularly in the oil industry, for several decades, although this activity has not been as widely recognized as in the public sector. The use of geographic information technology is growing in companies involved in marketing, retail, and wholesale sales, resource development, delivery services, real estate, site selection, development, construction, and land title assurance. However, with the exception of the resource development industries, private-sector use of GIS appears to be recent and still fairly limited.

Oil and Gas

Geographic information technology helps the oil and gas industry organize data and develop new prospects for exploration and production in a more efficient manner. GIS is used to support geological, geophysical, and drafting data management, as well as transportation and production facilities management, pipeline and platform design, and land work. Most major and small oil companies have at least one automated mapping or geographic information system.

"Everything an exploration geologist or geophysicist does relates to a position on the globe . . . The whole act of exploration is essentially a mapping operation," said Phillip B. Lelyveld, Director of Exploration Computing Services at ARCO International Oil and Gas Company. "We are in the process of developing a new generation of geographic-based database and project file systems to store and retrieve

data for exploration software tools. This effort is not unique to ARCO. Exxon, Shell, Mobil, BHP, Amoco, and many others are also in the midst of developing this type of system."

A GIS database used by an oil company contains planimetric and geopolitical data, lease boundaries, and seismic line and well locations. It also manages positionally linked nongraphic information such as text of legal descriptions, land ownership information, and well history. Noninterpretive GIS data are used to track lease obligations so that companies can adequately prepare for upcoming mineral-rights lease or concession offerings. GIS is also used to prepare management presentations showing the relationships of facilities, hydrocarbon prospects, and development scenarios. Some companies are moving toward a fully integrated GIS database system to serve the multiple needs of exploration, production, legal, land management departments, and senior management (Figs. C-33, C-34).

"While geographic-based computer systems are employed by most oil companies, few are topologically rigorous or capable of the sophisticated analysis commonly performed with GIS," said Mark Settle, ARCO manager for integrated exploration research. "Many oil industry systems use batch processing, rather than interactive workstations, for production of maps."

"It's been my perception from going to professional meetings," said Phillip Lelyveld, "that the oil industry views a GIS as a small system, while the industry's need is for large systems. From what I've [learned about GIS], that probably is a misperception. New off-the-shelf database software products appear to be able to overcome the problems facing the industry. The quality of data dealt with by large municipalities is probably of the same magnitude as oil industry concerns. When GIS firms and oil companies reexamine each other, they will probably conclude that their products and goals are more similar than they currently believe."

The use of geographic information technology is growing in companies involved in marketing, retail, and wholesale sales, resource development, delivery services, real estate, site selection, development, construction, and land title assurance.

BOX 3-5   The Oil Industry Develops New GIS Applications

"The oil industry as a whole began efforts to develop geographic information systems as early as . . . the mid sixties. Concepts were drawn up and the first attempts were developed on mainframes. They were of questionable success because they tried to solve all things for all people, and therefore adequately addressed few, if any, specific needs," said Philip B. Lelyveld, Director of Exploration Consulting Services at ARCO International Oil and Gas Company. This moderate but low-key in-house development effort continued through the early seventies.

The rapid oil price rises of the seventies triggered a frenzy of geological and geophysical data gathering, and a corresponding flood of data poured into exploration offices. Many major oil companies reevaluated, redefined, and reinvigorated their prototype GIS efforts in order to handle the data influx. They invested manpower and research/development time directed toward integrated systems development—GIS plus computer-based exploration data-interpretation tools.

The late 1970s and early 1980s were years of rapid evolution and improvement in the fields of computer hardware and databases. The microcomputer, new types of peripherals, and new ways of working with computers all came into being. By the early to mid-1980s, oil company management realized that if they did not have a centralized database, they needed one; and if they did have one established, it was awkward and inadequate. Clearly, more efficient packages for posting, contouring, and manipulating data on a map were needed.

A new generation of systems is now in development or in use within the oil industry. These systems have the concept of a GIS at their core, but because they include data manipulation, interpretation, and other capabilities, they are called Integrated Exploration Workstations/Systems or some variant of this name.

Today, the development effort has to some degree moved away from strictly in-house programming and toward a blend of in-house efforts and modification of off-the-shelf systems. However, the basic system envisioned in the early sixties is now finally beginning to appear in many forms throughout the industry. The American Association of Petroleum Geologists (AAPG) is working to standardize data structures and data exchange formats.

Forestry

Private timber companies use geographic information technology to assist foresters with mapping, planning, and managing areas of timber ownership, harvest history, timber type, soils, silvicultural treatment, and other features.

In 1981, Scott Paper Company began looking for a more efficient way to update its manual maps and records. A rapidly expanding silvicultural program and epidemic of spruce budworm in Maine provided the initial impetus to investigate GIS. In 1985, after a trial project, Scott purchased two geographic information systems running GeoBased Systems software for its northeast timberlands and southern timberlands divisions. The systems enable analysis previously considered impossible because of the very large amounts of data involved (Manogue and Hatch 1987).

Transportation

GIS is emerging as a tool in the transportation industry, as well as in industries that involve logistics or use transportation services. Applications include dispatching and vehicle fleet management.

According to Donald F. Cooke, founder and president of Geographic Data Technology, Inc., logistics are a $0.5-trillion portion of our gross national product. Tens of thousands of companies operate fleets of twenty-five or more delivery vehicles. Studies of route optimizing and scheduling systems show that they save between 5 and 15 percent of fleet operating costs. Although these systems operate in many different ways, all require that delivery addresses be geocoded or address-matched before the routing algorithm can be run. Some systems also require a complete or partial street network so they can estimate actual driving time along the route.

Dispatching applications allow companies such as home-delivery pizzerias and taxi companies to define fixed districts and send out vehicles efficiently in response to calls. Dynamic dispatching—a system that allows the dispatcher to make judgments based on recent knowledge of where the resources are—is used for dispatching emergency vehicles, plumbers, repairmen, and parcel pickup services. In its simplest form, spatial information system technology helps by displaying the call for service graphically on a map.

Commercial delivery services like Federal Express Corporation and United Parcel Service, Inc. (UPS), are developing and testing geographic information technology for various applications.

Federal Express developed an in-house GIS system to aid dispatch decision making, route planning, estimation of labor requirements, facility planning, service evaluation, and other functions. But the company discontinued use of the system after trial runs at four test sites during 1989–1990.

The goal of the Federal Express prototype, which tracked the location of every vehicle in the test areas, was to decrease the number of miles driven, increase the number of stops, and provide better customer service. However, the system was somewhat cumbersome and required more resources than the company had available to make it operational throughout the distribution network, said Jeff McLeod, a senior dispatch management specialist. Federal Express is now developing and testing a digital mapping system for tracking long-haul, over-the-road routes, he said.

Roadnet Technologies, Inc., sells GIS vehicle routing and scheduling software to food, soft drink, wine and spirits distributors, and other firms that make large deliveries. Roadnet, which was acquired by UPS, in 1986, is developing a nationwide hardware and software network to support vehicle dispatch and routing.

Vehicle navigation systems are emerging that may someday help unsnarl traffic; these will be used in passenger, emergency service, and utility repair vehicles.

Etak, Inc., of Menlo Park, California, developed the Navigator, a vehicle tracking and navigation system that works in conjunction with a digital street map. It has built a data base of street centerlines that covers the United States and a significant portion of the developed world. The company has sold about 3,000 Navigators that are now used in vehicles in the United States and an additional 2,000 are used internationally. Distribution of the Navigator is now through Robert Bosch of Germany.

The Federal Highway Administration is conducting a pilot project with the California Department of Transportation and General Motors (GM) that is testing a linked car–road computer navigation system on fifteen miles of Santa Monica freeway. The joint project, called Pathfinder, will equip twenty-five GM cars with Navigators. The project will provide GM with information about the marketability and cost-effectiveness of electronic navigational systems.

Generally, efforts to apply geographic

information technology to the problem of unsnarling traffic congestion are not moving as quickly in the United States as in Europe, where the European governments, fourteen automakers, and seventy universities and research institutes are co-operating to test traffic navigation systems. A seven-year, $1-billion research project is analyzing the feasibility of building "smart highways," wherein sensors monitor information about delays or hazards and relay it to a computer center, which then transmits the information to computer screens in cars. A similar joint effort is being undertaken by Japan's Ministry of International Trade and Industry and fifty Japanese electronics firms (Beck 1988).

Marketing and Sales

Companies involved in selling products and services use automated census and other demographic data, and some use geographic information technology to map and analyze the data (Figs. C-35–37). Site location studies are a common private-sector use of demographic data. Customers and potential customers are evaluated using demographic and lifestyle data to identify new markets and to assess store performance.

Retail and wholesale product sales and service organizations make use of geographic information technology for telephone referral and territory management, as well as transportation applications.

A simple example of such an application involves referral of customers or prospects to the nearest facility, such as a retail outlet, hotel, repair facility, or medical center. Using a system that calculates proximity through addressing or zip code matching, a telephone operator can select the nearest facility for referral.

Management of sales and service territories is another GIS application. Territories are usually defined as groups of counties or zip codes, but often they are not compact or contiguous geographic zones and may not have balanced sales potential. According to Don Cooke of Geographic

Data Technology, this problem, which calls for a solution very similar to political gerrymandering, is ideally suited to a new class of geographic information system software called the *spatial spreadsheet*.

A spatial spreadsheet consists of two parts: a spreadsheet of statistics about zip codes or counties—the building blocks of territories—and a computerized map of zip codes or counties. The map and statistics are linked in a way that allows the user to design a sales territory by simply pointing to its constituent zip codes. As the system assigns the zip codes to a territory, it automatically computes the spreadsheet statistics. This lets the user know immediately whether sales potential is balanced from territory to territory, while keeping track of the size and shape of each territory to minimize travel time.

A large industry of private data vendors has evolved since the 1970 census—the first census to be released on magnetic tape—to meet the demand for digital census and market data. Demographic digital data—about sex, race, income level, housing, occupation, education, and many other variables—is available by metropolitan statistical areas, census tracts, zip codes, and other measures. Much data is based on the decennial census, and may be updated or enhanced by private vendors. For example, some vendors have segmented markets into "lifestyle" categories and common buying units based on demographic and spending patterns. Demographic digital data also is available from public sources.

A survey of twenty major retailers in the grocery, general merchandise, and speciality goods sectors revealed that Census Bureau data are the primary sources of information, augmented by data from proprietary companies and local planning agencies. Residential building permits are among the most important sources of local data, especially for structures built since the 1980 census. The survey found wide use of integrated spreadsheets, but scant use of automated cartography and of the more complex mathematical models (Dakan 1987).

A large industry of private data vendors has evolved since the 1970 census—the first census to be released on magnetic tape—to meet the demand for digital census and market data.

Some political party organizations and campaigns are using geographic information technology to raise money and to promote their candidates through targeted direct-mail campaigns. By using census data and voter registration files, groups of voters or contributors can be identified and specific messages tailored to them. The technique also is used by political action committees, labor unions, public and private institutions, grass-roots movements, and others (McDonald 1988).

Real Estate

"The long-standing adage about real estate is that the three most important components of value are 'location, location, and location.' The potential relevance of a geographic information system is obvious even to a novice," said Gilbert H. Castle III, Senior Manager for Deloitte & Touche, Real Estate Consulting Group.

Real estate development historically has been the province of a very large number of small organizations. That circumstance is now changing. An ever larger portion of new development is managed by large, sophisticated domestic and international organizations. Organizations like Japanese banks, pension funds, insurance companies, and Wall Street investment houses, are beginning to dominate capital flows in real estate. These players are in general more sophisticated, comprehensive, data-intensive—and, hence, computer literate—than previous generations of developers. Especially given the highly competitive environments in which they operate and the large dollar (or yen, or deutschmark) amounts they are putting at risk, they want to be as sure as possible of the factors that could affect the return on their investments.

Real estate applications of GIS technology promise tremendous potential that to date is largely unexplored. In the private sector, such applications will no doubt focus on national, regional, and local screening of investment opportunities. In the public sector, the applications most likely will focus on asset management opportunities, both for reducing expenditures on real estate needed for certain governmental functions and to maximize revenues from publicly owned lands and facilities.

REFERENCES

The AM/FM Scribe (1985, February). San Diego Gas & Electric DFIS Project: A long-term information investment.

1990 *AM/FM International Referral Directory* (1990), p. 225.

Anderson D. R., Wendt D., Christensen B. (1987). CAMS: An integrated system for geographically viewing criminal activity. *Microcomputers/Information Resources Management/Systems Integration/New Technology: Papers From the 1987 Annual Conference of the Urban and Regional Information Systems Association 3:* 126–136.

Antenucci J. C. (1982). A GIS Generation GAP: MAGI and KNRIS. *Computers, Environment and Urban Systems 7:* 269, 272.

Beck M. (1988, December 5). Smart cars, smart streets. *Newsweek:* 86–87.

Brown R. T., Reel J. S. (1987). The beginning of a regional/local geographic information network in south Florida. *Geoprocessing and Geographic Information Systems: Papers From the 1987 Annual Conference of the Urban and Regional Information Systems Association 2:* 266–276.

Chambers J. (1987, June). HL&P installs automated mapping system. *Transmission & Distribution:* 36–38.

Cooke D. F. (1988, Summer). From our president. *GDT News 4:* 1.

Dakan A. W. (1987). The private sector and geographic information; A survey of end-users. *Geoprocessing and Geographic Information Systems: Papers From the 1987 Annual Conference of the Urban and Regional Information Systems Association 2:* 184–193.

Design Graphics World (1986, January). Automated mapping technology triples productivity for CGTC. Copyright Communication Channels, Inc., Atlanta, GA.

Federal Digital Cartography Newsletter (1987/88, Winter). A demonstration GIS project to assist the Chesapeake Bay "clean-up" in the Elizabeth River drainage basin. Federal Interagency Coordinating Committee on Digital Cartography 7: 4.

Federal Digital Cartography Newsletter (1988, Summer). Western Oregon Digital Data Base. Federal Interagency Coordinating Committee on Digital Cartography 3: 6.

Federal Digital Cartography Newsletter (1987/88, Winter). Bureau of Mines' inventory of land use restraints program. Federal Interagency Coordinating Committee on Digital Cartography 7: 3.

Federal Interagency Coordinating Committee on Digital Cartography (1988, August; 1987, July). Reports Working Group. *A Summary of GIS Activities in the Federal Government.*

GDT News (1988, Summer) Coming down to neighborhoods with AIDS mapping.

Gibson D. A. (1986, June). TransAlta Utilities Corporation—An electric utility company's AM/FM conversion program. *The AM/FM Scribe:* 1–2.

Gilbrook M. J., Sheldon P. K. (1987). Coping with Florida's new growth management legislation: A first time application of a geographic information system. *Geoprocessing and Geographic Information Systems: Papers From the 1987 Annual Conference of the Urban and Regional Information Systems Association* 2: 254–265.

GIS News, Photogrammetric Engineering & Remote Sensing (1988). Corps of engineers apply GIS.

GRASSCLIPPINGS, Geographic Resources Analysis Support System Newsletter 2 (1988, Spring). Army uses GRASS in INF treaty assessment. GRASS Interagency Steering Committee, United States Army Corps of Engineer Research Laboratories, Champaigne, IL.

Hanigan F. L. (1988). GIS recognized as valuable tool for decision makers. *The GIS FORUM 1:* 4, 5, 8.

Hull W. V. (1987, July). NOAA map and chart products: Present activities and trends. *The American Cartographer 14:* 213, 214.

IBM Update (1987, August). Computers—Silent partners in making state and local governments work efficiently.

Jannace R., Ogrosky C. (1987, July). Cartographic programs and products of the U.S. Geological Survey. *The American Cartographer 14:* 197.

Johnson G. O. (1987). Toward an emergency preparedness planning and operations system (EPPOS). *Geoprocessing and Geographic Information Systems: Papers From the 1987 Annual Conference of the Urban and Regional Information Systems Association* 2: 171–183.

Juhl G. M., Wallick P. (1986, May). New York: A map for all reasons. *Planning:* 26–27.

Licht J. C. (1987). A comparison of Minnesota's centralized and Wisconsin's decentralized approach to modernized land records management. *Land Records/Natural Resources: Papers From the 1987 Annual Conference of the Urban and Regional Information Systems Association 1:* 7.

The Louisville Courier-Journal (1986, August 9). Plans for "super map" of city rescaled after 2 utilities back out.

Maizel M. S., White D. Experts in land information systems development. *Lands Systems Technology.* Lincoln Institute of Land Policy, monograph 86-11, pp. 19–32.

Manogue C., Hatch J. (1987, May/June). Scott Paper—Two years with a GIS . . . in the Southeast . . . in the Northeast. *the Compiler 5:* 17–19.

McDonald E. C. (1988, April). The message is the medium. *Government Data Systems 17:* 6–8.

Metropolitan Council Data Center, Regional Mapping Consortium (1988, June). *1988 Minnesota Inventory of Computer Mapping Systems: Software, Applications, Graphic Data Files and Expertise:* 45, 47, 48, 79.

Reid W. B. (1987, March). United Telephone Company of Florida—Among the first telephone companies to complete records conversion. *AM/FM International Scribe 3:* 1–2.

School of Information Studies (1989, August). Managing Information Resources: New Directions in State Government. Syracuse University.

Urweider K. (1986, February). Map database sheds new light on maintenance in New York. *Computer Graphics Today:* 12.

Chapter 4

Benefits and Costs

Investment in geographic information technology will not occur unless an organization's leaders believe that the benefits will exceed the costs of implementation. This chapter gives practical advice on the preparation of benefit–cost analyses and discusses the subjective nature of such analyses.

Although in many instances the net benefit of a geographic information system is self-evident to technical and managerial personnel, a persuasive business management case in support of investment in a GIS requires objective demonstration as well as intuitive assumption of benefit.

The benefit–cost analysis is often necessary to demonstrate a system's net benefit to elected and appointed officials, senior executives, regulatory and rate-setting bodies in the utility world, stockholders, and others.

The nature and magnitude of the benefits will vary among organizations, as will the costs. Various organizations assess and weigh benefits differently when they make the critical decision whether to design, install, and operate a GIS. Generally, the information that supports such decisions is unbalanced. Explicit recognition of benefits and costs, and quantification when possible, is an important step leading to the decision to implement a GIS. Organizations often have a more thorough knowledge of the costs involved, but must rely on a subjective appreciation of the potential benefits. Recognizing this disparity, this chapter presents an overview of techniques used to structure a GIS benefit–cost analysis.

BENEFITS

The potential benefits of geographic information technologies are as diverse as the organizations that put them to use. For example, as early as 1968, the Texas Electronic Service Company documented $1 million in savings in four years over its grid-based transformer load management system.

The city of Long Beach, California, uses an automated mapping system to draft maps at twice the speed of manual mapping. Updates to these maps now are accomplished four times more quickly than was possible with traditional drafting techniques.

The Denver Water Department implemented engineering and planning applications on an AM/FM system and reduced the labor and calendar time needed to create cross-sectional drawings of underground pipes. Before automation, drafters spent two months to produce drawings for

When you talk about cost justification, it's more a matter of, "Can you afford not to?" I know that doesn't give management a warm, fuzzy feeling, but at least in our world, with the demands on our time, and with fewer people to do more, and more departments asking for more information, automated mapping provides a cost-effective answer to increasing demands.

Patricia Mitchell, manager of field support group, Public Service Indiana

65

There's a certain amount of risk in being a leader, but there's a lot of risk in not being a leader. If you don't want to keep your job very long, start getting a lot of sewer backups that you don't do anything with.

Because I had no inventory of my systems, not only was I unable to make good executive decisions, I was unable to meet regulatory obligations that I saw coming in the future until we had IMAGES. So it's the ability to comply with the law—whether you're going to [meet] future demands with a new bureaucracy that sits around and files, or whether you're going to do it with an automated mapping system.

Barbara Gole, former Director, Indianapolis Department of Public Works

100 cross-sectional maps. With automation, a similar effort was accomplished in less than two days.

Ducks Unlimited, a private nonprofit corporation in Chicago, is using Landsat satellite and GIS technology to assess where to invest millions of private dollars to rehabilitate duck habitats.

The City of New York, using its Planning Department's GIS, helped the FBI plan the logistics of President Reagan's attendance at the Statue of Liberty Centennial celebration.

The examples illustrate the potential range of benefits offered by GIS—from productivity and timeliness to protection of life. They are not the same kinds of benefits, and it is not always possible to measure their value.

Benefit Types

An anecdotal recital of benefits experienced by various organizations is not adequate to characterize the scope of the technology's impact on operational environments. Benefits that are defined and categorized casually will not make a persuasive case for implementing a GIS in a particular setting.

A more formal structure for examining benefits is required. Benefits either are or are not quantifiable. They can be either direct or indirect, predictable or sporadic. The method of examining potential benefits must accommodate these traits.

In this chapter, five distinct benefits are discussed:

- Type 1—Quantifiable efficiences in current practices, or benefits that reflect improvements to existing practices.
- Type 2—Quantifiable expanded capabilities, or benefits that offer added capabilities.
- Type 3—Quantifiable unpredictable events, or benefits that result from unpredictable events.
- Type 4—Intangible benefits, or benefits that produce intangible advantages.
- Type 5–Quantifiable sale of information, or benefits that result from the sale of information services.

Types 1, 2, 3, and 5 are quantifiable benefits; Type 4 benefits are not easily quantified. For all types, the benefits can be either direct or indirect.

Direct benefits are defined as those that accrue to the organization or unit sponsoring the GIS. Productivity improvements in drafting that reduce the workload and labor costs of the agency are an example of Type 1 benefits.

Indirect benefits are defined as those that accrue to organizations or individuals who are not the sponsors of a GIS. For example, an indirect benefit could be improved access to more timely and accurate maps for the general public.

Consider another example: The Corporation of Burnaby, British Columbia, a municipal government near Vancouver, Canada, began implementing a GIS in 1977. During the early stages, the system was used to evaluate a proposed subdivision of municipal land. The initial plan proposed forty five building sites at a cost of $100,000 (Canadian) per tract. Using the GIS, the subdivision was redesigned to include fifty lots, all meeting the initial design criteria.

Two direct benefits accrued to the corporation because of the GIS. First, the additional five lots raised land-sale income to the corporation by $500,000 (Canadian). Second, property tax income was increased by $1000 (Canadian) per lot per year. The benefits were both direct and indirect—the corporation received additional income, and the residents of Burnaby enjoyed the benefits of an enlarged tax base. Though the same design could have been achieved without the system, the availability of the GIS provided an impetus to reexamine the initial design.

Wisconsin Gas Company implemented an AM/FM system and reduced the time required to draft as-built engineering drawings. Decreased labor and costs of updated drawings are direct benefits to the utility, as are the higher quality and timely updates of the drawings, and the utility's improved management. Improved utility service to customers is an indirect benefit of the system. Quantifying indirect benefits can range from easy to impossibly difficult. In the Burnaby example, each of the approximately 120,000 residents of the city received an indirect benefit from the modified subdivision equivalent to $3.34 (Canadian) exclusive of the long-term tax revenue. Similarly, utilities frequently assess the value of projects in terms of savings and costs to their customers.

Type 1 Benefits

Automation of geographic data capture, handling, archiving, and manipulation improves upon many existing practices and makes some obsolete. The benefits of automation are generally the easiest to identify and quantify.

Cost savings achieved through the reduction or elimination of redundant activities, improvements in map drafting and maintenance, more efficient engineering, drafting, and design, and reductions in the time required to locate and transport geographic information illustrate Type 1 direct benefits. Much of the savings or benefits are associated with enhanced productivity.

Productivity improvement can be measured in units of "time saved" by automating some parts of a particular task. Note that productivity improvements are limited by the extent to which a task may be automated. Most tasks require some level of manual intervention.

The process of updating an existing map or engineering drawing typically includes these steps:

1. Collecting and confirming information documenting the change
2. Locating the original map or engineering drawing
3. Setting up the original drawing or new material on a drafting surface
4. Manually redrafting some or all of the map or engineering drawing that has not changed
5. Manually drafting all additions and modifications
6. Editing all portions of the redrafted and added material
7. Correcting errors

8. Distributing the updated document and replacing all outdated documents

Automation of the drafting function using a computer-aided drafting (CAD) system is likely to reduce the level of manual effort thus:

- Electronic setup of map sheet, borders, and boilerplate is accomplished more quickly because the items are extracted from existing files or templates (step 3).
- Redrafting unchanged information is all but eliminated in an automated environment (step 4). The existing digital file is used to the extent practical. Although redrafting may not be required in a manual drafting operation for changes that affect small areas, redrafting is required when a drawing must be of high quality or the cumulative number of changes begins to deteriorate the drawing surface.
- The drafting process is accelerated by various commands that draft and add dimensions to lines, define arcs and other geometric forms, place previously designed symbols, and set annotation with the ease of typing, among others (steps 5 and 7).
- Certain types of errors (not all) can be detected using system commands, such as checking whether intersecting lines physically intersect (step 6).

In a system designed to perform more than a drafting function, several additional benefits accrue from automation, including these:

- The collection of information to be added to the drawing may be augmented by electronic transfer of as-built drawings, subdivision plats, and other files. The reductions in "walk around" time can be considerable if this step can be automated to some degree (step 1).
- A digital file represents the original to be modified, and efforts to locate it are minimal in an automated environment (step 2).
- Distribution of the updated document is accomplished by placing it in the system with an access code that allows users to examine and print copies as required (step 8).

BOX 4-1 Findings of Nordic Kvantif Economics and User Experiences in North America

"Research reports published in Norway and Sweden show that the benefit–cost ratio for [automating] conventional maps is greater than 3 : 1."

"A digital system only used for mapping and updating gives you your money back (B/C 1 : 1)."

"If the system is also used for planning and engineering purposes in your organization, your money will be doubled (B/C 2 : 1)."

"If you manage to create a common system where information can be shared among the different relevant organizations, you will have your investment back four times. (B/C 4 : 1)"

"For organizations . . . [with] . . . a poor system for manual map production, the automated system has given . . . benefit–cost ratios up to 7 : 1. As an average, we found a reduction in time of 50 percent in map production."

"For special effects, it [automation] has derived an efficiency rate of 20 : 1 compared to manual production."

Digital Map Data Bases, Report 3, pp. 5, 13, 21, Nordic Kvantif, March 1987.

The productivity increases offered by automation of drafting and update activities are potentially substantial. The city of Edmonton, Alberta, Canada, experienced productivity improvements at a ratio of 3 : 1 for drafting functions and in the range of 5–20 : 1 for updating functions using GIS. The experiences of many other organization confirm that productivity enhancements of this magnitude are the rule. Efficiencies in tasks other than map

compilation are common. The reduction of redundant data collection, maintenance, and use are key benefits for several reasons, the most obvious the immediate reduction of labor and direct costs as the repetition of each activity is eliminated.

The maintenance and use of address files in public- and private-service organizations is a simple and common example. A key geographic component for many functions, addresses identify the location of people and equipment, enable the billing of commodities such as water and electricity, and indicate where services such as waste collection and emergency responses are needed.

Municipal and utility organizations typically maintain multiple address files, all tended by different operating departments. These files are often incomplete, inconsistent, and of varying form and content. Moreover, each organization usually is convinced that the file in its domain is the "best," considering its criteria and responsibilities. Each file requires—but may not receive—the same level of support to assure that new addresses and changes are promptly entered in the database, that addresses assigned are actually used, and that all other files referencing an address are updated concurrently.

Consider the example of a large Midwest metropolitan area that encompasses a county government, a major city, more than a dozen smaller incorporated jurisdictions, three utilities, and an organization responsible for tax assessment. No fewer than sixty five address files were identified, not including those maintained by the smaller incorporated jurisdictions. Of those files, forty six were automated and nineteen were manual. Only seven of the automated files had a true parent–child relationship in which the addresses were maintained in parallel with concurrent updates. Although the files generally were of uneven quality and were dissimilar in their level of accuracy and timeliness, considerable cumulative effort was required to maintain them. If the jurisdictions were to pool their efforts to build and maintain a centralized address file from which each organization would extract information and updates, the resultant savings (benefits) would be considerable, simply in terms of the labor effort required.

Other benefits accrue from the elimination of redundant activities: Accuracy and efficiency improve. It is difficult, if not impossible, to assure that all updates of redundant records are made promptly. Even if the updates are accomplished, some of the multiple records will not be updated as quickly, introducing a temporary but perhaps critical discrepancy. The confusion in comparing temporally different records causes misinformed judgments and time lost in identifying and rectifying discrepancies.

Redundant generation and maintenance of maps and engineering drawings is common in government and private industry. Generally, there is a rationale for the multiple maintenance activities. In some cases, the drawings contain more or less content than the "base" drawing. Sometimes the scales or symbology are different. In other instances, institutional factors—turf wars or lack of trust in data accuracy between organizations—influence the decision to preserve redundancy. Sometimes the reason is simply reluctance to stray from tradition. Regardless of cause, the costs of redundancy can be reduced or eliminated with geographic information technology.

The cost of mistrust is frequently high. A United States telephone utility maintained a series of maps manually for a number of years. A backlog of 30–45 days in posting changes to the schematics that show cable throws and use of underground ducts was claimed by the drafting manager, but field crews claimed it was as much as six months. Although the drafting manager could accomplish the posting in 45 days, generation and distribution of hard-copy products to the field crews took much longer.

As a result, when a work order for underground connections or expansions was received by the outside plant engineering department, its supervisor dispatched a field crew to open the vault and review the actual circuits against the most recent schematic. This required that each of two people spend two to four hours, depending on weather, circuit layout, and other factors, to accomplish the review. The on-site observations were noted freehand on the schematic, which was returned to the engineering office for manual redrafting. The engineering was completed, noted on the revised schematic, and sent to field construction. Field construction then sent a crew (two people for another two to four hours) to the field to confirm that the schematic and engineering drawings were correct. If the design was confirmed, a third crew was then dispatched to execute the change; if not, the entire process was repeated. As a routine matter, four to sixteen hours were diverted to checking records for fear that the original schematic was not accurate.

Could automation stem this cost? Perhaps, if only because it would reduce backlog and inspire confidence in the accuracy of the schematics and engineering drawings.

Many organizations have demonstrated that automation of "base" drawings creates an opportunity for the generation of multiple specialized products without redundant development and maintenance of the drawing's common elements. In these cases, benefits can be pinpointed by measuring the labor and direct costs required to develop and update each product manually, then subtracting the labor and direct costs required to develop and maintain each product with an automated database.

Another telephone utility, Southwestern Bell, projected nonlabor cost savings from an automated mapping system by eliminating or reducing external costs for drafting services, commercial map products, blueprinting, and mailing. Those former expenses were supplanted by electronic access to a centralized file of key engineering documents and base maps.

In summary, Type 1 benefits are quantifiable and are computed by comparing existing practices with the potential efficiencies, productivity enhancements, and

Type 1 benefits are quantifiable and are computed by comparing existing practices with the potential efficiencies, productivity enhancements, and reduced nonlabor direct costs expected to result from automation.

reduced nonlabor direct costs expected to result from automation. Each specific function of current operations must be defined and analyzed to account for the impact that automation may have on it. In addition, automated operations must be examined to identify what additional functions will be required that are not part of the current manual effort.

Type 2 Benefits

In every organization some tasks are not performed routinely because of the extraordinary level of effort required to accomplish them. Regardless of how interesting, useful, or beneficial those functions may be, the labor to do them routinely is not available. Geographic information systems often yield expanded capabilities— Type 2 benefits—that make it feasible to accomplish such tasks.

Police and school organizations may redefine service districts reluctantly because of the difficulty in calculating the balance of officers and equipment or pupils and classrooms required in each district. School bus routing frequency is planned after several weeks of actual (bad) experience during the fall term, rather than calculating in advance the location of the pupils and the most efficient routes. Utility organizations evaluate electric load balancing and sanitary sewer routing less frequently than desired because these networks are complex and undergo frequent changes. Even perfunctory tasks requiring significant labor in drafting and drawing, such as developing illustrations or maps for public hearings, may go unaccomplished for lack of time and labor.

Some of these functions are rendered more practical or less onerous in an automated environment. The introduction of geographic information technology, perhaps tied to the appropriate analytical tool, may eliminate the need for additional labor to perform these functions. The GIS or mapping system can be viewed as an added complement of staff.

Although also related to productivity, Type 2 benefits differ from Type 1 in a sim-

ple way. Type 1 benefits accrue from increasing the productivity of staff. Type 2 benefits are the equivalent of additional staff.

A system developed by the Public Works Department of the Municipality of Anchorage, which received the URISA 1987 Exemplary System in Government Award, produces many Type 2 benefits. The agency's automated mapping system was developed to store and maintain approximately 100,000 land records. After implementation, however, the system found many other uses. Tasks formerly left undone due to the lack of staff and support services were accomplished routinely with the help of the new system. Tracking permits, planning routes for snowplow crews, creating street sweeping districts, and producing and maintaining storm drain logs for use by maintenance crews are examples of some of the expanded capabilities.

Carolina Power and Light started planning its automated mapping system in the late 1970s and initiated a pilot in 1979. The mapping system began production in 1983. Coincidentally, based on a project evaluation by senior executives, the mapping system evolved into facility management the same year. The data base is scheduled for completion in 1992. Major benefits of the system will come from many functions in addition to drafting, including distribution trouble-reporting, transformer load management, calculation of tax deductions for line repairs, and line and service crew efficiencies (Table 4-1).

These benefits are quantifiable. They are generally measured in labor equivalences and nonlabor costs that would be incurred if the practice were routine and accomplished manually.

Type 3 Benefits

Some of the benefits of a GIS cannot be anticipated, although they can be measured after the fact. Despite rigorous planning before system implementation, applications with considerable benefits are typically "discovered" after the system is

You buy a system, and all of sudden there are so many potential applications and new capabilities developing these takes dollars. The benefits are difficult to quantify. How do you measure saving the kid who did not get hit by the car, or the value of storm drains not being clogged because we now know where they are? The biggest benefit of the system is [having] better information for making better decisions

Kimball Forrest, former GIS Manager, Department of Public Works, Municipality of Anchorage, Alaska

in place. Frequently, these application opportunities emerge as responses to events that could not have been predicted.

For example, a system might expedite response to a natural disaster or assist a major development proposal for a new airport or office complex. The equivalent level of manual effort can be estimated with some precision after the event.

Consider another example: Most utilities that have underground facilities track the number of outages, or service interruptions due to accidental breaks in water or gas lines and cuts in power or telephone cables. Typically to blame are inaccurate maps or an equipment operator's failure to use available information, perhaps because records are unreliable or difficult to obtain.

Assuming geographic information technology can improve the availability and quality of information, it stands to reason that, by combining information from multiple utilities on a single map, the number of service disruptions and breakages should decline. The projected degree and extent of the decline compared to current experience establishes a benefit produced by the GIS. One can measure the benefit after observing the operation for a period of time. The difficult lies in estimating these benefits in advance.

Wisconsin Public Service (WPS), a gas and electric utility, found that its AM/FM system produced the unexpected benefits of improving the utility's credibility with regulators, which expedited a rate case. WPS project manager Jack Bernard told a 1987 industry conference that in the utility's last rate case, the cost-of-service analysis presented was prepared on the system in only twenty two minutes and was accepted by the commission without argument. "It gives you credibility. A benefit we never expected in a million years was expediting a rate case," Bernard said.

Sometimes unexpected events prompt organizations to recognize the benefits of geographic information technology. A gas line explosion in Indiana that killed two elderly grocery-store owners precipitated an attempt to identify the potential for

TABLE 4-1 Distributed Information System Estimated Annual Benefits, Carolina Power & Light Company

Engineering labor	$ 850,000
Line crews	150,000
Servicemen	95,000
Distribution operations	50,000
Line loss reduction	945,000
Planning	180,000
Distribution automation	100,000
Transformer load management	50,000
Trouble reporting	125,000
Fixed asset accounting	75,000
Rates and services	200,000
Street light inventory	65,000
Joint use billing	150,000
Equipment inventories	50,000
Conservation projects	25,000
Right-of-way management	250,000
Tax renewal allowance	1,000,000

Benefits of the Distribution Information System, Robert E. Tripp III, P.E., Carolina Power & Light, undated, briefing paper.

similar accidents. When a manual search of the gas utility's records was estimated to require five person-years of work, management made the commitment to develop an AM/FM system.

Type 4 Benefits

Some benefits of the technology defy quantification in traditional monetary terms. Type 4 benefits are diverse, and they vary in type and significance. At times, intangible benefits are the critical factor in system justification. Charles Litecky (1981), writing in the *Journal of Systems Management*, noted that decision makers "tend to assume extreme positions toward intangibles, either ignoring them or assigning dominant roles to them."

Certainly, intangible benefits should

Perhaps the most important benefit of the AIMS project to our agencies and departments has been the atmosphere of sharing and working together. This new atmosphere is not just confined to the project participants. It is carrying over into other projects and functions performed by the county government.

Ken Brelsford, Project Manager, Automated Information Management Systems (AIMS), Johnson County, Kansas

not be ignored in the development of systems; they do exist. On the other hand, the benefits must have an upper limit since development of GIS will not occur despite all costs. The difficulty lies in giving dimension to subjective values. What value should be assigned to better decisions that are based on information that is more timely, more accurate, more consistent with other data sets, and more readily available?

Although all GIS users enjoy the benefit of improved decision making, improved service to customers and constituents is another potential benefit important to most organizations. The ability to produce an answer or product more quickly, more accurately, in a readily usable form, and with specific content has significant albeit unquantifiable value.

Intangible benefits also might include the reduction of tedious functions, improved workforce morale, and higher self-esteem for employees given the opportunity to work with "high-tech" tools. Benefits associated with reductions in staff turnover and absenteeism may exceed those that can be quantified. In some cases, the sharing of data and analytical tools may trigger better working relationships within and between organizational units.

Type 5 Benefits

As a result of automation, opportunities emerge for the sale of information services and digital data. Geographical information resources, once captive to analog forms and disorganized filing systems, have become readily accessible and transferable. What was once an operational expenditure becomes an organizational asset leveraged by the GIS.

Only recently has the value of digital geographic data based on its resale potential been recognized by public and private organizations alike. Spurred initially by efforts to develop cost-sharing arrangements in anticipation of GIS development, many organizations discovered latent interest in both "basic" data and in its manipulation into standard and custom products.

In 1988, the city of Virginia Beach, Virginia, completed a planimetric, topographic, and parcel database. The city then recouped 50 percent of its investment by selling the digital base map, parcels, planimetric detail, and rights-of-way to Virginia Power. In addition, the city's system generates a small but steady revenue through over-the-counter sales of tax map sheets, stock planimetric and topographic maps and by selling area maps to engineering consultants.

The resale potential of licensed databases, database maintenance services, and services leading to the production of GIS products has whetted many organizations' appetites. Still, there are many obstacles. Discussed later in this volume, the legal issues relevant to the sale of products and services, particularly by government organizations, are a critical consideration. Moreover, organizations attempting to market information resources must prepare to function as entrepreneurs—a most uncharacteristic role.

The benefit–cost model, though, cannot be complete without considering cost sharing, income generated by the sale of products, services and rights to the resultant databases, and the value of inkind services contributed in lieu of payments.

COSTS

The cost of geographic information technology is not a trivial matter, and it can range over several orders of magnitude. Systems vary in size, configuration, and level of sophistication; each element influences the eventual costs.

Some general rules persist. The costs of hardware and software, often the primary focus of attention, rarely exceed 20 percent of the total system cost. Software maintenance and upgrade fees quickly exceed their initial cost. Data base development makes up the largest component of total system cost, though internal development of the data base may disguise its true cost. Unless the data base is maintained in a rigorous manner, the initial investment can be substantially lost, requiring major reinvestments to restore data base integrity.

Although the costs of GIS development are easier to quantify than the benefits, a number of cost components are not immediately apparent. A life-cycle approach to cost accounting is preferred because it addresses the ongoing costs of maintenance and anticipates periodic expenditures for updates and planned enhancement. Such an approach also establishes a foundation for comparative benefit–cost analysis.

Cost Model

The cost model developed in this section accounts for a large number of components typically associated with implementation of a GIS and yet is not intended to be comprehensive (Fig. 4-1). Its utility spans a variety of systems from uncomplicated to sophisticated in design and content and its function during the life cycle of GIS development and operation evolves.

Cost accounting varies among organizations. As a result, cost components can be included, excluded, or otherwise modified to accommodate those differences. In addition, the degree to which specific components are treated as capital assets will influence the model's content.

Given the particular needs for cost analysis, the model can be used with greater or lesser degrees of detail. Implementation of a small system for a single department may not require the same level of cost analysis as would a system meeting the needs of multiple organizations. Although cost modeling is a necessary component of benefit–cost analysis, the model also can serve other purposes (Fig. 4-1).

The most straightforward use of a cost model is a cash-flow analysis that describes funding requirements for system development and operation over a given time frame. The model provides a consistent framework for comparing alternative system designs and implementations. The model can also document costs as a pre-

A life-cycle approach to GIS cost accounting is preferred because it addresses the ongoing costs of maintenance and anticipates periodic expenditures for updates and planned enhancement.

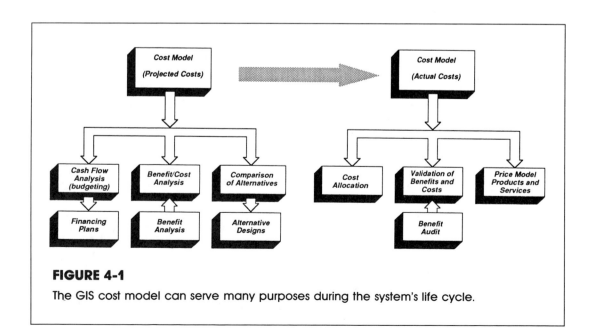

FIGURE 4-1

The GIS cost model can serve many purposes during the system's life cycle.

cursor to determining where to allocate costs and charges for the services and products the GIS supports.

Capital and Operating Costs

The choice of the term *capital* here refers to expenditures for items that have a durable life of at least several years. Whether or not an organization chooses to treat these components formally as capital assets is less material to the model. However, a portion of the model does account for depreciation, financing costs, and sinking fund charges. Components with an extended life include the initial data base, software and hardware, and implementation costs.

Operating costs include expenditures that are typically "expensed" or paid from operating revenues. Charges may include personnel, overhead, maintenance fees, and other direct costs such as utilities and supplies. Certain expenditures for data base maintenance and software development also may be treated as operating costs.

Data Base

Initial investments in the data base can be substantial. For systems using planimetric and topographic data constructed through photogrammetric techniques, the cost components will include ground surveys, aerial photography and photographic products, analytical triangulation, and planimetric and topographic mapping. Each of these cost components can be subdivided further as necessary.

As an example, detailed cost estimates for photogrammetric mapping need to account for researching the existing survey control network; locating or reestablishing survey monuments; undertaking field work with traditional or satellite survey techniques to establish both horizontal and vertical survey control; establishing and maintaining control panels marking survey monuments for visibility on the aerial photographs; obtaining aerial photographs; and purchasing a variety of photo-

graphic products such as mosaics, contact prints, and diapositives (transparent positive images). Following the acquisition of the aerial photographs and survey control, various photogrammetric procedures are used to accurately associate the survey control with features or points on the diapositives. The vertical and horizontal coordinates of the survey control are computed after adjusting their associated networks using a technique called *analytical triangulation*. The results are used to develop a digital planimetric image using photogrammetric equipment and techniques. A parallel process is followed for constructing topographic contours or digital elevation models.

The mapping process is labor-intensive and the production of digital files is generally accompanied by "hard-copy" output. Both the hard-copy products and the digital files are edited and corrected before loading into the GIS. Additional quality-control efforts are required at many points in the process.

Other databases normally associated with geographic information systems are constructed from source material, or are converted (transferred to a digital format) from existing drawings or maps. The preparation of property maps and facility records is a good example.

Property maps can be created by referring to deeds, subdivision plats, and other land records. Engineered facility designs can be created by referring to the initial design documents, work orders, and as-built drawings. Using source material is called *reconstruction*. Alternatively, existing property maps or engineering drawings can be automated by "trace digitizing" or, in some cases, by using a semi-automated technique called *scanning*. In either instance, the cost model can account in summary fashion for the costs of reconstruction and editing, or it can identify each of the primary functions in the process and account for costs in great detail.

A model of data conversion costs typically includes the efforts required to locate and assemble source and collateral data; correct that information manually (a process commonly called *scrubbing*); define the data structure for both graphic and nongraphic data; extract the nongraphic data from the source material; enter the text data into the nongraphic digital file; digitize the graphic data; and edit and perform independent quality-control measures for the cartographic and digital files. Ordinarily, the conversion of source and collateral data must be supplemented by field examination or advice from individuals who are knowledgeable about the data.

Data base development usually involves labor efforts both internal and external to the organization. Internal staff are used to organize material for conversion, develop specifications, select and manage contractors, mediate discrepancies in the data, fill information gaps, and provide quality control. Considering the magnitude of the conversion effort, the type of data, equipment requirements, and time and funds available, internal staff may be assigned to the conversion. Costs for labor, fringe benefits, overhead, and direct expenses should be included in cost accounting.

Photogrammetric mapping and data conversion often are accomplished by contracting with a firm that specializes in the conversion process. Consultants frequently are engaged to assist in database design, oversee conversion contractors, and provide independent quality assurance. The role of external sources of labor depends largely on the sponsoring organization's approach to system implementation, its internal resources, and the time available to accomplish the effort. The cost model must account for the external costs, although the level of detail may not be great.

Hardware and Software

Although they are frequently the undeserved focus of system planning and cost discussions, the cost components of hardware and software are less significant than might be perceived initially. The obvious components include the computer(s), file servers and other processors, data storage devices, terminals, printers and plotters, the operating system of the computer, and the geographic information management software. Beyond these, an actual installation includes a considerably more complex array of cost components (Table 4-2).

In addition to the processing, workstation, and output devices, configurations are likely to include data communication equipment to support transmissions between sites and systems, structural modifications, and environmental controls such as air conditioning, raised flooring, and power conditioning, among others.

Most of the equipment components have an annual maintenance cost in addition to their initial cost. This fee traditionally is charged by the supplier and covers repairs, preventive maintenance, documentation, and responses to customer queries. Standard annual equipment maintenance fees are about 8 to 15 percent of the initial cost, although they may be higher for equipment that is specialized or outdated.

Software costs can be divided into four types: operating system software, core geographic information management software, application software, and device software. The distinctions are somewhat arbitrary and are intended to account for the range of software cost components.

The *operating software* generally is provided with the computer. Other basic software packages that are host-specific, such as FORTRAN, COBOL, or C, as well as special performance routines that are not specifically required for the geographic information management software but may contribute to the usefulness of the overall system, may be purchased with the CPU.

Generally, software fees are associated with the purchase of a license to use the software under specific conditions, most of which reserve the ownership to the licensor and require confidentiality on the part of the licensee. In most cases, these software packages have a first cost and annual maintenance fee.

The *core geographic information management software* serves as an application pack-

TABLE 4-2 Representative Equipment Detail Hardware Procurement

Hardware: General	
Processor	Ethernet taps
Impact line printer	TCP/IP connector
Tape drive	Cable (feet)
Disk storage devices	Cable connectors
Cabinets	Ethernet cable
Plotters	**Hardware: Gateway**
Screen copiers	X/SNA server
Digitizers	Clocking cable
Alphanumeric terminals	Communication card
Workstations	SNA/API
Graphic terminals	**Hardware: External Communication**
Hardware: Internal Communications	Telephone circuits (shared and dedicated)
Host-based communication cards	**Hardware: System Environment**
Host controller	Security
LAN transceiver	Air conditioning
PVC drops	Backup Power (battery)
	Subflooring

age for the processing of geographic information. It may be a comprehensive code allowing the processing of graphic and nongraphic data; it may be a tightly integrated set of programs from one or more sources that manages the geographic information; or it may be a loosely bundled set of programs from different suppliers that operate in a coordinated fashion. Although the packaging may differ, the core software typically has a first cost and an annual maintenance cost. The core package also may have various options or features that may or may not be required for a specific installation. These, too, traditionally have a first cost and annual maintenance fee.

Application software performs specialized processes that are specific to a particular type of analysis or function, such as routing vehicles through a street network; calculating fill material in road design using topographic data; calculating property assessment values; tracing utility networks; or logging customer complaints. Application programs may be available from the supplier of the core GIS software, from third parties who have developed interfaces with a particular core package, or by designing a custom program for a particular system and application. Again, the software costs generally have a first cost and annual maintenance fee. In those cases when the product is customized for a particular site or when no maintenance is provided, the labor costs of maintaining the program internally or through third parties must be considered.

Licensing of workstations or "seats" is now the trend. Until recently, core and application software licenses were associated with the central processing unit. The costs sometimes increased with larger computers to account for the potential increase in the number of users. As workstations with local processing power have become more popular and available, software vendors have developed a range of pricing techniques to relate their sales to the number of users that may access the system. By licensing seats, the software vendors' revenues track the success of the system. Potential growth in system users and workstations, then, must be considered in the system's life-cycle cost for software. A life-cycle analysis also should include the cost of upgrading the hardware from a smaller to a larger CPU.

Software maintenance fees usually include modifications, corrections, and enhancements made to the software during the term of the maintenance agreements. However, vendors retain the ability to define substantial modifications or improvements as "new" products that require supplemental investments by user organizations that want the enhancements. While this occurs infrequently, a cost model examining a life cycle of more than a few years should account for major reinvestments in software every five to eight years.

The term *device software* encompasses the various software packages needed to communicate within and between systems. It varies from the operating, core, and application software in that it is not required specifically for these packages to function. However, it is essential to the operation of peripheral equipment such as plotters and printers, as well as to data communication functions such as Standard Network Architecture (SNA) gateways between different CPUs. Licensing and annual maintenance fees are again the norm.

Training and user support is a significant and frequently ignored cost of hardware and software. Training in the use and maintenance of hardware and software is usually provided for a fee by the respective supplier and may occur at the user site or at special facilities operated by the vendors. Training costs of the employee include tuition, travel, and per diem. In addition, direct costs may include supplemental equipment rental, supplies, and computer time, among others. An organization also may account for the costs of staff time devoted to training programs.

Personnel

Internal personnel costs are incurred when an organization's staff are used to design, develop, implement, and operate the system. Should another organization's personnel be retained to accomplish one or more functions, the costs are to some degree external. External sources may be commercial, such as a consulting or conversion firm, or they may be nonprofit, such as a university or other governmental unit where funds or "in-kind" services are exchanged.

Each source of personnel has several common cost components, although they may vary substantially in relative value. For labor costs, gross wages are measured in terms of hourly, daily, or annual salaries. To these costs the fringe benefits—such as leave, health and insurance benefits, and matching tax contributions—are added. Labor charges, particularly those

from external organizations, are typically burdened with overhead charges. These are fees that account for the indirect expenditures required to support an individual. In the simplest cases, overhead charges may include the costs of space, utilities, furnishings, and supplies. In more complex instances, the list of operating expenses recovered through overhead charges can be quite extensive. External labor costs, including fringe benefits and overhead, are easily accounted for through direct billings from external sources.

Internal accounting of labor costs is frequently more difficult for many organizations. Typically, overhead charges are not tracked or assigned to an individual's salary in nonprofit and public organizations. Frequently, these expenditures are handled as line items in a budget, and cost recovery is not a primary consideration.

Even if internal labor is available without direct charge to a project, burdened labor costs should be factored into the decision whether or not to accomplish a task internally or externally.

Moreover, internal labor costs are required to account for the total cost of system development. This is more important if one desires to recover some or all of the initial investment through allocation of costs to other organizations or from the sale of products and services following system implementation.

The duties split between internal and external sources of labor will vary, depending on a organization's personnel and financial resources, and its operating philosophies. The talent and expertise required to implement a GIS is specialized and, in many instances, not readily available within an organization. In other cases, either a particular expertise or large bodies of labor are required for relatively short periods of time, and it may not make practical sense to meet these needs with in-house staff.

Labor costs are associated with each step of system conceptualization, design, development, and implementation. Feasibility and definition of requirements, presentations to management, coordination,

Training and user support is a significant and frequently ignored cost of hardware and software.

project management, system and database procurement, system installation, training, operation, maintenance, and a wide range of collateral functions are labor-intensive. In any event, the model should account completely for labor sources and costs to ensure an accurate comparison with the labor savings and benefits.

Financing

If the hardware, software, or database is purchased with borrowed funds (e.g., bonds) or through a lease purchase agreement, financing costs arise. In other instances, an organization may wish to evaluate the cost of using available funds for system development rather than earmarking them for alternative projects. In cases where a group is taking the lead in developing a system on behalf of other organizations, either by agreement or speculatively, the organization may desire to recover its "cost of money" when the other users begin to make financial contributions.

Some organizations also choose to depreciate the equipment over a period of time, with the depreciations acting as a charge against assets like the value of the computer. In addition, some organizations establish a reserve account or sinking fund that accounts in advance for anticipated replacement of hardware and software. Sinking funds ranging from 20 percent (five-year replacement) to $12\frac{1}{2}$ percent (eight-year replacement) are not uncommon. The sinking fund should not be confused with the depreciation of initial capital investments.

Some financial officers argue that the costs of software and databases should not be depreciated. Their rationale suggests that continuing expenditures for software and database maintenance protect the initial investment. In fact, the value of the database actually may increase if it is rigorously maintained, particularly if records of change are retained, because the costs of replacement will be increasingly greater with time.

Intangibles

Change accompanies the introduction of new technology—the modification of existing practices, retrained personnel, and the diversion of attention to a new program. Change has a disquieting impact on all organizations, large and small, public and private, that causes subtle, intangible costs frequently unrecognized by those implementing a new system. Although subtle, these costs nonetheless may exert a substantial influence on the system's potential for success. An accounting of the intangible costs and their sources is an essential element of cost analysis.

BENEFIT–COST ANALYSIS

The comparison of benefits and costs serves many purposes. Foremost, the analysis is used to weigh the positive and negative aspects, both quantifiable and unquantifiable, of introducing geographic information technology into an organization. If the results are positive, one use of the analysis is to justify the introduction of the technology; if negative, it buttresses the argument for the status quo.

Benefit–cost analysis also can be used to compare the relative merits of implementation alternatives, to establish the priority of components when an implementation is to occur incrementally, and to aid in justifying enhancements as they occur during the life cycle of a system.

As mentioned earlier, benefit–cost analysis often influences the politics of system implementation, as well as managerial and technical considerations. The analysis can provide quantitative proof of the system's net benefit to elected and appointed officials, senior executives, utility regulatory and rate-setting bodies, stockholders, and others.

Techniques

The most frequently used technique for comparing quantitative benefits and costs is to compare the cumulative costs of managing geographic information in traditional ways over a period of time to the cumulative costs of implementing geographic information technology over the same time. In the simplest of presentations, the costs are set out in constant dollars. Benefits and costs that cannot be quantified are treated through discussion and anecdotes.

Figure 4-2 graphs a typical cumulative costs comparison of manual and automated geographic information management systems. The curve representing the manual system shows generally increasing costs over time. The curve reflects compounded costs such as increased numbers of staff, higher wages, and inflationary pressures. The curve characterizing the automated approach shows substantially higher costs in the earlier time periods. However, cost increases diminish over time. The higher costs in the early time periods reflect the investments in data automation, equipment, training, and added personnel, among others, while the organization continues to incur the manual operation's costs. The cost curve declines as investments diminish, as benefits (or reduced costs) are enjoyed, and as the system becomes operational. If GIS costs are cumulatively less than those incurred under the existing process, the curves will cross. The point of intersection represents the time when the additional investment in GIS technology is "paid off."

Table 4-3 summarizes an actual analysis for a geographic information system supporting a large urban area. The cost components selected for comparison, with and without automation, are listed in the rows of the matrix, and the costs projected for each of ten years are stated in the columns in constant dollars (inflation was included in the projection of wage costs). The table illustrates staged system implementation costs as well as staged benefits (reduced costs of operation), based on the extent of implementation during specific calendar periods.

Figure 4-3 graphically depicts the result of this analysis, expressed in terms of cumulative benefits. This figure illus-

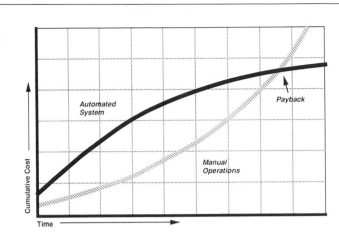

FIGURE 4-2

One technique for visualizing the benefits of GIS plots the cumulative cost of operations with and without investment in GIS once on extended time period.

trates the anticipated completion date for system implementation, although it is clear that benefits from the system will accrue earlier. The shaded area shows that the "payoff" in about the seventh year also can be viewed as occurring four years after the initial system is operating fully. This is a subtle difference, yet it may be important from a political or managerial viewpoint.

Another view of the same data analysis can take into account the time-cost of money. Project economics may warrant an account of the present worth of money to be spent in later years, as well as the present worth of the savings to be accrued in those later years. The technique effectively discounts both the costs and the benefits based on an estimate of the dollar's relative worth in the future. The present-worth analysis assumption regarding the reduced value of money frequently is based on an estimate of inflation or interest rates. Readily available interest tables that incorporate present-worth factors are used to adjust the costs of later expenditures by a factor that reflects the interest rate and the time frame measured in years. Present-worth factors also ac-

TABLE 4-3 Estimated Cost Savings Over 10-Year Period (In $1,000s)

	Year 1	Year 2	Year 3	Year 4	Year 5	Year 6	Year 7	Year 8	Year 9	Year 10	Total
Annual manual	2,836	3,068	3,318	3,589	4,198	4,541	4,911	5,312	5,746	5,746	41,400
Map acquisition cost	374	404	437	473	511	553	598	647	700	757	5,454
New base map	1,112	778	678	0	0	1,400	0	0	0	678	4,656
Total annual	4,332	4,250	4,433	4,062	4,392	6,151	5,139	5,558	6,690	6,503	51,510
Cumulative manual	4,332	8,582	13,015	17,077	21,469	27,620	32,759	38,317	45,007	51,510	
Personnel	2,945	2,973	2,890	2,773	2,941	3,120	3,310	3,511	3,725	3,951	32,139
Database develop	3,361	2,097	1,805	0	0	820	0	0	407	0	8,490
Hardware/ Software acquisition	1,310	0	941	0	0	0	0	0	0	0	2,251
Maintenance and communication	51	99	136	176	183	190	198	206	214	223	1,676
Training	20	30	10	10	10	10	10	10	10	10	130
Technical services	400	260	295	50	0	0	9	0	0	0	1,005
Map acquisition cost	374	192	81	0	0	0	9	0	0	0	647
Total	8,461	5,651	6,158	3,009	3,134	4,140	358	37,217	4,356	4,184	46,338
Cumulative total GIS	8,461	14,112	20,270	23,279	26,413	30,553	34,071	37,798	42,154	46,338	
Manual vs. automated (Type 1)	−4,129	−1,401	−1,725	1,053	1,258	2,011	1,621	1,831	2,334	2,319	5,172
Potential capability savings (Type 2)	0	0	198	419	452	479	508	540	573	607	3,776
Manual vs. automated (W/Savings)	−4,129	−1,401	−1,527	1,472	1,710	2,490	2,129	2,371	2,907	2,926	
Cumulative savings Exclusive of Type 3, 4, 5		−5,530	−7,057	−5,585	−3,875	−1,385	744	3,115	6,022	8,948	8,948

count for single expenditures, such as the purchase of equipment or conversion services, or for uniform series expenditures such as salaries, overhead, and maintenance fees.

The interplay of variables is far more evident in the benefit–cost tables than in the cost curves. Although not obvious from the graphics, a potent point can be extracted from the quantitative data: The return (or saving) begins at an early stage of implementation, even though the impact is not particularly visible in the cumulative figures. A labor-effectiveness factor can be derived from the data. That factor acknowledges the relative "savings" gained by the effective use of labor in the years preceding the payoff period, especially in those cases where the benefits are largely Type 1 and 2, as discussed above.

Many other techniques have been developed to express costs and benefits, such as returns on revenues (ROR), returns on investments (ROI), cost or savings per share (of stock), and a variety of other measures. An organization's chief financial officer frequently stipulates the format in which the benefits and costs are to be expressed.

Standard management and engineering economics texts provide a range of techniques as well as detailed explanations of both constant-dollar and present-worth comparisons. A growing body of literature in the systems management and business management press describes techniques for incorporating intangible benefits and costs into management information system decision making.

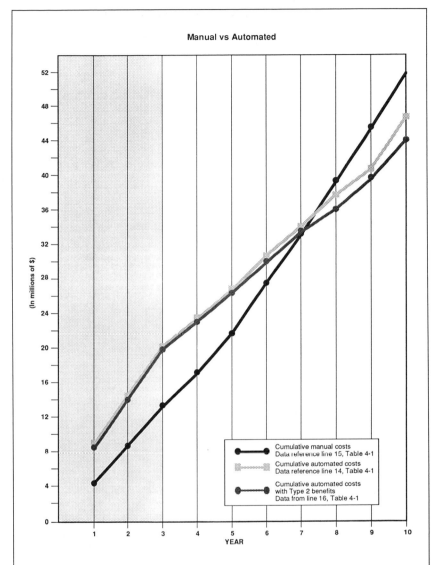

FIGURE 4-3

Cumulative costs of operations, with and without GIS and computed with and without consideration of the benefits attributable to increased capabilities, i.e., estimated in labor hour equivalents, are plotted against time.

REFERENCES

Joint Nordic Project; Nordisk Kvantif, Community Benefit of Digital Spatial Information; Viak A/S, Norway, ed; Svein Tveitval, project manager; *Digital Map Data Bases: Economics and User Experiences in North Amer-* *ica*; report 3 (Helsinki, Finland: Printed at Publications Division of the National Board of Survey, March 1987), p 5, 13, and 21.

Litecky C. R. (1981, February). Intangibles in cost/benefits analysis. *Journal of Systems Management*, 15–17.

SUGGESTED READINGS

Emery H. (1988, Spring). In the beginning. *Emery DataGraphic Quarterly*: 32.

Josephson M. (1986, June). Mapping water. *Plan and Print*, International Reprographic Association.

Lay P. M. (1985, June). Beware of the cost benefit model for IS project evaluation. *Journal of Systems Management 36:* 30–35.

Treworgy C., Bargh M. et al. (1988) Costs and benefits of GIS data management: a case study of database managed by a state agency. *Conference Proceedings, Vol. 2, Cartography*, 1988 ACSM-ASPRS Conference, March 1988, pp. 186–194.

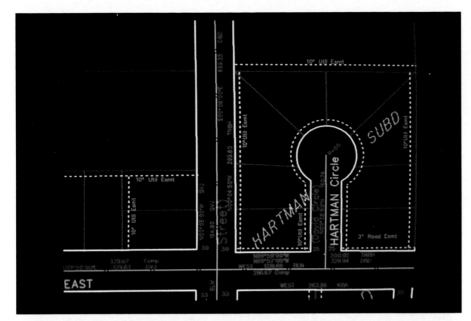

FIGURE C-1

Municipalities maintain parcel boundaries and public easements through automated mapping tools; colors differentiate the various types of the lines and the boundaries they represent (courtesy of Municipality of Anchorage, Alaska).

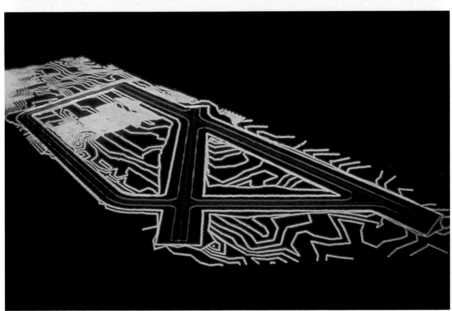

FIGURE C-2

Computer-aided mapping (CAM) plays an increasingly important role in engineering large facilities, such as runway placement within a local terrain map on a screen display (courtesy of McDonnell Douglas).

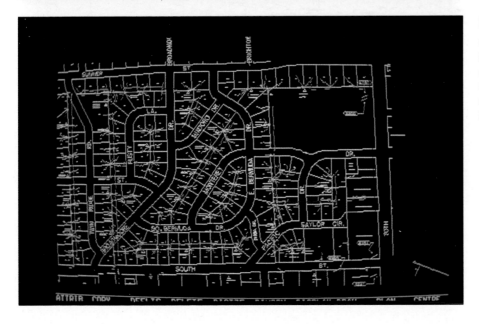

FIGURE C-3

Computer-aided drafting (CAD) tools increase the efficiency and consistency of drafting personnel: here is a screen display of subdivision and utility layout drawings before plotting (courtesy of Enghouse-CableCad).

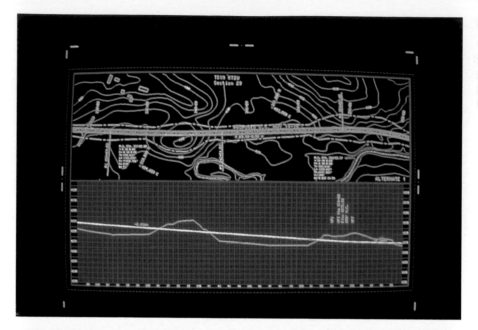

FIGURE C-4

Computer-aided design and drafting (CAD/D) facilities allow translation of planimetric and topographic detail into profiles of selected cross sections (courtesy of McDonnell Douglas).

FIGURE C-5

Geographic information systems (GIS) have traditionally emphasized spatial analysis and modeling: this screen display correlating population distribution to the location of contaminated wells and hazardous material generation facilities relies on the information overlay and proximity calculation functions of a GIS (courtesy of Environmental Systems Research Institute [ESRI]).

FIGURE C-6

Automated Mapping and Facilities Management (AM/FM) is used predominantly by utilities: a display shows proposed subdivision and electric facilities, e.g., poles, distribution lines, and power drops (courtesy of Synercom Technology, Inc.).

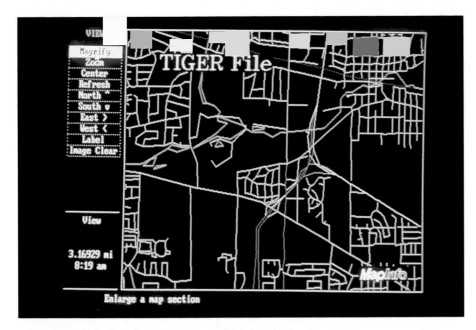

FIGURE C-7

The U.S. Bureau of the Census developed a second generation of street centerline files, or TIGER files. The centerline files are associated with address ranges of street segments for use in a wide range of locational analysis, sometimes called geoprocessing (courtesy of MapInfo).

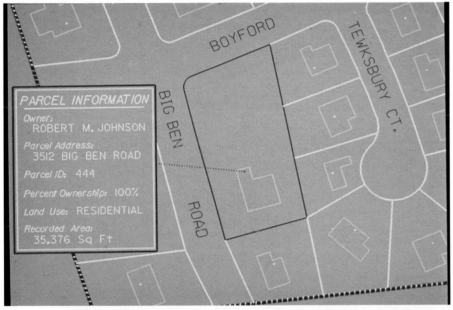

FIGURE C-8

Land information systems (LIS) are an increasingly widespread application of GIS technology: here, a screen displays information for a selected parcel in a land information data base (courtesy of Intergraph).

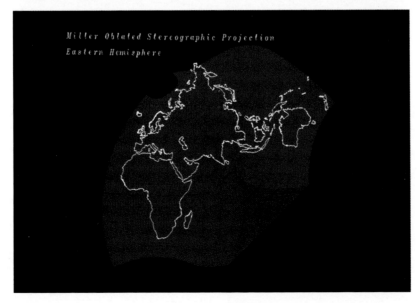

FIGURE C-9

The representation of the earth's sphere in two dimensions is accomplished through many "projections"; one cartographic tool of GIS is the transfer of data between various projections (courtesy of ESRI).

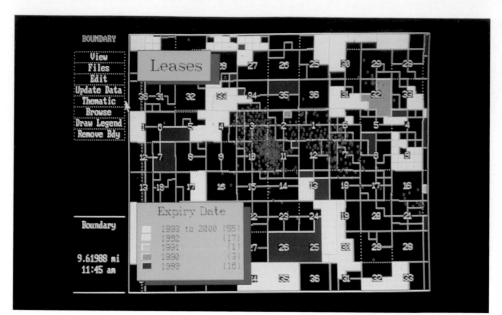

FIGURE C-10a

The linkage of nongraphic characteristics (attributes) with graphic representations of legal or physical features supports the analysis of databases and dual displays of the results (courtesy of MapInfo).

FIGURE C-10b

The interpretation of soils and slope characteristics from the nongraphic data base distinguishes their relative suitability for use as septic fields (courtesy of Minnesota Land Management Information Center [LMIC]).

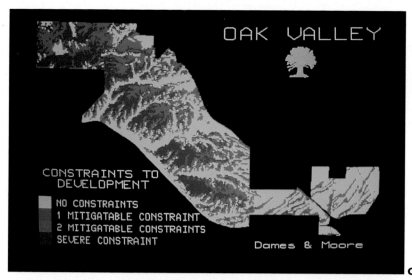

a

b

FIGURE C-11

The analytical capability of GIS supports many uses. The assessment of physical and cultural characteristics leads to the identification of constraints on development and subsequently, a land use plan for the areas (courtesy of Dames & Moore).

FIGURE C-12

The analysis of fire hydrant location and coverage supported the elimination and relocation of hydrants (courtesy of Burnaby British Columbia).

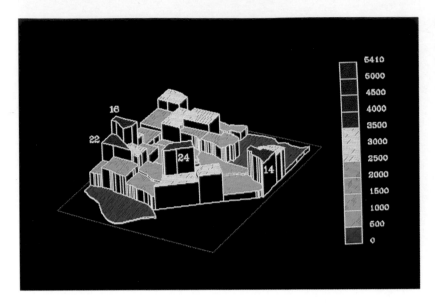

FIGURE C-13

Representation of population within service districts provides an immediate and graphic interpretation of statistical information (courtesy of Burnaby British Columbia).

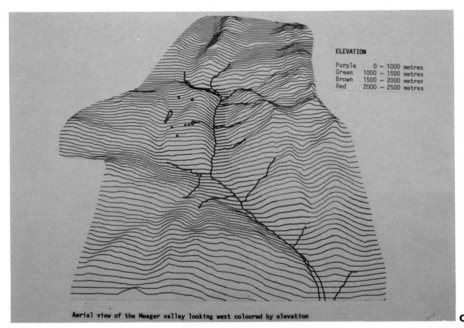

ELEVATION

Purple 0 – 1000 metres
Green   1000 – 1500 metres
Brown   1500 – 2000 metres
Red   2000 – 2500 metres

Aerial view of the Meager valley looking west coloured by elevation

a

FIGURE C-14

GIS and allied imaging technologies are used to model the impact of forest clearcutting: (a) digital elevation model depicting four classes of elevation and stream network; (b) computer simulated view of forest terrain with harvest area depicted in green; (c) digital photo image of site; (d) contour lines overlain on photo image with harvest area depicted in green; (e) digital photo image altered to depict harvest area; (f) computer simulated aerial view of clear cut area (courtesy of Clark University).

b

c

d

e

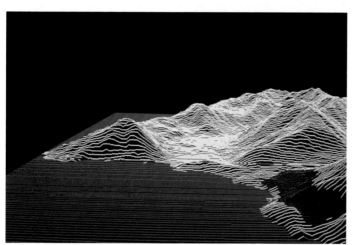

f

a

FIGURE C-15

Computations of distance between childcare centers provides human service delivery planners a foundation for determining accessibility and service effectiveness: (a) land use inventory; (b) proximity definition with zones differentiated by the use of color; (c) service availability in the context of building location (courtesy of Catherine Berris Associates).

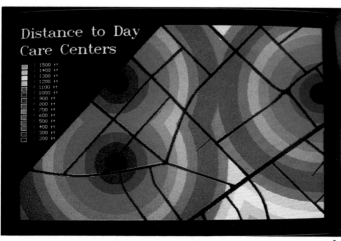

b

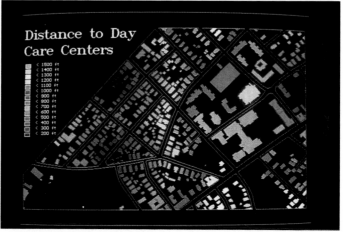

c

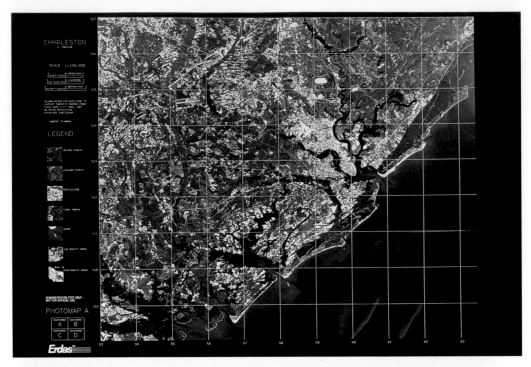

FIGURE C-16

Orthophotomap of Charleston, South Carolina, prepared from Thematic Mapper satellite imagery, providing a single synoptic view of 1987 land cover and land/water boundaries (courtesy of ERDAS, Inc.).

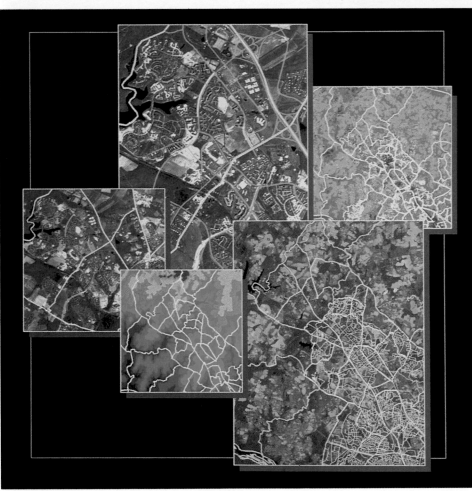

FIGURE C-17

Raster imagery serving as base maps to vector data. Clockwise from top: SPOT panchromatic satellite imagery overlain with census tracts and major roads; land cover classification from EOSAT Thematic Mapper; false color infrared EOSAT TM image with census overlay; digital elevation data with census tract overlay; and SPOT panchromatic images showing changes in one-year period.

FIGURE C-18

The results of a recreational suitability model, circa 1976, of the Maryland Automated
Geographic Information (MAGI) system produced on a DICOMED film recorder (legend
text was added subsequently).

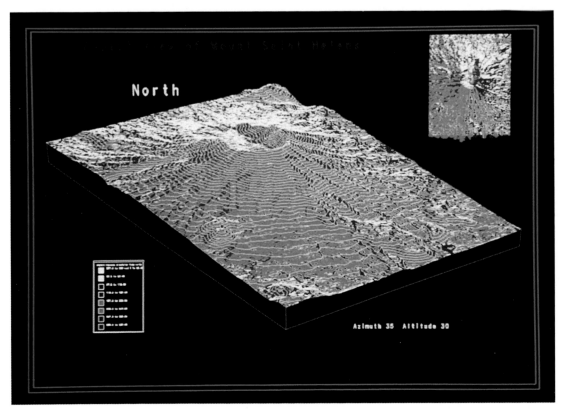

FIGURE C-19

Planimetric and oblique
digital views of Mt. Saint
Helens where
directional (compass)
aspect of slope is
computed and
differentiated in color
(courtesy of ESRI).

FIGURE C-23

U.S. Fish and Wildlife Service track polar bears with satellite telemetry and analyze population distribution and international ranging of bears that inhabit Alaskan coastal areas (courtesy of U.S. Department of Interior, Fish & Wildlife Service).

FIGURE C-24

Models of land use suitability and development by the Minnesota Land Management Information Center (LMIC) identify lands available for development (courtesy of LMIC).

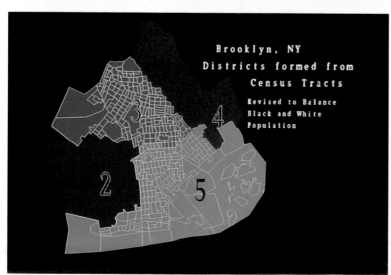

FIGURE C-25

GIS will support more than 75 percent of state-level redistricting activities following the 1990 Census while local governments place increasing reliance on it for defining both electorial and service districts (courtesy of ESRI).

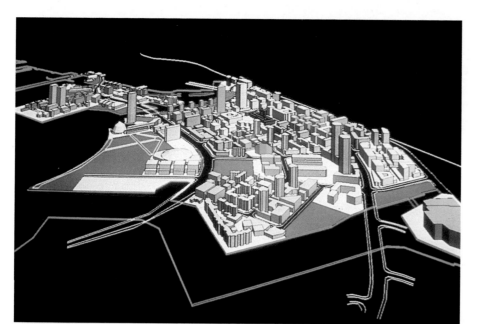

FIGURE C-26

Three-dimensional perspectives provide planners, elected officials, and their constituencies an opportunity to review and communicate proposals in a non-technical media (courtesy of McDonnell Douglas).

1986/1988 SPOT Panchromatic Change Detection with 1986 Landuse Arcs

Development Started Between 1986 and 1988
■ Development Started Prior to 1986, finished before 1988
— Landuse Arcs

DATA PROVIDED BY:
• NASA/EOCAP

arc-info LIVE LINK

FIGURE C-27

Use of SPOT Thematic Mapper data and GIS land use data for change detection analysis of Del Mar, California. Red represents development started between 1986 and 1988, and blue reflects development started prior to 1986 and completed prior to 1988 (courtesy of ERDAS).

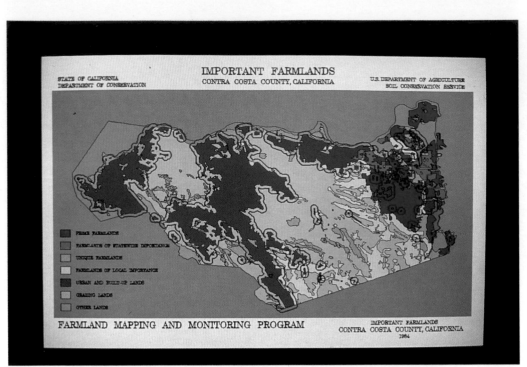

FIGURE C-28

GIS has long supported the development and monitoring of land use policies, such as the preservation of prime agriculture land in Contra Costa, California (courtesy of Intergraph).

FIGURE C-29

The distribution of liquor licenses within the City of Milwaukee, mapped in conjunction with the location of Aldermanic Districts (courtesy of City of Milwaukee).

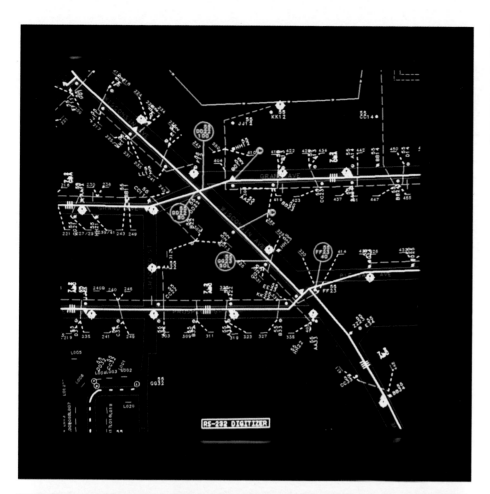

FIGURE C-30

Distribution network planning and engineering are common applications of AM/FM systems within electric utilities; real time linkages to System Control for load management may soon produce even larger benefits (courtesy of Geographic Systems Corporation).

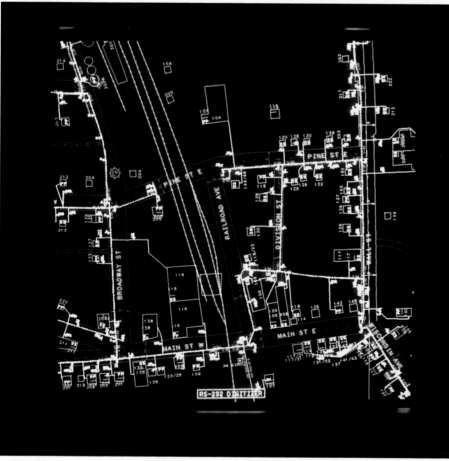

FIGURE C-31

AM/FM offers natural gas utilities opportunities to integrate engineering functions, map production, regulatory compliance, and facilities accounting by linking graphic and nongraphic databases throughout the organization (courtesy of Geographic Systems Corporation).

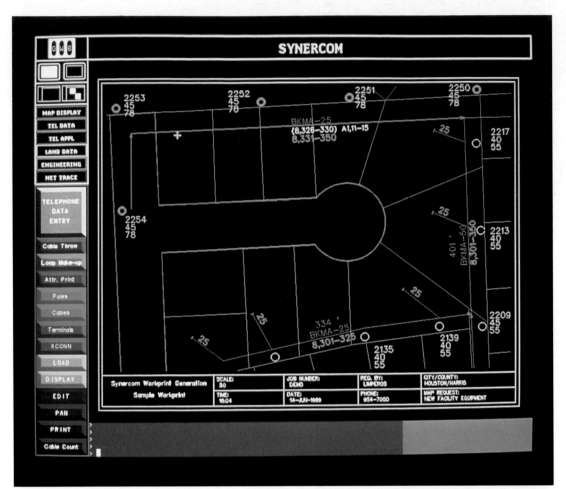

FIGURE C-32

Electric utility drops and poles are located against parcel boundary files provided by local government as part of the cooperative program being implemented by Nashville-Davidson County (Tennessee) and Nashville Electric Service (courtesy of Synercom and Nashville MAGIC).

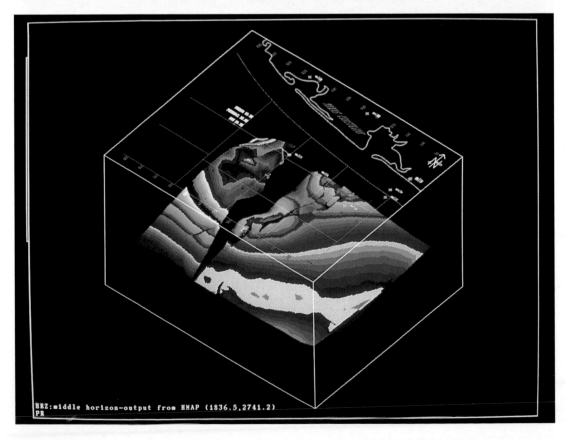

FIGURE C-33

Seismic interpretation software allows petroleum explorers to analyze huge amounts of seismic data to visualize the geological strata beneath the earth's surface. The isometric view from an oil reservoir in the Gulf of Mexico can be rotated in real time to visualize fault geometrics as well as structural highs and lows (courtesy of Landmark Graphics Corporation).

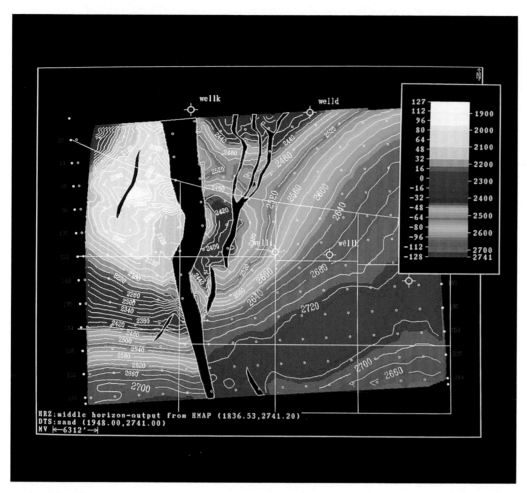

FIGURE C-34

A broad color range helps highlight subtle changes in depth and enables seismic interpreters to analyze complex geologies (courtesy of Landmark Graphics Corporation).

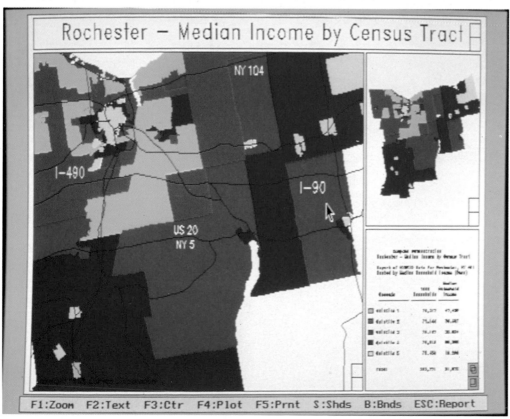

FIGURE C-35

The combination of demographic and economic data relative to known geographies, such as census tracks, postal areas, telephone exchanges, and street segments, provide diverse potentials for market analysis and marketing of products and services (courtesy of Claritas Corporation).

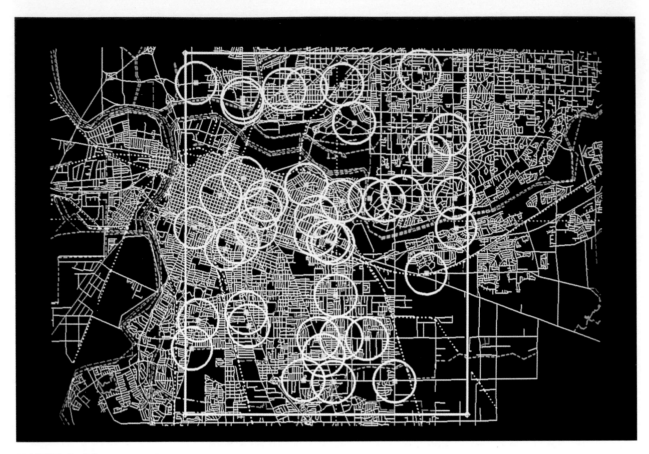

FIGURE C-36

Drawing upon the 1990 census geographies and TIGER files, market analysis at the street segment and census block level will become a routine function of GIS (courtesy of Intergraph).

Adult Entertainment Zone

Residentially Zoned Areas
Public Schools
Churches
Public Lands and Institutions(PLI) Zoned Areas
1000 foot Residential, School, and Church Buffers
Adult Entertainment Zone

FIGURE C-37

Municipality of Anchorage uses GIS to analyze the compatibility of land uses in proximity to adult entertainment enterprises (courtesy of Municipality of Anchorage).

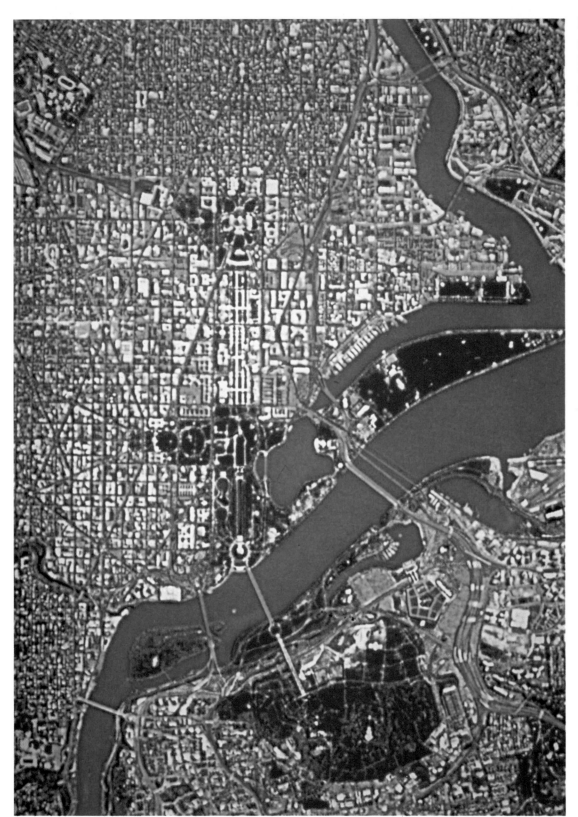

FIGURE C-38

Multispectral Satellite (MS) Imagery (SPOT) of Washington, DC, and environs typifies raster data.

FIGURE C-39

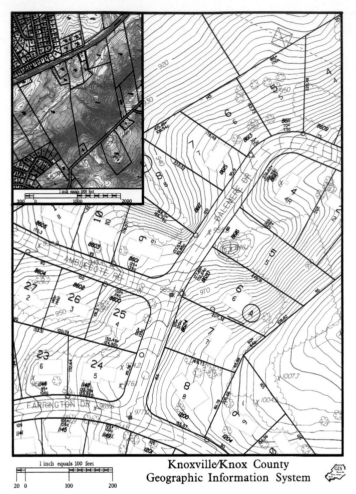

Map depicting subdivision detail at 1″ = 100′ and an overview of the area at 1″ = 1,000′ (courtesy of Knox County, Tennesse [KGIS])

Knoxville/Knox County
Geographic Information System

FIGURE C-40

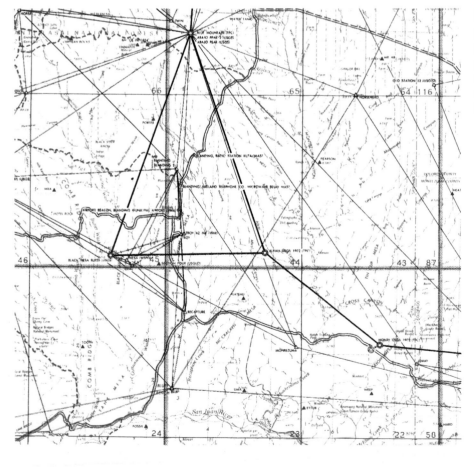

Geodetic control diagrams, prepared by the National Geodetic Survey, provide an index to the location of horizontal and vertical control monuments and related information (courtesy of National Geodetic Survey).

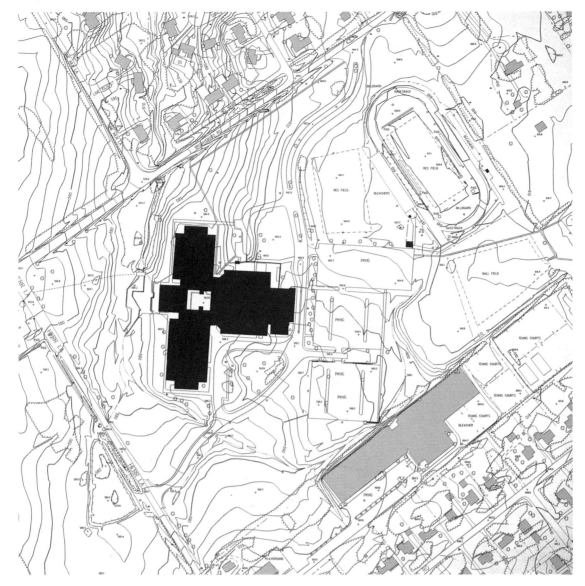

FIGURE C-41

Contours of equal elevation surrounding a high school (courtesy of
Louisville-Jefferson County [Kentucky] Information Consortium, LOJIC).

FIGURE C-42

Digital elevation model (courtesy of
LOJIC).

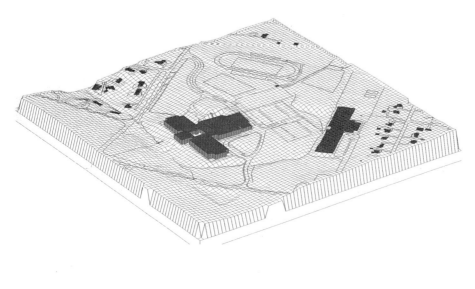

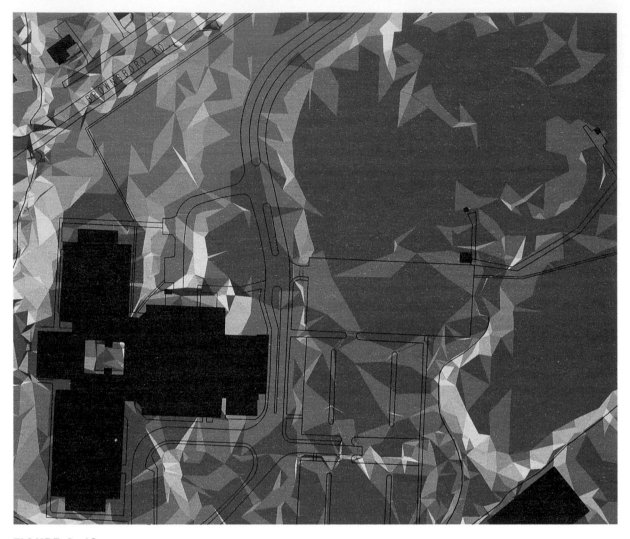

FIGURE C-43

Digital elevation model interpreted to indicate the relative steepness of the land surface's slope (courtesy of LOJIC).

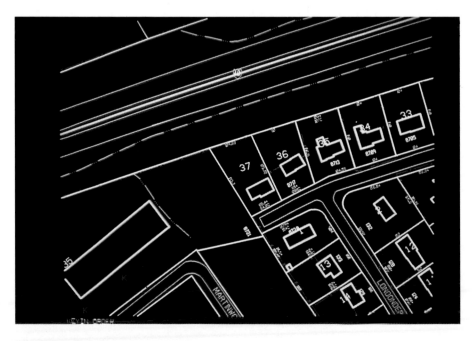

FIGURE C-44

Parcel maps showing private lands and easements (white lines), street edges (pink), building outlines (yellow), and subsurface utilities (blue) (courtesy of KGIS).

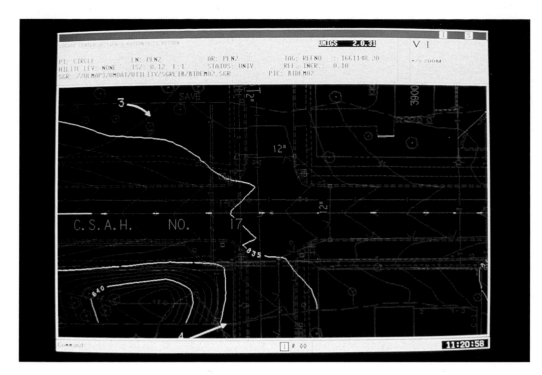

FIGURE C-45

Congested engineering drawing depicting both surface and subsurface engineered facilities (courtesy of Ultimap Corporation).

FIGURE C-46

Pen plotter planimetric maps of Columbus, Ohio (courtesy of Franklin County Auditor).

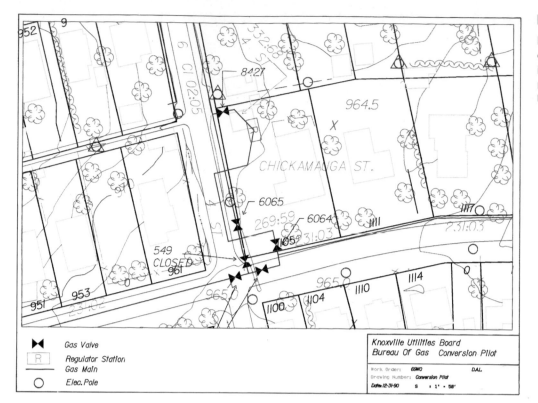

Figure C-47

Electrostatic plots are composed of many individual print elements (dots). (Courtesy of Knoxville Utility Board [KGIS]).

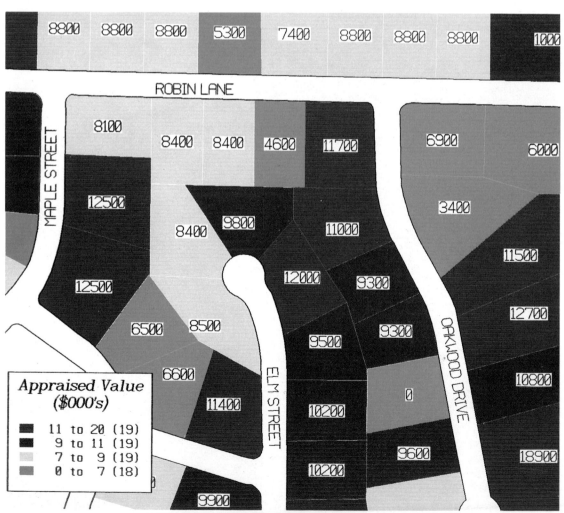

FIGURE C-48

An inkjet (HP Paintjet) screen image of property assessment records (courtesy of MapInfo).

Part II

SYSTEM COMPONENTS

Chapter 5

Data Base Concepts

Building a GIS data base is typically the most expensive, time-consuming, and problematic aspect of implementing GIS technology. Because of the significance of the data base, the next two chapters are devoted to it. This chapter discusses the concepts of GIS data and data management. Chapter 6 reviews the various classes of geographic data typically found in a GIS data base. See Chapter 8 for a discussion of data management software.

Some disagreement exists over the meaning of the term *data base*. As two separate words, the term refers to a large collection of data in a computer, organized so that it can be expanded, updated, and retrieved rapidly for various uses (*Webster's New World*, 2nd Ed.). The data may be organized as a single file or as multiple files or sets. Written as a single word, the term *database* as used in this book means a specific grouping of data within the structure of a data base management software system. The data base of a GIS typically is composed of multiple sets of graphic and nongraphic data managed by the software of a GIS (Fig. 5-1).

The data in a GIS are varied and complex. They may include digital descriptions of map features, logical geographic relationships among features, and nongraphic data that describe characteristics of the features and phenomena that occur at specific geographic locations. The contents of the data base are determined by the varied applications that make use of a GIS in a specific situation.

GRAPHIC AND NONGRAPHIC DATA

A GIS data base can be divided into two basic types of data: graphic and nongraphic. Each of these types has specific characteristics, and each has different requirements for efficient data storage, processing, and display (Figs. 5-2, 5-3).

Graphic data are digital descriptions of map features. They may include the coordinates, rules, and symbols that define specific cartographic elements on a map. The GIS uses graphic data to generate a map or cartographic "picture" on a display device (computer screen), on paper, or through other media.

Nongraphic data are representations of the characteristics, qualities, or relationships of map features and geographic locations. They are stored in conventional alphanumeric formats, though document management systems that manage data as graphic images in raster format are beginning to be linked with GIS technology.

The term *nongraphic* is used here to differentiate those data that do not describe

A GIS data base is a structured collection of digital graphic and nongraphic data that describe map features, spatial relationships, and characteristics of an area or feature.

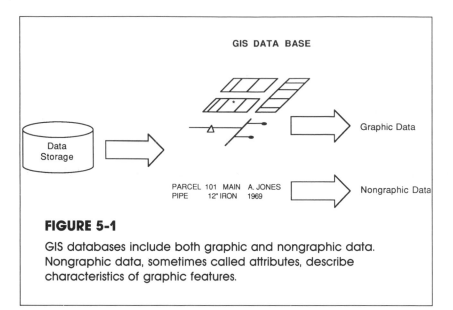

FIGURE 5-1

GIS databases include both graphic and nongraphic data. Nongraphic data, sometimes called attributes, describe characteristics of graphic features.

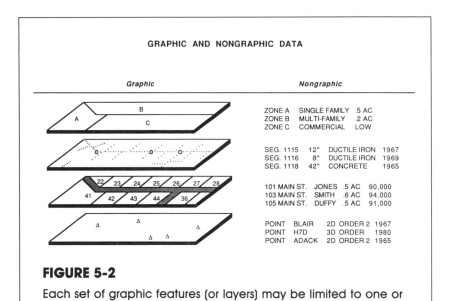

FIGURE 5-2

Each set of graphic features (or layers) may be limited to one or more sets of associated nongraphic data.

ten, nongraphic data are managed separately from the graphic data due to their different characteristics or their maintenance and use in other systems.

Nongraphic data may be further divided into attributes, geographically referenced data, geographic indexes, and spatial relationships (Fig. 5-4). *Attributes* describe in words and numerals the entities represented by the graphic elements. *Geographically referenced data* describe physical phenomena, man-made features, and events that occur at a specific geographic location. *Geographic indexes* help locate map features and data based on their geographic identifiers. *Spatial relationships* are descriptions of the proximity, adjacency, and connectivity of map features.

Graphic Data

Graphic data represent map images in computer-usable form. Graphic data use six types of graphic elements to depict map features and annotation (i.e., points, lines, areas, grid cells, pixels, symbols) (Figs. 5-5, 5-6).

As defined in the Proposed Standards for Digital Cartographic Data, a *point* is a zero-dimensional object that specifies a geometric location through a set of coordinates (The American Cartographer, Jan 1988).

A *node* is a special type of point, also a zero-dimensional object, that is a topological junction or end point and may specify a geometric location.

A *line* is a one-dimensional object. A *line segment* is a direct line between two points. Special forms of lines include these:

- *String*, a series of line segments
- *Arc*, a locus of points forming a curve defined by a mathematical function
- *Chain*, a directed sequence of nonintersecting line segments or arcs with nodes at each end

An *area* is a bounded, continuous two-dimensional object that may or may not include its boundary; individual areas are represented as polygons.

the graphic images of the map features. Nongraphic data are often called *textual* data or *attributes*. They are related to geographic locations or graphic elements and are linked to them in the GIS through common identifiers or other mechanisms. Of-

A *pixel* is a two-dimensional picture element that is the smallest indivisible element of an image.

A *grid cell* is a two-dimensional object that represents a single element of a continuous surface.

Symbols are graphic elements that represent features at points on a map.

Annotation is the text or labels plotted graphically on a map and includes such items as street names, place names, identification numbers, and dimensions.

Graphic images can be stored as vectors or as a raster of uniform grid cells or pixels (Fig. 5-7). Vector data are represented by horizontal (i.e., *x* and *y*) coordinates of point and line locations or as rules for computing the coordinates and connecting the points as lines or areas. Vector data define polygons, objects, and other complex entities that can be manipulated or displayed on the basis of their attributes.

Coordinate systems are of two types: The Cartesian system is most convenient for map plotting, and the angular system (geographic latitude/longitude coordinates) is used for terrestrial measurement. Map coordinate values are expressed in one of numerous potential map projections, which is the transformation of positions on the curved surface of the earth onto a flat map surface. Although any coordinate system may be used, the most commonly used in GIS technology in the United States are the State Plane Coordinate (SPC) System and Universal Transverse Mercator (UTM) projections (Fig. 5-8). Conversion of data from one map projection to another generally can be accomplished without significant loss of accuracy through specific transformation algorithms.

State Plane Coordinate System. The SPC system in the United States is based on a nationwide structure maintained by the National Geodetic Survey. Usually one of two projections—Transverse Mercator (for states with a north–south orientation) or Lambert Conformal (for states with an east–west orientation)—defines one or

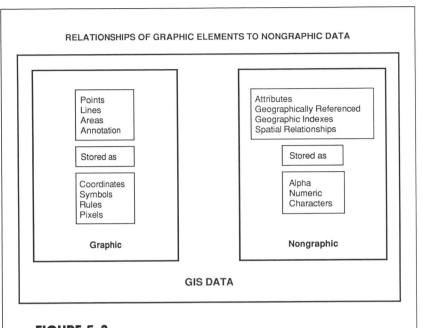

FIGURE 5-3

Graphic and nongraphic data are stored and manipulated in different formats for efficient processing.

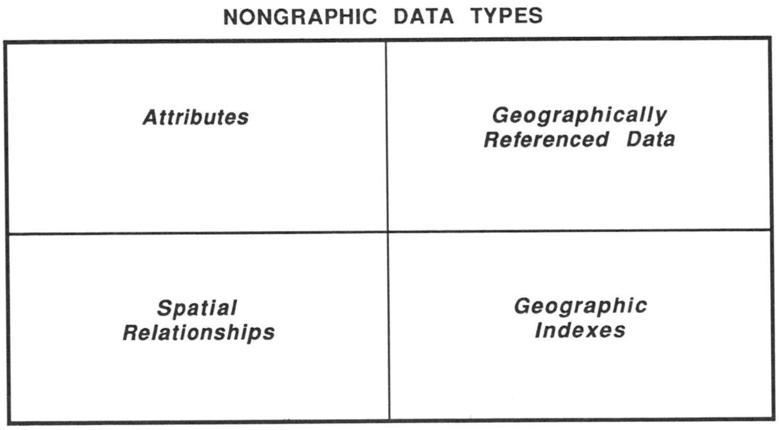

NONGRAPHIC DATA TYPES

Attributes	Geographically Referenced Data
Spatial Relationships	Geographic Indexes

FIGURE 5-4

There are four basic types of nongraphic data.

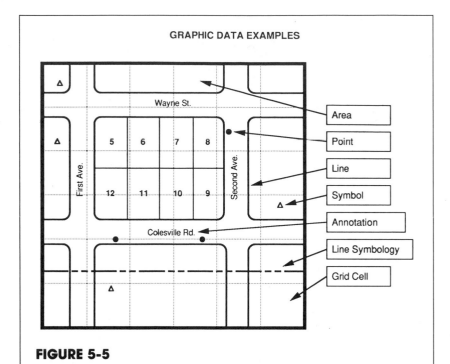

GRAPHIC DATA EXAMPLES

Area
Point
Line
Symbol
Annotation
Line Symbology
Grid Cell

FIGURE 5-5

Maps and engineering drawings are composed of six types of graphic elements and annotation.

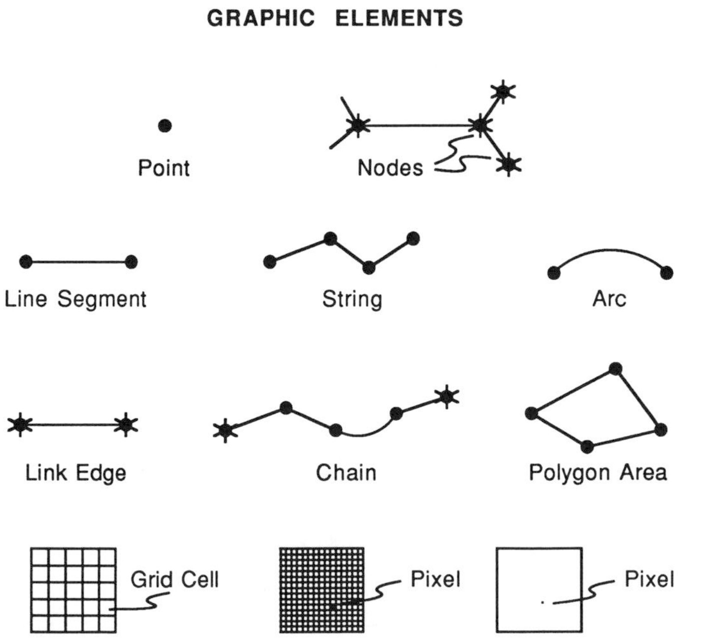

GRAPHIC ELEMENTS

Point   Nodes

Line Segment   String   Arc

Link Edge   Chain   Polygon Area

Grid Cell   Pixel   Pixel

FIGURE 5-6

Six graphic elements (points, nodes, line segments, link edges, grid cells and pixels) are combined to depict maps and engineering drawings.

more specific zones for each state. The geographic extent of the zones was originally limited to a distance of 158 miles (east and west in the Transverse Mercator or north and south in the Lambert Conformal) to limit the amount of distortion that accumulates due to the curvature of the earth. Modern techniques and equipment have allowed some zones to be extended. The SPC is tied to the National Geodetic Reference System (NGRS) and provides the accuracy and compatibility of that geodetic framework. For a GIS covering an SPC zone or smaller area, as a plane projection, the SPC is easy to use and relates to a national structure that can be transformed to other projections with relative ease (Fig. 5-9).

Universal Transverse Mercator Projections. The UTM is an internationally accepted projection that covers the entire globe as one continuous system. This worldwide system is composed of zones 6 degrees wide that are based on the Transverse Mercator projection. UTM values are generally recorded in meters. The projection provides sufficient accuracy for surveying and other detailed purposes.

Latitude-Longitude. Compatible throughout the world, the latitude and longitude coordinate system is also commonly used. This system of geographical coordinates (the geodetic latitude and geodetic longitude) records angular measurements relative to the equator and prime meridian.

Raster Data. Raster data are represented by uniform grid cells of specified resolution. Many different types of organizational schemes exist for raster data, including image data, layers of grids, and other grid cell-oriented data (Fig. C-36).

Data thus can be stored spatially as values for a matrix of grid cells. With this approach, a uniform grid structure is defined and a value for an attribute is assigned to each cell. The resolution of this approach depends on the size of the grid cell, so it is important to select an optimal grid size to suit the requirements. . The

VECTOR/RASTER REPRESENTATION

Vector   *Raster*

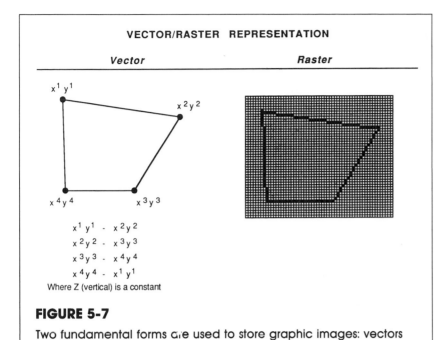

$x^1 y^1$ - $x^2 y^2$
$x^2 y^2$ - $x^3 y^3$
$x^3 y^3$ - $x^4 y^4$
$x^4 y^4$ - $x^1 y^1$

Where Z (vertical) is a constant

FIGURE 5-7

Two fundamental forms are used to store graphic images: vectors and rasters.

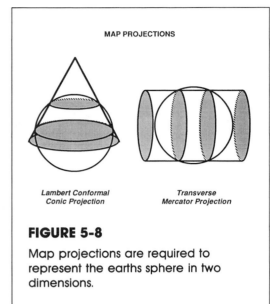

MAP PROJECTIONS

*Lambert Conformal
Conic Projection*

*Transverse
Mercator Projection*

FIGURE 5-8

Map projections are required to represent the earths sphere in two dimensions.

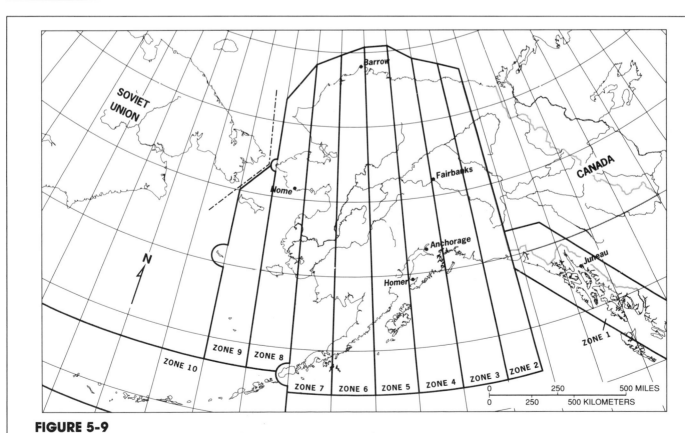

FIGURE 5-9

State plane coordinate systems are frequently the base of large-area GIS. Alaska requires ten zones for complete coverage; most states are covered by one or two zones.

larger the size, the less precise the information; the smaller the grid size, the larger the data base.

In image processing, which typically is applied to satellite data and document management, the grid cells are called *pixels*, the smallest indivisible picture element. In this case, the values represent image characteristics rather than attributes (Fig. C-38).

Symbology. A GIS is equipped to display a wide variety of symbols for points, lines, and areas (Fig. 5-10). Vendors of systems often provide a symbols library with the initial installation. The system users can create new symbols by digitizing desired shapes or elements. Those images can be assigned a name or identifier and stored in a library to be recalled. The symbols can then be placed into map locations as needed during data display. Different symbols may be applied to the same element in different maps. For example, the symbol for a pole may be used for all poles in a general purpose map. But in another map, the same locations may be identified with separate symbols for electric or telephone poles if a distinction is required.

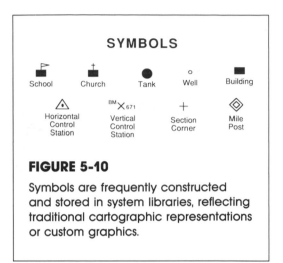

SYMBOLS

FIGURE 5-10

Symbols are frequently constructed and stored in system libraries, reflecting traditional cartographic representations or custom graphics.

Lines of different types also can be represented by a variety of symbols, such as solid, dashed, and dotted lines (Fig. 5-11). These too can be modified as needed, based on attribute values for the lines.

Annotation. Annotation is the alphanumeric text or labeling placed on maps. It includes items such as the names of streets, lakes, rivers, towns, and other en-

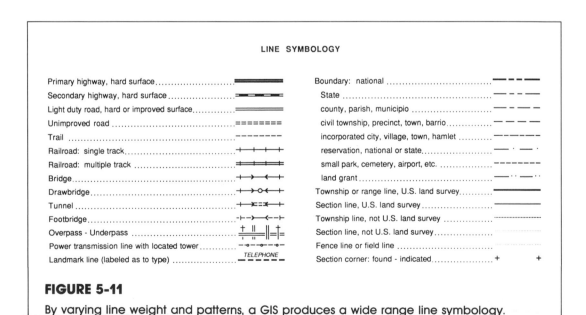

FIGURE 5-11

By varying line weight and patterns, a GIS produces a wide range line symbology.

SAMPLE FONTS

ABCDEFGHIJKLMnopqrstuvwxyz
ABCDEFGHIJKLMnopqrstuvwxyz
ABCDEFGHIJKLMnopqrstuvwxyz
ABCDEFGHIJKLMnopqrstuvwxyz
ABCDEFGHIJKLMnopqrstuvwxyz
ABCDEFGHIJKLMnopqrstuvwxyz
ABCDEFGHIJKLMnopqrstuvwxyz
ABCDEFGHIJKLMnopqrstuvwxyz
ΑΒΧΔΕΦΓΗΙϑΚΛΜνοπθρστυϖωξψζ
ABCDEFGHIJKLMnopqrstuvwxyz
ABCDEFGHIJKLMnopqrstuvwxyz
ABCDEFGHIJKLMnopqrstuvwxyz
ABCDEFGHIJKLMnopqrstuvwxyz
ABCDEFGHIJKLMnopqrstuvwxyz
✿✛⊹✚✦◇★✩☉☆✷★■□▢□▢▲▼◆◈▶▌▐

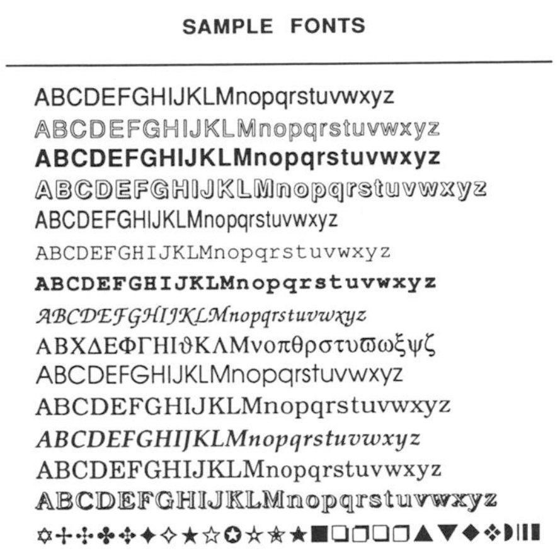

FIGURE 5-12

Standard fonts are provided for displaying and plotting text and many GIS provide facilities for designing custom fonts.

tities. It also includes numeric values such as elevations, dimensions, identifiers, and addresses. Annotation is handled differently from other map features in most systems. It is not stored as strings of coordinates but often rather as a combination of location coordinates, standard fonts, and placement or orientation rules.

A GIS typically provides a standard set of text fonts with descriptions for displaying or plotting each alphanumeric and special character (Fig. 5-12). The font descriptions often are stored in a table or library. Annotation is recorded and displayed using referenced character descriptions from the standard table or library. All characters that make up a label generally are manipulated and displayed together as a single string of graphics.

Three other items of information frequently are stored for annotation:

The *location* of annotation placement can be stored as a single x and y coordinate pair that indicates the beginning, centering, or ending of the annotation element or string (Fig. 5-13). The location can be indicated by two pairs of x and y coordinates, a beginning and an ending point, and the text height. Two pairs of coordinates, a lower left and upper right, indicating the beginning, ending, and height, also may be used.

The *orientation* of the annotation can be specified by the orientation of the two coordinate pairs or by a rule stored with the annotation. The rule can specify an angle of orientation, such as 45 degrees diagonal, or it can specify a relationship to another feature (for example, parallel to an identified line).

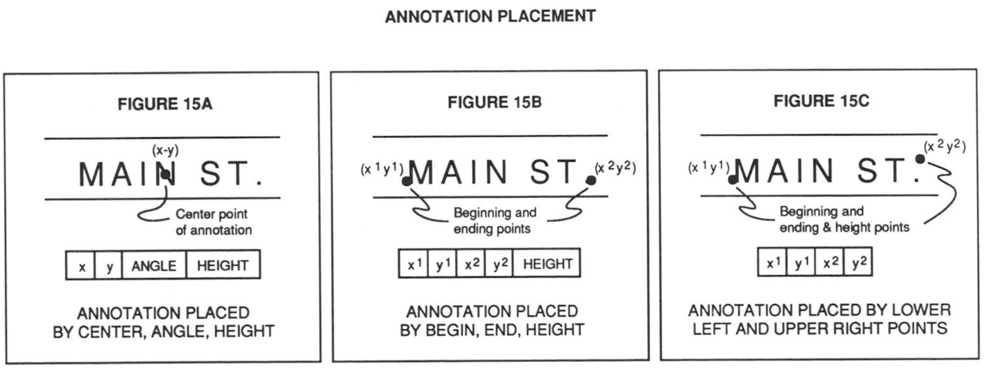

ANNOTATION PLACEMENT

FIGURE 15A	FIGURE 15B	FIGURE 15C

MAIN ST.
(x-y) Center point of annotation

| x | y | ANGLE | HEIGHT |

ANNOTATION PLACED BY CENTER, ANGLE, HEIGHT

(x¹y¹) MAIN ST. (x²y²)
Beginning and ending points

| x¹ | y¹ | x² | y² | HEIGHT |

ANNOTATION PLACED BY BEGIN, END, HEIGHT

(x¹y¹) MAIN ST. (x²y²)
Beginning and ending & height points

| x¹ | y¹ | x² | y² |

ANNOTATION PLACED BY LOWER LEFT AND UPPER RIGHT POINTS

FIGURE 5-13

Placement of annotation is dependent on location, orientation, and size.

The graphic component of the GIS data base is often described as a series of layers, each of which contains map features that are related functionally. Each layer is a set of homogenous features that is registered positionally to the other data base layers through the common coordinate system.

The *size* of the annotation can be indicated by the relative positions of lower left and upper right coordinates, by a height dimension, or by a scale factor.

Most systems also display and plot nongraphic attributes as annotation, though the capability for flexible placement may be more limited in some systems. Often, these data are displayed in a specified area or window or in a standard scale and orientation at a predefined label point for the feature.

Layers. The graphic component of the GIS data base is often described as a series of *layers*, each of which contains map features that are related functionally (Fig. 5-14). Each layer is a set of homogeneous fea-

tures that is registered positionally to the other data base layers through the common coordinate system. The separation into layers is based on logical relationships and the graphic portrayal of sets of features. One major purpose of the segregation is to simplify the combination of features for display.

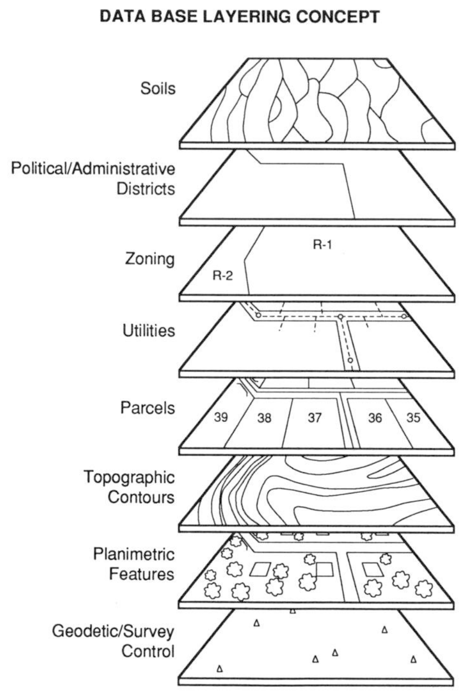

A geodetic control framework is commonly used to register the various map feature layers, although other reference schemes may serve the same purpose. This electronic layering scheme is comparable to a series of overlays in a manual mapping system.

Although the data base may be most often conceived of as a series of layers, in reality, GIS vendors use various techniques to differentiate between features. Some store sets of features as digital layers, while others assign data types to the elements or use attributes to separate or combine features.

The sequence of layers usually begins with the reference grid and base features. The base features typically include geodetic control points that use coordinates to record locations monumented on the earth's surface and planimetric features that are compiled from aerial photographs or field surveys. Numerous other feature layers, such as topography, parcel boundaries, administrative and legal boundaries, utility features, and natural physical conditions, are subsequently registered to the base.

Objects. Some vendors offer an alternative approach to graphic data organization in which the individual features or groups of features are defined as objects. In this case the relationships among features are defined in the data base specification and the software operates on the objects as entities. The conceptual layers can still be formed with this approach through the definition of objects.

Nongraphic Data

Nongraphic data describe the characteristics of the graphic images and the incidents or other phenomena that occur at

DATA BASE LAYERING CONCEPT

Soils
Political/Administrative Districts
Zoning
Utilities
Parcels
Topographic Contours
Planimetric Features
Geodetic/Survey Control

FIGURE 5-14

Though software systems vary in how graphic data are physically handled, the concept of data organization in layers is useful for communicating data structure.

BOX 5-1 Representative Nongraphic Attributes

For an Ownership Parcel—Owner name, size, land use, value

For a Water Line—Size, material, year installed

For a Street Segment—Pavement width, number of lanes, material, resurface date

For a Soil Polygon—Soil type, erosion and wetness identifiers

---

separately from the graphic data files and are linked through the common identifier.

A GIS is able to query or analyze the attributes separately and to generate a map based on nongraphic values (for example, it could display all 12-inch pipe segments installed in the last fifteen years). Most systems also can display attributes as annotation on a map or as control parameters for the selection of map display symbols.

Geographically Referenced Data. Geographically referenced data describe inci-

specific geographic locations. A GIS data base has four classes of nongraphic data, namely:

- Nongraphic attributes
- Geographically referenced data
- Geographic indexes
- Spatial relationships

Each class comprises alphanumeric data that are entered, stored, managed, processed, and reported in the same manner as data in conventional alphanumeric data processing systems. Nongraphic data used by a GIS can be acquired as needed from (and managed by) a related, separate data processing system, or they can be managed by the GIS directly.

Nongraphic Attributes. Nongraphic attributes provide descriptive information about the characteristics of map features. They are linked to the graphic elements through common identifiers, often called *geocodes*, that are stored in both graphic and nongraphic records.

Nongraphic attributes can include qualitative and quantitative data that describe a point, line, polygon, or network feature stored in the data base. For example, a map layer depicting pipe segments in a water system could have a file of attributes that includes the segment number, diameter, material, and date installed (Fig. 5-15). Attributes usually are stored

MAP FEATURES AND ATTRIBUTES

For a pipe segment in a water system:

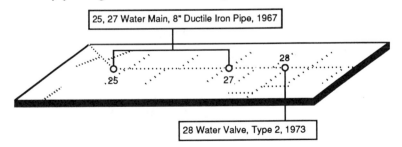

25, 27 Water Main, 8" Ductile Iron Pipe, 1967

28 Water Valve, Type 2, 1973

For a land parcel:

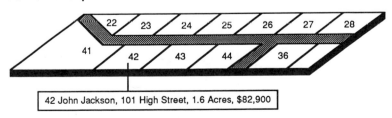

42 John Jackson, 101 High Street, 1.6 Acres, $82,900

FIGURE 5-15

Nongraphic data are linked to graphic elements, e.g., points, lines, and polygons, providing quantitative and qualitative characterization of the features.

Nongraphic attributes provide qualitative and quantitative information about the characteristics of map features, such as points, lines, polygons, or network features stored in the GIS data base. They are usually stored separately from the graphic data and are linked by a common identifier.

dents or phenomena that occur at a specific location. Unlike the attributes, they do not describe the map feature itself. Instead, this type of data describes items or actions (such as building permits, accident reports, or health inspections) that can be related to specific geographic locations (such as addresses, parcels, or intersections). Geographically referenced data typically are stored and managed in separate files and systems that are not directly associated with the graphic features in a GIS data base. However, these records do contain elements that identify the location of the event or phenomenon. In the examples above, the building permits and health inspections can be related to a geographic location through a street address; the accident reports can be related through an intersection or street name and milepost.

Geographic Indexes. Geographic indexes are maintained in a GIS to select, relate, and retrieve data based on geographic locations that have been described by geographic identifiers. An index may include multiple identifiers for entities used by different organizations as well as listings of geocodes that identify the spatial relationships between locations or between features and geographic entities. The index may, for example, contain the various identifiers for individual parcels, such as street address, mailing address, parcel

number, and account number. A very common index is the Geographic Base File (GBF), which contains names, addresses, blocks, census tracts, and other geocodes related to each street segment.

Spatial Relationships. The spatial relationships of entities at particular geographic locations are important to many GIS processes (Fig. 5-16). The relationships can be simple and logical, such as the fact that 101 Main Street is next in sequence and therefore adjacent to 103 Main Street, or that both are odd numbers and therefore fall on the same side of Main Street in the 100 block. These relationships are useful in numerous applications; for example, they can help determine how to dispatch the proper emergency vehicle or assign the proper census tract values.

The topological relationships (i.e., connectivity and adjacency) among graphic elements or geographic entities are more complex. Examples of such elements are the nodes (intersections), links (street segments), and areas (blocks) of the urban geography (Fig. 5-17). Spatial relationships that can be recorded for these elements include the facts that a specific pair of nodes terminates a specific link, and that a particular link is a boundary of a specific area.

Some relationships also can be interpreted or calculated from the geographic coordinates of graphic elements. For example, the number of telephone poles or other features within 1 mile of a specific

---

BOX 5-2   Sample Geographic Index

Street name, type, direction, from node, to node

Left address range, block, census tract, jurisdiction, fire district, police beat, traffic zone, plan areas

Right address range, block, census tract, jurisdiction, fire district, police beat, traffic zone, plan areas

---

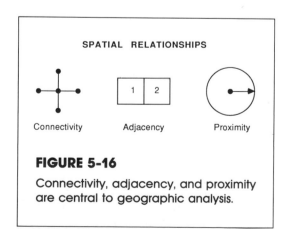

SPATIAL RELATIONSHIPS

Connectivity   Adjacency   Proximity

FIGURE 5-16

Connectivity, adjacency, and proximity are central to geographic analysis.

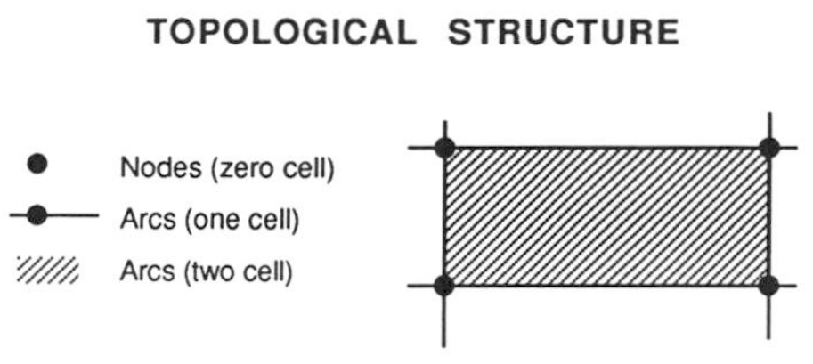

TOPOLOGICAL STRUCTURE

- ● Nodes (zero cell)
- ●—● Arcs (one cell)
- ▨ Arcs (two cell)

FIGURE 5-17

The ability of a system to develop and utilize topological relationships differentiate GIS from automated drafting and imaging systems.

most common method of linking the two is to store identifiers simultaneously with each set of graphic and nongraphic elements. The identifier can be simply a unique number assigned sequentially, randomly, or arbitrarily, or it can indicate the geographic or data storage location. The identifier for an entity can contain its location coordinates, map sheet number, area description, township and section, or a pointer to the storage location of related data (Fig. 5-19). The identifier is stored with the coordinates or other values describing the graphic image and with related nongraphic data.

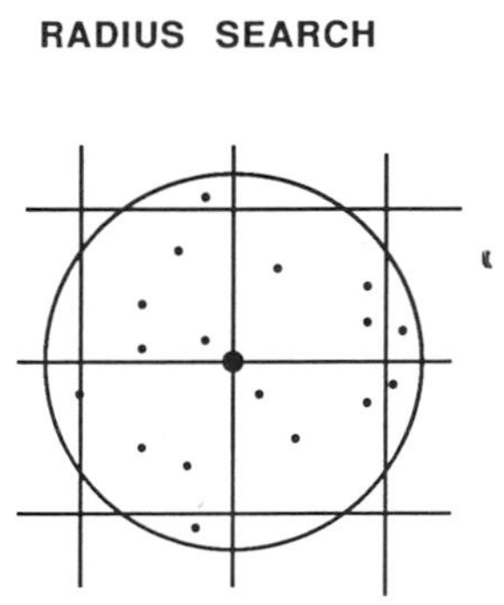

To use nongraphic and geographically referenced data in geographic analysis, the GIS may interface with other computers, typically found in business or administrative environments, to obtain or share nongraphic data.

intersection can be determined by a radius search analyzing the proximity of the features' coordinate values (Fig. 5-18). Spatial relationships can be encoded as nongraphic data or implied through coordinate values.

Relationships of Graphic and Nongraphic Data

The GIS maintains the relationship between graphic and nongraphic data. The

Relationships With Other Computer Systems. To use nongraphic attributes and geographically referenced data in geographic analysis applications, the GIS may interface with other systems to obtain or share nongraphic data. These other systems typically are found in administrative or business environments, and although they have different purposes, they are nonetheless a source for spatially referenced digital data. Similarly, the GIS sometimes is a source of geographic data for other systems.

There may be several levels of interface between systems. In the simplest form, data may be passed from one system

RADIUS SEARCH

FIGURE 5-18

Search techniques are used to define a particular area by specifying a radius or by calculating a radius (or distance between two features) required to satisfy a set of specified conditions.

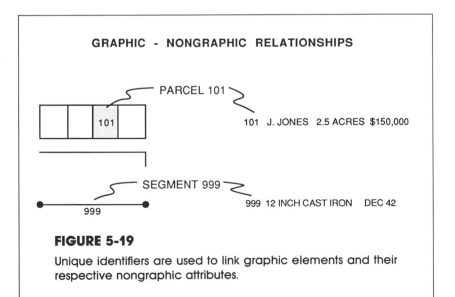

GRAPHIC - NONGRAPHIC RELATIONSHIPS

PARCEL 101

101   101 J. JONES 2.5 ACRES $150,000

SEGMENT 999

999   999 12 INCH CAST IRON   DEC 42

FIGURE 5-19

Unique identifiers are used to link graphic elements and their respective nongraphic attributes.

to the other by copying onto a magnetic tape or disk that is loaded into the other system. The next level is electronic linkage and movement of data in batches. In this approach, one system becomes a remote entry device for the other. This level of interface may be used for periodic updates from one system to the other or for ad-hoc data acquisition by one system. The most sophisticated level is direct on-line, real-time exchange, wherein an inquiry is entered in one system and data for the response are retrieved directly from another.

The appropriate level of interchange varies by installation and application. The difficulty of accomplishing an interchange also varies. In the simplest case, both the GIS and the related system operate on the same processor, or on compatible processors of the same vendor, and use the same operating and data management systems. At the other extreme, two separate processors of different vendors that use different operating systems require a gateway and translation between protocols, formats, and software structures. This technique is relatively new, especially to the GIS world, but rapid advances have been made in multivendor computer networks and it is becoming increasingly popular.

DATA MANAGEMENT

Graphic and nongraphic data have distinct characteristics and thus different optimum techniques for their management. One notable characteristic of graphic data is their high volume, which requires efficient find-and-retrieval procedures. Nongraphic data generally are processed using standard alphanumeric data management software and structures.

The GIS system vendors use varying design approaches to the diverse requirements for handling graphic and nongraphic data. Some systems separate graphic and nongraphic data and manage each type with software best suited to its characteristics. Other systems handle both types with a single software and management structure.

Approaches

Since the storage structures can be tailored to the characteristics of the various data types, a GIS could have a graphic structure for coordinates, a topological structure for spatial relationships, and a conventional data base management system for attributes. The basic data models in use include hierarchical, network, relational, and flat file models (Fig. 5-20).

The *hierarchical* model stores data in a structure based on parent-child, one-to-many relationships. If those relationships can be clearly defined, and if queries are standard, can be predefined, and follow those relationships, the model is very efficient. Many early geographic systems used the hierarchical model.

In the *network* model, records of the same type are grouped in conceptual files. They also can be grouped into owner-coupled sets reflecting one-to-many relationships between the files. A network system can model one-to-many or many-to-many relationships.

The *relational* model, as defined by E. F. Codd, the well-known data base model innovator, is based on a matrix structure of rows and columns of data. This approach, whereby tables of matrices store the data, facilitates ad-hoc inquiry and is particularly useful in situations where nonstandard inquiries are common. The model allows great flexibility in the range of queries that can be handled. Virtually any combination of values can be used as selection criteria; there is no need to predefine selection keys. This model has become very common in the GIS world (Codd, 1976).

Not all systems are exact implementations of a specific model. Hybrids and variations that offer functions of more than one model also exist. A GIS usually has a data base management system, either uniquely developed for the GIS or a standard commercial package, that will manage at least the nongraphic data.

A fourth model, the *sequential* or *flat file*, is not a true data base management system. However, this approach is commonly found as the data structure for existing nongraphic data when a GIS is being implemented. These files are managed by specifically developed systems or programs written in COBOL, BASIC, or other languages and are tailored to individual applications.

Graphic Data Storage

The graphic data require special data storage techniques to support efficient processing. The techniques must accomodate the very large quantity of data necessary to describe graphic images. They must support the location and retrieval of data from the large set with reasonable response times. This will require efficient organizing and structuring of the data on the storage media to minimize requirements for transferring data between internal and external memory devices. The techniques must also support the linking of the graphic data with their related nongraphic characteristics. Several alternative approaches are found among the GIS vendors. Those that have evolved from a drafting or CAD background may store the graphic elements as graphic objects that are specifically defined combinations of graphic elements. Spatial indexing techniques may also be employed to improve access efficiency.

A single graphic element may represent multiple features or may be defined as a part of multiple features (Fig. 5-21). For example, a graphic line element representing the edge of a road also may represent one side of a parcel boundary. Simultaneously, it can be an element in a census tract boundary, a city boundary, and a fire district boundary. This element can be

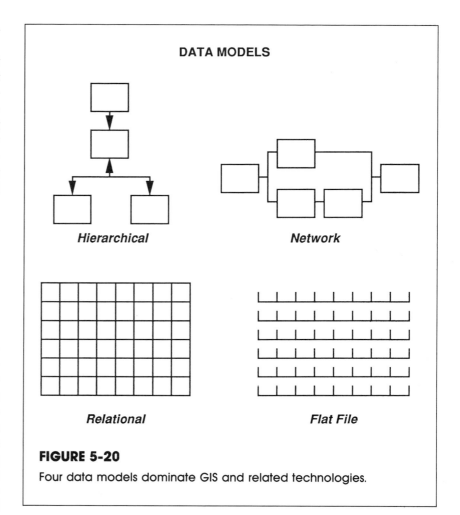

DATA MODELS

Hierarchical

Network

Relational

Flat File

FIGURE 5-20

Four data models dominate GIS and related technologies.

handled in a number of ways. One option is to define it as a single graphic element with indicators that it serves as multiple features for each entity of which it is a part. Alternatively, it can be defined as five separate elements that appear in each of the descriptions of the road, parcel, census tract, city, and fire district boundaries.

Assigning multiple definitions to single graphic elements is typically part of a topological structure, although other structures may support it. The second approach defines separate elements for each feature and is used in graphically oriented management systems. Each approach has its advantages and disadvantages. The topological structure is efficient in storage,

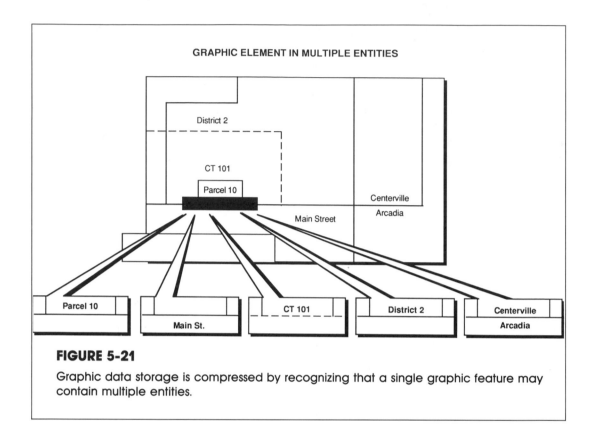

GRAPHIC ELEMENT IN MULTIPLE ENTITIES

FIGURE 5-21

Graphic data storage is compressed by recognizing that a single graphic feature may contain multiple entities.

since the element appears only once in the data base; however, it requires certain memory overhead to describe the various entities of which an element is a part and to handle those relationships. If an element is not stored on the disk near the other elements to be displayed with it, the system may require additional search-and-retrieval time to locate adjacent elements. The single-definition approach minimizes retrieval overhead but requires redundant storage of the graphic element itself, enlarging the size of the data base and requiring additional capacity. The second approach also can result in different representations of elements that should be identical.

Topology

Topology is a branch of mathematics dealing with two types of objects—points (called *nodes*) and lines (called *edges*)—and one type of basic relation between them (called *incidence*) (Frank 1984). Topology is

a technique used to record and manipulate the logical relationships of map features and geographic information in a GIS. Many GISs use a topological structure to define the relationships between nodes, lines, and areas.

In a topological approach, the end points and intersections of lines, such as street or water-line segments are recorded as nodes or 0 (zero) cells. The lines themselves between intersecting lines (or from node to node) are links or 1 (one) cells (Fig. 5-22). The closed areas bounded by a chain of lines are polygons or 2 (two) cells. Each of the topological elements is defined in a specific manner, as are the relationships between elements (Fig. 5-23). For example, a street segment is a specific 1 cell as it traverses from one node (0 cell) to another node (0 cell). These bounding nodes are recorded in the description of the 1 cell in some form. The node (0 cell) definitions also refer to each 1 cell to which the 0 cell is related (e.g., all street segments intersecting at a specific node). The descriptions of the areas (2 cells) such as city blocks, refer to all bounding lines (1 cells). The nodes bounding an area may be encoded explicitly in the area definition or may be identified indirectly through the bounding lines that are recorded with the area. All connectivity and adjacency relationships among features in the GIS data base thus can be identified through the topological information.

Topological information is used for several purposes in a GIS. Initially it may be used in editing and quality control. By applying logical tests to topological data, the system can identify errors such as missing or extraneous lines (Fig. 5-24).

Other uses include retrieval and analytical functions. For example, line and node connectivity relationships are used to model utility and other networks and to trace through these networks for analysis of flows. Routing, districting, and other algorithms use topological information to select or evaluate routes and to allocate, and evaluate proximity to resources (Figs. 5-25, 5-26). Additionally, topological information supports efficient polygon overlaying procedures.

Data Directory

A data directory is essential to effective management of a GIS data base. The directory defines the entities, their attributes, and associated domain values and conveys those meanings to the data base administrator and GIS users. A directory allows system users to identify the map features and attribute data available for specific applications, as well as the definitions, quality, and other characteristics of those data.

Data Base Administration

GIS data can be shared among multiple organizations that update, maintain, and use specific sets of data. This requires effective administration to protect the data base's integrity. The data base administration function uses and maintains the data directory described above. Access controls such as passwords restrict the opportunity

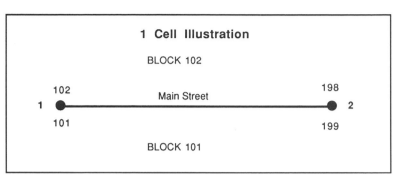

TOPOLOGICAL ELEMENTS

1 Cell Illustration

BLOCK 102

102   Main Street   198
1   •————————————————•   **2**
101   199

BLOCK 101

- **From Node 1 to Node 2**
- **Street name - Main Street**
- **Left address range - 102-198**
- **Right address range - 101-199**
- **Left block 102**
- **Right block 101**

FIGURE 5-22

Topological structures record the logical relationships inherent in many geographic features.

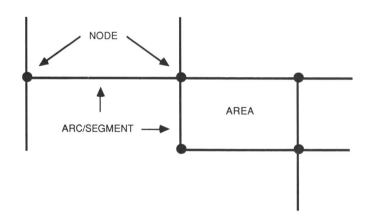

TOPOLOGICAL STRUCTURE

NODE

ARC/SEGMENT →

AREA

O Cell - Node - End point or intersection of arc
1 Cell - Arc/Segment - Line between two nodes
2 Cell - Area/Polygon - Bounded by chain of arcs and nodes

FIGURE 5-23

Topological structures become complex when networks of line segments bound areas; each line segment has both connectivity and adjacency with enclosed areas on either side.

TOPOLOGICAL EDIT

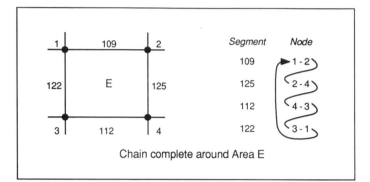

Chain complete around Area E

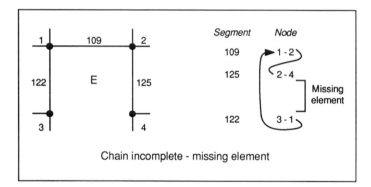

Chain incomplete - missing element

FIGURE 5-24

Logical tests of connectivity are used to test topological integrity.

EXAMPLE OF DISTRICTING

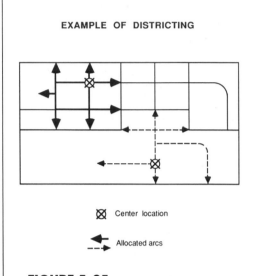

⊗ Center location

←--→ Allocated arcs

FIGURE 5-25

Radial searches along topologically structured line segments are used to construct districts meeting specified parameters.

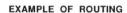

EXAMPLE OF ROUTING

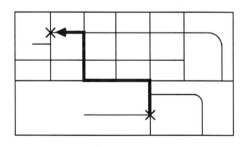

✕ Beginning and end nodes of route

← Optimum path between selected nodes

FIGURE 5-26

Routing along pathways requires a topologically structured network, which is tested against a series of optimization factors.

to view, enter, and modify the data base to authorized users.

Data base administration establishes standards including definitions, quality, and timing, among others. It also monitors the data base continuously to ensure that the quality controls and other standards are observed by all users.

Data base administration maintains backup copies of the data base and system software to allow restoration should a mechanical, electrical, or human failure occur. A specific "backup" cycle is established to duplicate parts of and the whole data base at regular intervals. Copies of the backup data often are stored at a separate site in case of a natural disaster at the system facility.

BOX 5-3   Sample Database Dictionary

Feature Name: GEODETIC CONTROL Feature Set: GEOGRAPHIC REFERENCE
Feature Source: SURVEY MONUMENT LIST
Source Organization Name: NORTH CAROLINA GEODETIC SURVEY
Source Organization Type: STATE
Source Format: TABULAR Source Quality:
Source Medium: DIGITAL Scale: Format: ASCII
GIS Input Method: DIGITAL TRANSFER
GIS Format: SYMBOL Geocode:
Description: Taken initially from the NCGS diskette of 1983 datum. Contact: Charles Friddle 467-0184.

Feature Name: ORTHOPHOTO REFERENCE GRID Feature Set: GEOGRAPHIC REFERENCE
Feature Source: WAKE COUNTY PROPERTY MAPS
Source Organization Name: LANDMARK GIS
Source Organization Type: PRIVATE
Source Format: MODEL Source Quality:
Source Medium: DIGITAL Scale: Format: GDS
GIS Input Method: DIGITAL TRANSFER
GIS Format: TOPOLOGIC Geocode:
Description: Grid generated through computer program using State Plane Coordinate values.

Feature Name: PROPERTY MAP GRID-MANUAL SYSTEM Feature Set: GEOGRAPHIC REFERENCE
Feature Source: WAKE COUNTY PROPERTY MAPS
Source Organization Name: LANDMARK GIS
Source Organization Type: PRIVATE
Source Format: MODEL Source Quality:
Source Medium: DIGITAL Scale: Format: GDS
GIS Input Method: DIGITAL TRANSFER
GIS Format: TOPOLOGIC Geocode:
Description:

Feature Name: BUILDING FOOTPRINTS Feature Set: PLANIMETRIC
Feature Source: AERIAL PHOTOGRAPHY (1988–89 FLIGHTS)
Source Organization Name: GEOGRAPHIC INFORMATION SVCS
Source Organization Type: COUNTY
Source Format: PHOTOGRAPHY Source Quality:
Source Medium: MANUAL Scale: 6,000 Format:
GIS Input Method: DIGITIZED
GIS Format: LINEAR Geocode:
Description: Other map scale 12.000.

Data base administration also advises system users of the content, characteristics, quality, and availability of GIS data. The administration helps users gain access to and use the data base, offers information on the most appropriate data for specific applications, and warns of potential problems or limitations inherent in the data base or source.

DATA QUALITY

Measures of Quality

The most commonly used measure for the quality of graphic map features is *positional accuracy*. Accuracy is the closeness of results of observations, computations, or estimates to the true values. Positional accuracy is a measurement of the variance of the position of a map feature from the true position of the entity. Two classes of positional accuracy are used: relative and absolute. *Relative accuracy* is a measure of the accuracy of individual features on a map when compared to other features on the same map. *Absolute accuracy* is a measure of the location of features on a map compared to their true position on the face of the earth.

Accuracy standards generally are stated in terms of an acceptable tolerance that must be achieved and the proportion of measured features that must meet the criteria. The scale of the map is also an important factor in evaluating map accuracy. For example, the United States National Map Accuracy Standards specify a horizontal accuracy for maps of scales larger than 1 : 20,000. To satisfy the standard, not more than 10 percent of points tested can be in error by more than $\frac{1}{30}$ of an inch. These tolerances are applied to well-defined points selected for evaluation (Thompson, 1979).

Positional accuracy can be tested at four levels: deductive estimates, internal evidence, comparison to the source, and comparison to an independent source of higher accuracy. *Deductive estimates* assess the potential errors that may occur in each production step and the propagation of those errors. *Internal evidence* uses tests based on repeated measurement and redundancy in measurement. *Comparison to the source*, the most common technique, generally involves producing "check plots" that are overlaid on and registered to source documents. Variances are noted and measured. Geometric tolerances relative to defined standards are then evaluated. *Comparison to a source of higher accuracy* is the preferred test of positional accuracy. This approach involves comparison of well-defined points using a set of rules such as the Spatial Accuracy Standards of the American Society for Photogrammetry and Remote Sensing.

Other data quality characteristics include lineage, attribute accuracy, logical consistency, completeness, and timeliness. *Lineage* addresses the source material from which the data are derived and the method of derivation. If data result from merging separate sources, the lineage identifies the original sources. References to specific geodetic control and transformations of coordinates also are part of data lineage.

Attribute accuracy is measured similarly to positional accuracy. Several tests can be performed to evaluate the accuracy of attributes, including deductive estimates, comparison with independent samples, and polygon overlay verification.

Logical consistency is a measure of valid or permissible values or conditions. For graphic data, logical consistency is tested by measuring characteristics, such as line intersections where intended, duplicate lines, over- or undershoots in intersections, polygons too small, or lines too close together. Topological tests can be made for chains intersecting at nodes and for complete closure of chains bounding all polygons.

Completeness is an evaluation of the existence of all necessary graphic and nongraphic data in the data base. Completeness can be evaluated through topological

The most commonly used measure of the quality of graphic map features is positional accuracy. Positional accuracy is a measurement of the variance of the position of a map feature from the true position of the entity.

BOX 5-4 United States National Map Accuracy Standards

With a view to the utmost economy and expedition in producing maps that fulfill not only the broad needs for standard or principal maps, but also the reasonable particular needs of individual agencies, standards of accuracy for published maps are defined as follows:

1. *Horizontal accuracy.* For maps on publication scales larger than 1:20,000, not more than 10 percent of the points tested shall be in error by more than $\frac{1}{30}$ inch, measured on the publication scale; for maps on publication scales of 1:20,000 or smaller, $\frac{1}{50}$ inch. These limits of accuracy shall apply in all cases to positions of well-defined points only. Well-defined points are those that are easily visible or recoverable on the ground, such as the following: monuments or markers, such as bench marks, property boundary monuments; intersections of roads, railroads, etc; corners of large buildings or structures (or center points of small buildings); etc. In general, what is well defined will also be determined by what is plottable on the scale of the map within $\frac{1}{100}$ inch. Thus while the intersection of two road or property lines meeting at right angles would come within a sensible interpretation, identification of the intersection of such lines meeting at an acute angle would obviously not be practicable within $\frac{1}{100}$ inch. Similarly, features not identifiable upon the ground within close limits are not to be considered as test points within the limits quoted, even though their positions may be scaled closely upon the map. In this class would come timber lines, soil boundaries, etc.

2. *Vertical accuracy,* as applied to contour maps on all publication scales, shall be such that not more than 10 percent of the elevations tested shall be in error more than one-half the contour interval. In checking elevations taken from the map, the apparent vertical error may be decreased by assuming a horizontal displacement within the permissible horizontal error for a map of that scale.

3. *The accuracy of any map may be tested* by comparing the positions of points whose locations or elevations are shown upon it with corresponding positions as determined by surveys of a higher accuracy. Tests shall be made by the producing agency, which shall also determine which of its maps are to be tested, and the extent of such testing.

4. *Published maps meeting these accuracy requirements* shall note this fact on their legends, as follows: "This map complies with National Map Accuracy Standards."

5. *Published maps whose errors exceed those aforestated* shall omit from their legends all mention of standard accuracy.

6. *When a published map is a considerable enlargement* of a map drawing (manuscript) or of a published map, that fact shall be stated in the legend. For example, "This map is an enlargement of a 1:20,000-scale map drawing," or "This map is an enlargement of a 1:24,000-scale published map."

7. *To facilitate ready interchange and use of basic information for map construction* among all federal mapmaking agencies, manuscript maps, and published maps, wherever economically feasible and consistent with the uses to which the map is to be put, shall conform to latitude and longitude boundaries, being 15 minutes of latitude and longitude, or 7.5 minutes, or $3\frac{3}{4}$ minutes in size.

Issued June 10, 1941; revised April 26, 1943; revised June 17, 1947 by the United States Bureau of the Budget.

testing, comparing the number of features with the universe of features, and testing for entries in all necessary data fields or cells.

Timeliness addresses the currency of feature representation and their respective attributes. Tracking the age of information aids the user in evaluating the data's legitimacy and any changes that may have occurred since its inclusion into the data base.

Development of a GIS data base usually involves integration of data from multiple sources with varying accuracies, scales, geometric structures, spatial resolutions, and other characteristics. The quality of GIS data should be measured periodically as the data base is developed and updated. Quality descriptions can be entered in the data base as attributes of the map features and related data. This aspect of the GIS data base too often is overlooked, but its importance is gaining appreciation. The presence of quality characteristics allows users to make valid decisions about the usefulness of data for specific applications.

Map Scale

A map is an analog (physical model) that represents some portion of the earth. Each map is produced at a specific scale (or is designated "not to scale"). Map scale is the ratio of units of measurement on the map to units of measurement on the earth. Scale often is stated as uniform parts such as 1 : 2,000, where one part or unit of measurement on the map is equal to 2,000 parts or identical units of measurement on the earth. Scale also can be stated in specific units such as 1" = 100', where 1 inch on the map is equivalent to 100 feet on the ground.

The concept of large and small scale often is difficult to perceive because a large-scale map generally covers a relatively small area. The large- and small-scale terms refer to the size of the scale and not to the area covered or to the sheet size. For a map of 1 : 1,200 (one part in 1,200 or 1" = 100'), the ratio of map units to ground units is rather large (thus a "large"-scale map); on a map of 1 : 12,000 (1" = 1,000'), the map ratio is relatively small.

Scale is important to the measurement of positional accuracy in a map. As the scale gets smaller, it becomes increasingly difficult to identify positional errors, and any notable positional errors will be relatively large. At a scale of 1" = 2,000', the thickness of an ink line on the map may be 25 to 40 feet of ground distance. Thus, the positional accuracy that may be achieved and verified at that scale is relatively gross (Fig. 5-27).

Scale also is an important factor in the display of maps. A GIS can display or plot maps at any desired scale. However, as map scales become smaller, the density of data per unit of area becomes greater. As significant changes in scale (e.g., from 1" = 100' down to 1" = 1,000') occur, it is typically necessary to thin or remove detail to produce a map that is aesthetically pleasing or cartographically readable. If all boundaries, annotation, and symbols were retained on a typical county tax map during the above change in scale, the resulting 1" = 1,000' scale map would be a mass of lines and text features (Fig. 5-28). The converse also is true: Increasing the scale dramatically either greatly reduces the display area or results in a document of unwieldy size (Fig. 5-29). A significant increase in

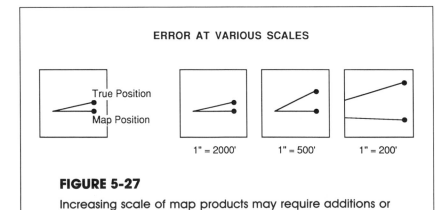

ERROR AT VARIOUS SCALES

True Position
Map Position

1" = 2000'   1" = 500'   1" = 200'

FIGURE 5-27

Increasing scale of map products may require additions or movement of annotation.

scale may require added detail. For example, at 1″ = 1,000′ scale, it may be sufficient and desirable to plot street names only once in ten blocks on a city map, while a 1″ = 100′ map might not cover ten blocks and might not display the street names at all unless annotation is added.

DATA BASE DEVELOPMENT

A GIS requires the availability of at least a substantial portion of the digital data before production operations can begin. As a result, an extensive data base development effort is frequently needed. This often means converting a full set of existing maps to digital form or compiling a new set of maps in digital form from aerial photographs or field data. The expression "data conversion," referring to the creation of digital data from analog data, or the translation of digital data, has become a popular term to describe these activities.

The extent of the data conversion effort depends on many factors, including the number of feature sets to be entered into the data base, the availability of maps to be digitized, the need for compilation, the scale of the source maps, the accuracy to be achieved, the condition of the source materials, the extent and number of sources of attribute data, and the resources available for data base development. At this time only relatively small amounts of existing digital map data are available from government or private sources to help create a GIS data base. Nongraphic attribute data are more likely to be available in digital form, although they still may require extensive reformatting or restructuring.

Source Preparation

The first stage in data base development is to prepare the source materials. The necessary maps and other source materials must be acquired and may have to be copied.

FIGURE 5-28
Manual entry of graphic data is accomplished by tracing lines or digitizing using special data entry tools.

The source materials may have ambiguities or feature representations that require clarification, especially if, as is often the case, the data base is developed by a contractor unfamiliar with the materials. Usually it is necessary to prepare the source materials by reviewing and marking them to enable rapid, high-volume entry of data by persons not familiar with the sources.

An inventory of the source materials should be taken and checked throughout the data base development process for control over the effort and to ensure that all necessary information has been entered into the system. The prepared and logged materials then are given to the contractor for conversion to digital form.

Data Entry

The conversion of existing paper, mylar, or other manually produced maps and ma-

Regardless of the technique used to develop the data base, the quality of the data must be controlled carefully. The positional accuracy of the map features, the completeness of the data entered, the proper definition (such an assignment to the proper layer or feature type of all elements), the topological integrity of spatial relationships, and the logical consistency of data values must be verified.

FIGURE 5-29

Analytical stereoplotters are used to compile digital planimetric and topographic data from aerial photography.

terials to digital form can be accomplished using one or more of at least six techniques or sources:

- Manual or semiautomated digitizing
- Photogrammetric digitizing
- Scanning
- Coordinate geometry
- Key entry
- Translation of existing digital files

Manual Digitizing. Until now, the most common technique for creating digital map data bases has been manual or semi-automated digitizing to measure and store coordinate values. This approach is used when existing maps are available as source materials but cannot be scanned. The technique often is termed *manual* because, although a GIS or computer-aided drafting (CAD) system is used, the maps must be retraced by a person using a digitizing ta-

ble and cursor. In this procedure, the map to be digitized is mounted on a digitizing table (Fig. 5-30) and the scale, angle of rotation, and area coverage are registered with the computer system. Then, each of the map features is digitized by pointing the cursor and pushing appropriate buttons to activate commands (see below).

Each system provides a repertoire of commands for digitizing. Several modes of entry are available, such as indicating a point in the crosshair of the cursor and pushing a "point" recording button, indicating beginning and ending points of a line and using an "enter a line" command, or tracing a line with the cursor using a continuous or "stream" digitizing mode. Digitizing software includes commands to draw a curve, snap to an existing point or line, and place predefined symbols. Annotation is put into the system by key entry of the text and by specifying with the cursor, its location and angle of rotation or reference feature. The capability of a digitizing workstation is discussed in more detail in Chapter 8.

Photogrammetric Digitizing. A similar method, the photogrammetric digitizing technique, is used to compile new maps from aerial photographs. This approach also is labor-intensive, requiring considerable manual effort in the digitizing process. In this case, however, the table digitizer is replaced by a photogrammetric instrument such as an analytical stereoplotter (Fig. 5-31). Photogrammetric digitizing most often is used to record very precise, accurate digital planimetric features and elevation data from stereophotographs. Elevation data can be recorded either as continuous-contour lines of a specific interval or as point data in this procedure.

Scanning. Manual or semiautomated digitizing is a labor-intensive, time-consuming, and costly endeavor for a GIS with a sizable data base. Scanning technology has been emerging for many years as a solution to these constraints. This approach uses an optical laser or other electronic device to "scan" an existing map and convert its images to digital format. Several variations of scanning technology have been developed, some of which are described in Chapter 8. Most scanners produce digital data in raster format, recording a value of dark (e.g., representing a line or symbol) or light (no line or symbol) for each grid cell or pixel of the scan. Compared to manual digitizing, scanning is very quick and less costly.

Unfortunately, because most GIS applications require a vector version of the data and attachment of attributes, the usefulness of scanning has been limited. Algorithms have been developed to convert raster data to the vector format and to recognize line, symbol, and annotation patterns that supply the necessary intelligence to the scanned data. To date, this approach has not been cost-effective for most GIS applications, but the technology is being used increasingly. As new pat-tern-recognition, edit, and translation software becomes available, the cost-effectiveness may improve dramatically. Source materials that lend themselves to scanning are relatively simple maps, maps for which separations or overlays differentiate the various feature types, and maps for which the applications are satisfied by raster-format data.

Coordinate Geometry. Coordinate geometry procedures represent an entirely different approach to creating digital map data. Here the geometric descriptions of map features are keyed into the computer. Mathematical algorithms are employed to compute the resulting coordinates, which are then stored and used to generate graphic displays of the map images (Fig. C-39). The most common software used for this function is COGO, a term that has become the common shorthand name for the coordinate geometry technique. This approach requires the definition of a point of origin through digitizing or entry of coordinate values. Bearings and distances or other geometric descriptions of the features to be mapped are entered on a keyboard.

The coordinate geometry technique often is used to enter survey data. Recent developments support the direct entry of digital data collected in the field from modern electronic survey instruments such as total stations and electronic distance measurement (EDM) devices. The coordinate geometry approach can produce very accurate cartographic data, more accurate than is practical with conventional manual digitizing of existing maps.

Key Entry. Nongraphic data and map annotation commonly are entered through a key-entry procedure. The data may be keyed directly into a GIS workstation or nongraphic data may be keyed into other less-expensive terminals. Data entered separately can be bulk-loaded later into the GIS data base. Annotation generally is key entered and then positioned on the map image through the interactive graphic capabilities of the GIS.

Existing Digital Data. Existing digital data files are a common source of attribute and geographically referenced data for the GIS data base. Less common at present, but growing in availability, are digital map files. Potential sources for digital data are organizations that use maps of the same geographic area and have automated operations, commercial firms that create and market digital cartographic data, and government agencies, particularly the United States Geological Survey (USGS). Digital map files, especially of road systems and statistical and administrative boundaries, are becoming available commercially. The USGS has developed a 1:100,000 digital line graph that has been incorporated in the Bureau of Census TIGER map set for the entire United States and is converting its standard 1:24,000 topographic maps to digital format in a long-term project. USGS operates the National Cartographic Information Center, which disseminates available digital map data bases nationwide.

Use of existing digital files often requires translation from the format and structure of the source system to that of the receiving GIS. These translations have caused serious problems in the past. The development of translation programs between the more common systems and the increasing use of digital cartographic standards are facilitating data translation and exchange. The National Committee for Digital Cartographic Data Standards in the United States has developed and published standards for definition, exchange, and accuracy for digital cartographic data (National Committee for Digital Cartographic Data Standards 1988).

Translation between GIS file formats may be accomplished in two ways. In some cases binary translators are available or may be developed. These translators are specifically designed to translate directly form one format to another. This is the most efficient approach, though it is not always practical due to the proprietary nature of the vendor formats. Where a direct translation is not possible the use of a "common" format in a two-step process is typical. First, the data are translated from the proprietary structure and format of the

source system into a neutral structure and format, often called a Standard Interchange Format (SIF) or Digital Exchange Format (DXF). Second, the data are transformed from the neutral or commonly available format into the structure and format of the receiving system. This procedure works reasonably well in translation between the more common systems, although problems are still encountered. Significant problems may result if the basic structures of the systems are different: For example, if one system uses a topological structure and the other a graphic structure for storing data, or if features such as curves, annotation, or text are stored as rules rather than coordinate strings.

Quality Control

Regardless of the technique used to develop the data base, the quality of the data must be controlled carefully. As map data are converted to digital form, the quality of each element must be verified before it is loaded into the permanent data base. Quality-control procedures must be designed to ensure that the specifications of the data base are met. The positional accuracy of the map features, the completeness of the data entered, the proper definition (such as assignment to the proper layer or feature type to all elements), the topological integrity of spatial relationships, and the logical consistency of data values are verified through these procedures.

Often, data conversion is performed by a mapping company that can offer quick and cost-effective mass-production procedures. To prepare the source materials, the client organization may need to remove ambiguities and clarify features and characteristics before delivery of source material to the conversion contractor. The data conversion contractor is expected to meet high quality-control standards before delivering the digital products to the client. The client also must verify the quality of the products before accepting and loading them into the permanent data base.

Typical quality-control procedures involve graphic and digital edits, including

---

Regardless of the technique used to develop the data base, the quality of the data must be controlled carefully.

edit plots; topological, consistency, and completeness edits; and others. Edit plots typically are produced on high-speed plotters and are overlaid on source maps to verify that all features have been digitized and that they are in the proper location within the tolerances delineated in the conversion specifications. It may be necessary to display or plot each layer or feature type separately to verify the proper segregation among layers or types has been accomplished.

Topological edits order and analyze the features in accordance with their logical relationships and verify completeness, proper identification, and correct orientation. For example, the bounding chains of each polygon are arrayed so they can be traced from a beginning point completely around the polygon and back to the beginning point. Such an edit will identify missing boundary elements and extraneous elements not properly related to a polygon. Other computer and manual edits are performed to ensure that all data fields are complete, proper codes have been assigned, and values fall within acceptable ranges.

DATA BASE MAINTENANCE

Ongoing maintenance of the data base is essential to a successful GIS. Most map data bases include information that changes with development, administrative activities, the evolution of natural processes, or other events. In addition, as requirements for the GIS change over time, the data base must evolve as well. As a result, data are added, modified, archived, or deleted as conditions change to maintain the data base's currency. As changes are made, procedures must be invoked to ensure that the integrity of the data base is protected.

Such changes may require additional elements, moves or modifications of the existing configuration of elements, or modifications of the attributes. Changes in the data base may initiate procedures such as production of updated maps or transactions with other systems.

Data base maintenance includes procedures to identify changing conditions and to ·capture or generate data about the changes, so that revised data (either in the form of new graphic features or attribute data) can be entered accurately into the correct locations. New or changed relationships among data also must be recorded. These and other maintenance procedures must operate continuously over the life of the GIS. Although maintenance may be shared among the various participants in the GIS, a system manager or data base administrator typically oversees these activities. Key issues in data base maintenance are these:

- Sources of change
- Authorization
- Positional registration
- Audit trails
- Protection of the permanent data base
- Generation of transactions

The GIS data base design must include ways to identify sources of updated information and procedures to capture and enter data from those sources. Success in acquiring updated information depends to a large extent on the interests of the parties responsible for data capture. For that reason, data base maintenance procedures should be incorporated into the operating activities of those who routinely use and manage the GIS and, to the greatest extent possible, should be part of the normal, day-to-day activities of the GIS users.

Because multiple organizations may contribute to the maintenance of the GIS data base, procedures controlling access and modifications must be established. Access control requires procedures limiting authorization to use the data base, passwords or access codes, verification of those codes by the system prior to a modification, and recordings of both the change transactions and the person making each change.

Similarly, it is important to ensure that new features and changes to existing features are placed in their accurate geographic locations. This task, called *positional registration*, often uses features already in the data base as references to place new ones. Again, procedures must account for placement and registration of new information, as well as rules governing the selection of existing features for registration.

As the data base is modified, information about those changes should be recorded to provide an audit trail for the data base. Ideally, all transactions changing the permanent data base should be recorded to allow later reconstruction. At a minimum, basic characteristics of the change, including the time and the person who made the change, should be recorded. This will allow the data base administrator to monitor changes, identify any unauthorized changes, and trace the origin of problems.

The GIS data base frequently is divided into temporary and permanent storage areas, although each system and installation has its specific approach and procedures. Temporary areas are used for initial data entry and temporary processing activities, such as copying for calculation or analysis. New data or modifications to the data base are entered into temporary storage areas before verification. As the data are verified, they move to the permanent storage area of the data base. Only data that meet the quality standards of the GIS are recorded in the permanent area. This procedure protects the data base from unauthorized or accidental deletion or modification of its contents.

Some updates to the GIS data base, or to its components, require corresponding changes in related systems. This may require transactions to abstract subsets of the data entered into the GIS, computation of new values based on the data entered, or other operations. The update transactions are formatted and transferred automatically to the receiving component after the quality of the updated data is verified.

REFERENCES

The American Cartographer (January, 1988). The proposal standards for digital cartographic data, Falls Church: American Congress on Surveying and Mapping: 15: 25–26.

Codd E. F. (1976, June). A relational model of data for large shared data banks. *Communications of the ACM (American Congress of Mapping) 13*, No. 6. Republished in *Communications of the ACM 26*, No. 1, (January 1983).

Frank A. (1984). Toward more intelligent systems: A general trend in computers. *FIG Commission 3: International Symposium on LIS*. Edmonton, Alberta.

National Committee for Digital Cartographic Data Standards (1988). *Issues in Digital Cartographic Data Standards, Report #7, A Report on Evaluation and Empirical Testing*. Harold Moellering, Editor.

Thompson Morris M. (1979). Maps for America. Washington, D.C.: U.S. Department of Interior, Geological Survey: p. 104.

Webster's New World Dictionary of the American Language, 2nd college ed., s.v. "data base."

SUGGESTED READING

American Society of Civil Engineers. Manual of map uses, scales and accuracies for engineering and associated purposes. ASCE Committee on Cartographic Surveying, Surveying and Mapping Division.

American Society of Photogrammetry (1985, February). Accuracy specifications for large scale line maps. Photogrammetric Engineering and Remote Sensing.

Croswell, Peter L. (1987). Map accuracy: What is it, who needs it, and how much is enough? *URISA Proceedings*, Vol. II. Ft. Lauderdale, FL: Urban and Regional Information System Association.

U.S. Geological Survey (1986). *Large Scale Mapping Guidelines*. Reston, VA: National Mapping Division.

U.S. Geological Survey (1986). *Map Accuracy Testing Procedures Manual*. Reston, VA: National Mapping Division.

Pearson, Fredrick II (1984). *Map Projection Methods*. Blacksburg, VA: Sigma Scientific Inc.

Manual of Photogrammetry Fourth Edition (1980). Falls Church, VA: American Society of Photogrammetry.

Chapter 6

Data Types

GIS data bases contain many types of graphic and nongraphic data, integrated and registered to form a comprehensive source of information. This chapter reviews the graphic data types most often found in a GIS data base. The graphic data derive their positions from one of two generic sources—ground position or legal description. These sources affect how data are integrated in the data base and how problems that arise in the spatial registration of features from separate sources are solved.

Accuracy and level of resolution are important issues in the development of a GIS data base, determined by the system's intended use. A GIS that supports engineering or legal activities typically requires a high level of accuracy, a large scale, and high resolution. Systems used for planning, policy analysis, or management applications generally do not require this high level of accuracy and detail. Cost is an important collateral issue as the prices of data base development and maintenance rise rapidly with higher accuracy and resolution (Fig. 6-1).

Even when a high level of accuracy is required, some organizations choose the expeditious data base development approach of entering readily available, lower-quality data to minimize initial costs. This strategy generally includes incremental improvement through the replacement of initial data with more accurate data as they become available. Several early automated mapping and GIS projects did this, but not as successfully as they had anticipated. The retrofit of more accurate data has proven very difficult in actual practice. For the best results, the initial data base

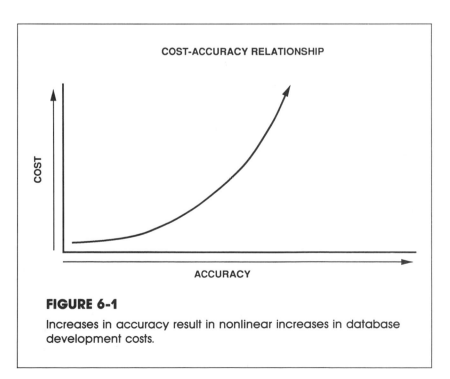

COST-ACCURACY RELATIONSHIP

FIGURE 6-1

Increases in accuracy result in nonlinear increases in database development costs.

should be established with the required level of accuracy and detail at the outset.

As illustrated in Figure 5-14, the data base requires a logical structure to support spatial integration. In the case of GIS data bases that are not intended to be purely schematic in addressing small and discrete areas, the structure begins with the geodetic control network, which provides an accurate positional framework and from which the locations of other features then can be measured. Next are planimetric and/or topographic features, which are generally compiled through very accurate photogrammetric procedures from aerial photographs. A common layer is the parcel or cadastral information, entered from the legal descriptions of parcel ownership and fitted to the control and planimetric features for positional accuracy. A GIS data base commonly contains several other types of data—administrative or jurisdictional boundaries, utility and infrastructure features, and natural and physical features.

Development of the initial data base involves the entry and integration of data from multiple sources. The various sources often have differing scales and formats that must be reconciled. A truly integrated data base requires a carefully planned structure of control and reference features to which other features must be precisely adjusted and positionally registered in the various layers. Each set of features is adjusted to features of higher positional accuracy, in accordance with specified rules of precedence.

Each of the layers and features has specific characteristics that are important in the initial data base development as well as in its maintenance and use in specific applications. These characteristics, describing the quality, entry scale, source, and date entered, should be entered into the data base as attributes of the graphic features and updated as changes are made in the data base. The following sections describe the characteristics of typical layers and features, their sources, and entry methods.

For best results, the initial data base should be established at the outset with the required level of accuracy and detail. A truly integrated data base requires a carefully planned structure of control and references features that must be precisely adjusted and positionally registered in the various layers to the control and reference structure.

CONTROL FRAMEWORK

The basic structure of the GIS map features must be tied to the geodetic framework of the area. The geodetic framework is a network of monumented geodetic control points whose positions have been measured very accurately in accordance with national survey standards and specifications (Fig. C-40). The local network, on which the GIS data base is developed, should be either a part of or tied into the National Geodetic Reference System (NGRS) in the United States, or a comparable geodetic framework in other countries. The horizontal, or x and y coordinate, measurements should be based on the horizontal control network. The elevation, or z values, should be related to the vertical control network.

In the United States, the Federal Geodetic Control Committee (FGCC) develops and publishes geodetic control standards. These standards describe the methods, equipment, and procedures used to achieve specified accuracy levels. The geodetic control points have been divided into three classes or orders according to the accuracy to be achieved (Table 6-1).

To support a GIS, a geodetic control network must have sufficient density (spacing) of points of a specific order to achieve the required data base accuracy levels. In a typical urban or suburban area where mapping is at $1'' = 100'$ and $1'' = 200'$ scales, the density of first-order points should be at 5- to 10-mile intervals, and second-order Class 1 and Class 2 points at 1-mile intervals.

The control points should be permanently monumented in accordance with FGCC standards, with a metal cap embedded in a concrete base (Fig. 6-2). Each point should be documented properly, and its description entered into the GIS data base. The documentation should identify the point and describe its location, a means to locate it, its quality class, the other control points to which it has been related, and the survey information for the point. The survey data also should be used

TABLE 6-1   Summary of FGCC Order Descriptions Standards for the Classification of Geodetic Control and Principal Recommended Uses

Horizontal Control

Classification	First-Order	Second-Order		Third-Order	
		Class I	Class II	Class I	Class II
Relative accuracy between directly connected adjacent points (at least)	1 part in 100,000	1 part in 500,000	1 part in 20,000	1 part in 10,000	1 part in 5,000
Recommended uses	Primary National Network. Metropolitan Area Surveys. Scientific Studies.	Area control which strengthens the National Network. Subsidiary metropolitan control.	Area control which contributes to, but is supplemental to, the National Network.	General control surveys referenced to the National Network. Local control surveys.	

Vertical Control

Classification	First-Order		Second-Order		Third-Order
	Class I	Class II	Class I	Class II	
Relative accuracy between directly connected points or benchmarks (standard error)	5.5 mm $\sqrt{K}$	0.7 mm $\sqrt{K}$	1.0 mm $\sqrt{K}$	1.3 mm $\sqrt{K}$	2.0 mm $\sqrt{K}$
Recommended uses	Basic framework of the National Network and Metropolitan Area Control. Regional crustal movement studies. Extensive engineering projects. Support for subsidiary surveys.		Secondary framework of the National Network and Metropolitan Area Control. Local crustal movement studies. Large engineering projects. Tidal boundary reference. Support for lower order surveys.	Densification within the National Network. Rapid subsidence studies. Local engineering projects. Topographic mapping.	Small-scale topographic mapping. Establishing gradients in mountainous areas. Small engineering projects. May or may not be adjusted to the National Network.

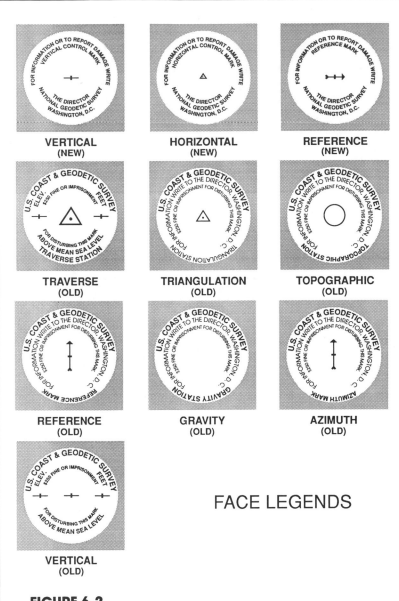

FACE LEGENDS

FIGURE 6-2

Survey control monuments are established for many purposes; metal caps document the differences.

ellite-based or conventional equipment. Recently, the Global Positioning System (GPS) has become available for geodetic control surveying. This technology uses Department of Defense navigation satellites, which transmit signals that can be decoded by specially designed receivers to determine positions precisely (Figs. 6-3, 6-4). GPS technology is able to measure positions within centimeters and with an error factor of one part in 1 million, well within the conventional first-order standards. These measurements can be made at a fraction of the cost of traditional first-order surveying techniques.

GPS is especially useful in extending a basic first- and second-order network into an area, which may then be densified using conventional survey instruments. The total station survey instrumentation now available is greatly improved, highly automated, and able to record data in a digital format for electronic transfer to a computer system (Fig. 6-5). These devices are used to conduct control surveys and in particular to densify areas between higher-order GPS-measured points.

The monumented geodetic control points are the bases for measurement of

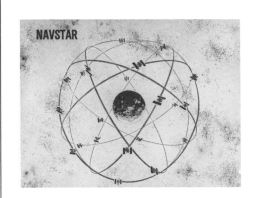

FIGURE 6-3

The Department of Defense's NAVSTAR constellation of satellites will provide worldwide real time positional information (courtesy of Trimble Navigation).

to adjust the point relative to the overall national network. Data that are documented according to the FGCC "Blue Book" format and submitted to the National Geodetic Survey (NGS) are adjusted to the national network and entered into the national geodetic data base.

The points may be surveyed with sat-

tion of data on existing geodetic control points in the area. Sources for that data include the NGS, state and local government agencies such as highway and public works departments, and the private survey community. Then these points are relocated and reoccupied (resurveyed) for verification. A control diagram is prepared and the adequacy of the control network for the area evaluated. If the existing control is not adequate, as is generally the case in the United States, a control densification program is required.

The geodetic control data are entered into the GIS data base in both graphic and attribute forms. The locations of the monumented points can be identified with symbols for the point type and class, and the relationships between points indicated by connecting lines to form a control diagram.

Recently, the satellite-based Global Positioning System (GPS) has become available for geodetic control surveying. GPS technology is able to measure positions within centimeters and with an error factor of one part in 1 million at a fraction of the cost of traditional first order surveying techniques.

FIGURE 6-4

Receivers for positional data from the NAVSTAR global positioning system (GPS) are becoming more portable and less expensive, increasing their utility for GIS database development efforts (courtesy of Trimble Navigation).

most other GIS feature locations. Before aerial photographs are taken, the control points are marked on the ground so they will be easily visible in the photographs. The marked monuments are used in the photogrammetric process to control relationships between air photo models and to enable highly accurate measurement and placement of other features.

Surveyors use these same points in the field for other purposes, such as parcel boundary locations and construction projects. By using geodetic control points, surveys can be referenced to their respective positions on the earth and individual surveys can be related to each other. Geodetic and survey control data form the registration base for other GIS feature data.

The process begins with the acquisi-

FIGURE 6-5

Electronic survey instruments such as this electronic distance measurement (EDM) device can provide direct digital readouts for GIS databases.

FIGURE 6-6

Aerial photography frequently serves as the primary source of base maps and thematic data (courtesy of Johnson County, Kansas).

The descriptions of each point should be entered into the GIS database as attribute data linked to the graphic symbols by point identifiers. The control data then can be used in the data base development to provide an accurate positional reference for the other map feature layers.

PLANIMETRIC FEATURES

Planimetric features are those natural and cultural physical entities such as roads, buildings, and water bodies that are visible and identifiable on aerial photographs and can be compiled into map features through photogrammetric or surveying procedures. Planimetric features are mapped in their horizontal dimensions, providing a "bird's-eye" or othogonal view of the landscape. These features generally are mapped at a relatively high degree of positional accuracy, achieved through reference to the geodetic control, use of aerial photographs and high-precision photogrammetric equipment and procedures, or precision surveying.

In most cases, planimetric maps are produced with photogrammetric mapping techniques that use aerial photography (Fig. 6-6). Acquisition of the aerial photographs begins with preparation of a flight plan for the project area that reflects the specifications of the photo mission (Fig. 6-7). The area to be photographed and the flight lines or paths of each strip of the photo mission are plotted on a map. The appropriate flight altitude needed to achieve the specified scale and accuracy, spacing and overlap, film and camera requirements, and other necessary specifications are decided in the plan.

The photographs are usually taken in the late winter, early spring, or in the fall when leaves are off the trees and snow is not on the ground, at a time of the day when the sun is high enough to minimize shadows. The geodetic control monuments are marked on the ground on the ground for clear visibility in the photographs. Special cameras are calibrated to photogrammetric specifications and mounted in the aircraft. Once exposed, the photographic film must be processed in accordance with specific procedures to achieve adequate quality.

Maps are compiled from aerial photographs using a stereoplotter and pairs of overlapping aerial photo diapositives. A diapositive is a positive transparency (the mirror image of a photograph negative). The photo pairs are oriented in the stereoplotter to match their relationships at the time they were exposed. This orientation forms a positively correct three-dimensional model of the area photographed. The control points are used to establish the orientation of the photographs in the stereoplotter for mapping. The mathematical relationships among control points within and among the models are computed through a process called *analytical triangulation*. Analytical triangulation involves a computer-aided extension of the control points in order to mathematically densify the control network and to provide a positionally accurate structure for the photographs.

After proper orientation, maps are produced by tracing feature outlines (i.e., compiling) from the three-dimensional photo image. High-precision photogrammetric instruments are used to compile the map features from the photographs. The most current devices are analytical stereo-

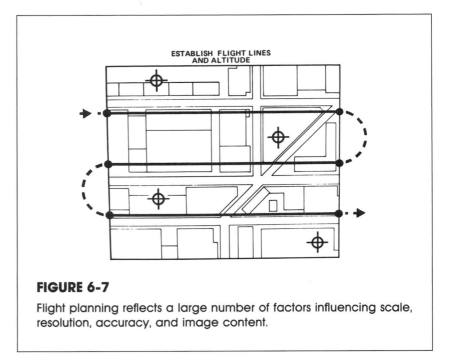

FIGURE 6-7

Flight planning reflects a large number of factors influencing scale, resolution, accuracy, and image content.

plotters (Fig. 6-8), which use a combination of operator controls and computer programs to orient the stereopair correctly and to automatically translate the compilation into digital data that can be edited, updated, and intensively manipulated.

In the compilation process, the stereoplotter is used to digitize the horizontal position of all selected points and features visible in the photographs. (The same instruments are used to capture elevation, as described in the next section.) The operator locates selected features on the stereophotograph pair in the stereoplotter and identifies points or traces lines along the photographic images. Some of these instruments enable superimposition of a digital image, such as lines and points, over the view of the aerial photograph in the stereoplotter to aid compilation.

Quality control is particularly important for planimetric features, since they serve as the reference for locating the positions of other map features. The content, accuracy, and other quality characteristics must be verified carefully. Accuracy tests are made using well-defined points that are easily visible or measurable on the

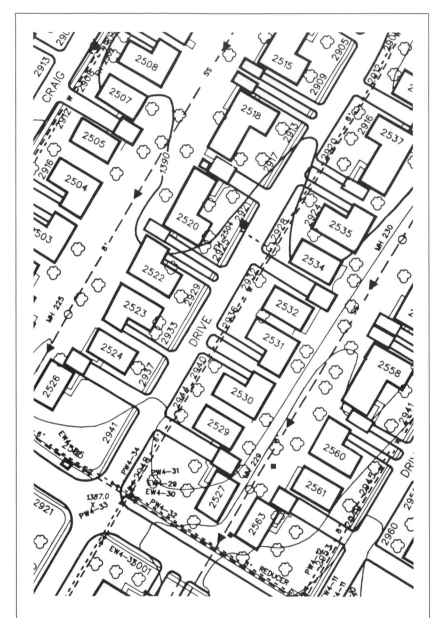

FIGURE 6-8

A planimetric and topographic map depicting building outlines, street edges, sidewalks, trees, and contour intervals, among other features (courtesy of KGIS).

analytical triangulation techniques. The results of this photogrammetric procedure are digital files containing the coordinates of the various planimetric features.

A typical listing of planimetric features includes the following:

airport runways
bridges
building outlines
catch basins
cemeteries
curbs
dams
fences and walls
hydrants
lakes and ponds
manholes
quarries
railroads
rivers and streams
road centerlines
road edges
sidewalks
storage tanks
storm water drainways
swamps and marshes
towers
trails
utility poles
wooded areas

TOPOGRAPHIC FEATURES

Topographic features define elevation information, often called *third-dimension* or z values. Like the planimetric features, the topographic data are most commonly compiled from aerial photographs. The same photography used for the planimetric mapping often is used to compile the topographic data, and is recorded in a separate process at the same time the planimetric mapping is compiled. In limited cases, often for small project areas, topographic data are compiled from field surveys.

Compilation of topographic data requires the availability of vertical survey control monuments, or *bench marks*. These monuments provide the control structure from which elevations can be measured.

ground, such as building corners, center-points of poles, and road intersections. The accuracy of the photogrammetric product should be verified by comparing coordinates of features as measured on the map with coordinates for the same features measured by field survey or by

When aerial photography is the source for compilation, these monuments or other vertical control points are marked on the ground for clear visibility on the photographs. Specific densities of vertical control points must be available to achieve particular levels of accuracy.

Topographic data are recorded in one of two basic modes. In the first mode, continuous contour lines connecting points of equal elevation are digitized or compiled by the operator at specific intervals from the stereophotography using the stereoplotter (Fig. C-41). This approach records thousands of point x and y coordinate values at a selected elevation value as the contour line curves along the surface of the land. Individual contour lines are traced at specified intervals selected to satisfy the functional requirements of the data base and within the precision standards of the photography specifications. This approach can involve the storage of very large volumes of data to represent areas of irregular terrain within a small contour interval. Normally, the elevation values are attached as attributes to each contour and labels stored for index contours.

The vertical accuracy of the contour lines is typically specified in relation to the contour interval. No more than one-half of a contour interval error may be present in at least 90 percent of the elevations and no elevation may be in error by more than a full contour interval to achieve United States National Map Accuracy Standards. For example, the accuracy of a contour line for a topographic map of 5-foot contour intervals would be plus-or-minus 2.5 feet.

The second mode of recording topographic data is as elevation values for individual points. In its simplest form, this involves recording spot elevation values at significant points on the map. An approach that is becoming increasingly popular in GIS system development is the production of a digital elevation model (DEM) (Fig. C-42) or digital terrain model (DTM) for an area. A DEM is the digital cartographic representation of the elevation of the land at regularly spaced intervals, while a DTM embodies the elevation of significant topographic features on the land and change points (breaklines), reflecting abrupt changes such as building edges, streams, or ridges. These data provide a pattern of elevation information that can be used for many purposes.

In their basic representation as a mathematical model of the terrain, the data can be used in various computer programs or models that analyze the slope (Fig. C-43) or angle of terrain. Examples of these are storm water runoff and viewshed models. A pattern of digital elevation data is needed by the models to represent the surfaces mathematically. The pattern may be either a rectangular or triangular grid, depending on the model being used. The elevation data thus can be passed directly from a DTM to such a model, whereas contour data would first have to be interpolated into an elevation point structure before entry.

Contour lines can be generated mathematically from a DTM through the use of a contour interpolation program (CIP). The CIP will compute the curves of the contour lines from the elevation values, the special values (e.g., breaklines), and the relationships among the point values. Very sophisticated CIP algorithms are available that will match or exceed the accuracy of stream-digitized contours from photogrammetric map compilation.

Both the DTM and DEM approaches have lower data storage requirements than required storing each contour because only a fraction of the number of x, y, and z values are recorded. When contours are required, they can be generated for the area needed, usually a very small portion of the overall area covered by the GIS.

Topographic data, like planimetric features, are generally compiled at relatively large scales and high levels of positional accuracy. The features recorded typically include contour lines, point elevations, and elevation values.

CADASTRAL FEATURES

The cadastral layer contains boundaries of property ownership and other rights to land. Many of the applications that use the cadastral layer involve overlaying other layers such as soils or flood plains to determine parcel characteristics or relationships to certain conditions. To support these types of applications, it is necessary to register accurately the spatial locations of the various layers to each other. It is equally important to place the parcel boundaries in their accurate spatial locations and to portray their proper geometric shape as found in legal descriptions.

Legal Aspects

The cadastral features are unlike the natural and cultural features discussed above in that they are based on legal descriptions rather than physical locations. They should be monumented on the land, although often the markers are not adequate. In some cases, the property corners were never marked, while in other instances the survey monuments have been lost or destroyed. In addition, surveys vary in levels of accuracy. Mapping of cadastral features often can be difficult. In many instances, confusing legal descriptions must be deciphered and conflicts between legal and physical aspects of location must be reconciled.

The ownership and rights to the land may be described in various ways. Most parcels created in recent years are described on subdivision plats that consist of a map, typically at a large scale of 1″ = 20′ to 1″ = 200′ for urban and suburban areas, with parcel identifiers and boundary dimensions. Subdivision plats should be tied to geodetic control points, although most often they are not.

Parcel boundaries are usually described in narrative form on a deed. The description records boundary measurements as metes and bounds or as bearings and distances. In much of the United States, the descriptions may refer to the Public Land Survey System (PLSS) township, range, section, and aliquot parts (half- or quarter-sections). The PLSS was laid out during the settlement of the country, dividing land areas into townships of thirty-six 1-square-mile sections. Land title was originally recorded or "patented" on

the basis of these units and has often been resold through some uniform division or combination of them. However, deed descriptions frequently use neither of these reference schemes and are vague and difficult to map.

In addition to fee-simple ownership of land, the cadastral layer may depict other boundary information, such as rights-of-way or easements, or may describe the locational extent of selected property rights such as minerals and livestock grazing. These too may be described on a surveyed plat or in narrative form.

Developing the Digital Cadastral Layer

Development of the digital version of the cadastral layer may use existing parcel maps or may compile new maps from legal and other source materials. Most property-taxing authorities maintain tax maps that show all legal parcels. These maps often are used merely as an index or inventory, and positional accuracy is not high. Other organizations, such as planning departments, public works, and utilities, also may maintain parcel maps. Before beginning a parcel mapping program, a review should be made of available parcel maps to determine if any will be satisfactory as a source. Factors influencing the decision include; the quality of existing maps, requirements for the cadastral layer, and resources available for the project.

If the maps require little or no modification, digitizing available maps generally is the least expensive and time-consuming approach to development of the cadastral layer. If the positional accuracy of current source maps is not adequate, the accuracy may be upgraded by adjusting the source maps to available survey control data and a planimetric base prior to digitizing. With this approach, the parcel boundaries are adjusted as necessary to match reference features on the base (Fig. C-44). This process can make a significant improvement in positional quality with relatively limited additional cost, although it does not affect the legal description, which still may contain errors.

The most thorough method of developing the cadastral layer, required when existing maps are unsatisfactory, is recompilation of parcel maps through research of plats, deeds, and legal records. Boundary lines are compiled from the descriptions as an overlay to the reference base that is used to locate the boundaries in their accurate spatial locations relative to the reference features. The reference base may be a planimetric map or an orthophoto. Again, this process will not affect the erroneous descriptions in the legal document, although procedures may be designed to record the conflicts.

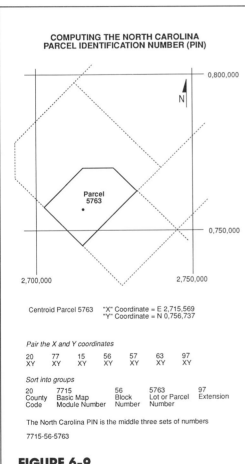

FIGURE 6-9

The North Carolina Parcel Identification Number (PIN) serves both as a unique identifier and a geographic index.

A variation on this approach is the use of coordinate geometry programs (COGO) to generate the parcel map features. The origin, bearings, and distances of parcel surveys or plat dimensions are entered into the COGO programs. The resulting coordinate values are computed and stored in the data base. The displayed parcel graphic images are adjusted to correct their positions interactively. Anomalies between adjacent parcels (where common boundary segments are described differently in adjacent parcel records) are also reconciled interactively to produce an aesthetically pleasing map. In this case, however, the true boundaries should be stored permanently on a separate layer to be available for applications that require depiction of the boundaries at the level of accuracy described in the survey documents.

Annotation

Several sets of annotation are generally used with parcel data. Typical parcel annotation includes parcel number, dimensions, lot number, block number, subdivision name, and address. Placement of the desired annotation poses a composition problem. Since the automated system can store annotation separately and display various combinations at different times, it is best to specify the locations for placement so that the individual sets will not overlay or interfere with each other.

Parcel Identification

Each parcel must be identified uniquely to allow the map features to be linked with the nongraphic attributes that describe the parcel characteristics. Parcel identification numbers (PIN) and numbering schemes have been debated since the early use of computer systems for parcel data processing. The large number of possible schemes ranges from a simple sequential number to a geocode that also defines location by incorporating x and y coordinates for the parcel. Each approach has its strengths and weaknesses. Sequential numbers are easy to assign, manage, and process, and

DIAGRAM OF MULTIPLE LOTS AS ONE PARCEL

FIGURE 6-10

Variances in parcel definition occur frequently even within a single government's jurisdiction.

terleaves x and y coordinates to form the PIN. As exhibited in Figure 6–9, the interleaving allows rapid identification of successively smaller rectangles in which the point (or parcel) lies. This approach also presents difficulties, since the relationships between numbers are difficult to recognize and adjacent parcels have widely different numbers.

Possibly the most commonly used scheme incorporates a map sheet number with a quadrant or grid number and an individual parcel number within the map and quadrant to form a unique parcel identifier. This approach is easier for people to understand but less efficient for the computer. One difficulty is the assignment of new numbers when a parcel is divided. Often, the next available parcel number is out of sequence for the area, again introducing the problem where the PIN numbers vary greatly between adjacent parcels.

Another problem with the assignment of parcel identifiers occurs when a parcel is divided but some part of the original parcel remains. In many jurisdictions, the remaining portion retains the original parcel number. This may cause confusion and difficulties over time, as records with different characteristics (e.g., acreage) but assigned the same parcel number will exist.

The GIS data base requires a clear definition of what constitutes a parcel before the parcel map is compiled. Various definitions are found among jurisdictions and even among agencies within a single jurisdiction. For example, the tax assessor may combine parcels for billing efficiency when adjacent parcels are owned by the same person. Similarly, a single parcel may be divided into multiple units when a leasehold represents a different use and tax responsibility (Fig. 6-10). Parcels may be defined as groupings of contiguous homogeneous land use or as the smallest legal entities, regardless of ownership or use. In general, the GIS data bases should record and manage the most basic unit and link or aggregate these units into various configurations to best serve the requirements of individual applications.

each number is unique. They convey no other information, however, and are somewhat difficult for people (as opposed to computers) to use. In a sequential numbering scheme, parcels with closely related identifiers may be distant from each other, and there will be no pattern to help locate a parcel by its identifier.

The coordinate-based identifier provides an indication of geographic location and is used for some spatial analyses such as proximity and point-in-polygon searches. Coordinates are determined for the center or another point within the boundaries of each parcel. The State of North Carolina has popularized an approach for the use of coordinates that in-

BOX 6-1 Construction of the Geo-Coded Parcel Identifier

EXAMPLE

x–coordinate 2,715,569
y–coordinate 0,756,737

The digits in each coordinate value are paired together by taking each digit separately from the x-coordinate and matching it with the corresponding digit of the y-coordinate:

20   77   15   56   57   63   97

With this arrangement, the above example of a parcel identifier may be sorted as follows:

20	7715	56	5763	97
Redundant lead number for any one county	Number of basic map module	Block Number	Lot or parcel number	Utilized only to extend the capacity of the system

Normal usage of the parcel identifier in any one county would involve these ten (10) digits.

7715   56   5763

The system of paired digits used above as a generator of a unique parcel identifier has some decided advantages over the simple statement of the x and y coordinate values. Each pair of digits relates graphically to a successively smaller square cell of the coordinate system. Thus, the first pair of digits on the left identifies a huge square area bounded west and east by $x = 2,000,000$ feet and $x = 3,000,000$ feet, respectively, and bounded south and north by $y =$ zero feet and $y = 1,000,000$ feet. The second pair of digits identifies a square cell bounded by $x = 2,700,000$ and $x = 2,800,000$; $y = 700,000$ and $y = 800,000$. Next, the third pair of digits identifies a square cell that is 10,000 feet on each side and it is, in fact, the area covered by a single map sheet in the statewide map series at the 1″ = 400′ scale. The fourth pair of digits relates to a 1,000 foot square that is called a "block." Succeeding pairs of digits work right down to a grid cell of 1 foot square. In this way, the parcel identifier provides a systematic sequence of successively smaller windows that leads the inquiry to an exact location without resort to key maps or other irregular patterns of an arbitrary nature.

Parcel Attributes

The parcel layer may include a wide variety of nongraphic data describing its legal, physical, economic, administrative, and natural characteristics.

In addition to parcel characteristics, other sets of nongraphic data are related to the parcels through geographic identifiers, such as the street address. These include incident and activity data taken from the day-to-day functions of various organiza-

---

BOX 6-2 Sample Nongraphic Parcel Data

Basic Parcel Characteristics

- Parcel identifier
- Topography class
- Area
- Road class
- Frontage
- Nondeveloped class
- Depth
- Mobile home code
- Land use(s)
- Zoning(s)
- Number of buildings
- Building identifiers
- Number of units
- Unit identifiers
- Utilities available
- Develement status
- Parking spaces
- Well
- Septic
- Swimming pool
- In flood plain
- Slope class

Parcel Tenure

- Legal description
- Owner name(s)
- Prior owner names
- Parent parcel number
- Child parcel numbers
- Sale dates
- Deed book/page numbers (current and past deeds)

Parcel Value Characteristics

- Commercial/residential codes
- Land values
- Improvement values
- Total value
- Prior land values
- Prior improvement values
- Sale price
- Sale date
- Prior sales
- Prior sales dates
- Agricultural program status
- Agricultural program value(s)
- Exemption

Administrative Characteristics

- Administrative case numbers
- Permits
- Active cases
- Active case status
- Development conditions/limitations
- Incident identifiers
- Development activity codes

Tax Status

- Tax district
- Tax bill amounts
- Tax accounts receivable
- Payment status
- Tax claim status

---

tions, then retrieved, manipulated, and displayed on the basis of geographic location or spatial relationships as defined by the parcels.

AREA BOUNDARY FEATURES

A GIS can support multiple sets of area boundaries, such as jurisdictions, administrative units, legal districts, and statistical units, among others. These features are characterized as enclosed areas established by administrative or legal definitions. They often are specified by legal survey descriptions or they may merely be drawn on reference maps or described narratively. Most area boundaries follow other physical or legal features such as streets, rivers, or parcel lines. Most are defined to be easily recognized on the ground.

Areas may be defined based on a desirable size or on physical conditions such as fire truck response time, on the aggregation of a maximum or desirable number of entities, such as voters, or on the outer bounds of a set of facilities or conditions, for example, a water pressure zone. Some area boundaries are dynamic and change as administrative or legal decisions are made or as development occurs. Others, such as political jurisdiction boundaries, may be very stable with little or no change over long periods of time.

Conversion to Digital Form

Conversion of area boundaries to digital form is generally less difficult and costly than conversion of other map feature sets. The number of areas and the volume of associated data is generally low. Areas are typically at least one block in size, often covering several blocks, and even many square miles. Defining boundaries accurately may be difficult, however. Often, the source materials are small-scale maps on which the boundaries have been drawn imprecisely. While there may be a legal or other specific description of the area, the source map may be very general. It may be unclear whether a boundary follows a street centerline or one edge, yet the difference may be important. Background research similar to that investigating the ownership parcel may be necessary to prepare materials for digitizing.

The actual digitizing often can use physical or other features as references while entering and positioning area boundaries. In many cases, the existing features themselves may be copied or referenced to enter area boundaries. In fact, an alternative way to define area boundaries is to enter area identifiers in the topological descriptions of other elements in the data base. In this case, the graphic presentation of the areas is generated from the topological definitions of features serving multiple purposes.

The area boundaries generally are displayed with specific line symbology to differentiate them from other features. Identifiers also support linkages to nongraphic data describing the bounded areas. Most areas themselves will have little or no attribute data. Rather, the identifiers will refer to incident and other data sets to allow aggregation of data for reporting and analytical purposes. The areas generally are defined as closed polygons for overlaying and other analytical functions and for graphic display.

A sampling of areas found in a GIS includes:

- City boundaries
- Election districts
- Precincts
- Zoning districts
- Police beats and reporting areas
- Fire response areas
- Emergency response areas
- Maintenance districts
- Pressure zones
- School attendance districts
- Tax districts
- Planning districts
- Traffic zones
- Transit districts
- Health districts
- Service districts
- Wire centers

FACILITIES FEATURES

The GIS (or AM/FM system, as it is often called when dealing with engineered facilities) may be used to support the management of many types of facilities, including roads, bridges, water, sanitary sewer, storm sewer, electric, gas, telephone, and cable-TV lines. Each facility has components, such as the lines defining pavement, transmission, distribution, collection, and other systems, and symbols for the various related appurtenances, such as manholes, hydrants, valves, culverts, transformers, switches, and poles.

Along with the graphic representations of these features, the GIS must store and manage related nongraphic data and the logical relationships among elements.

Most facilities data are interconnected into open (e.g., sewer) or closed (e.g., water) networks. The connectivity and characteristics of these networks are important to many retrieval and analytical functions. The connectivity describes the interchanges between elements in the systems—the flows from one segment to the next. Nongraphic data record the types, capacities, materials, installation, maintenance, operational status, and other characteristics of the individual elements. These attributes are attached to the graphic elements through unique identifiers. The combinations of the graphic representation, element relationships, and attributes support many analytical applications, such as tracing a reported breakdown back through the network to a particular device or connection.

Facilities such as roads, power lines, and water mains are located on, above, and below the ground, respectively. The mapping and data maintenance of these features must address these facts. While individual utilities generally are operated by separate organizations, their relationships often demand that they be mapped in the same system. In particular, the underground utilities often share the very limited space of a road right-of-way or easement and are vulnerable to damage during construction and maintenance.

The management of engineered facilities requires storage and maintenance of their map features, a wide range of attribute data describing the facilities and their maintenance characteristics, and the physical and logical relationships between and among individual devices.

The map features are entered into the GIS from multiple sources. The planimetric features described above can serve as the reference base for positional accuracy of the facilities. Some of the facilities themselves can be compiled from the aerial photographs as a part of the planimetric base construction. As examples, poles, manholes, and other features are visible in the aerial photographs at certain scales. Other features, and the connectivity among facilities, are compiled from exist-

ing maps, drawings, schematics, and other office records. In limited cases, field observations also may be necessary.

Symbology is a particularly important aspect of utility facility maps. Each utility has many types of devices, and each requires a unique symbol. The symbology within an individual organization or utility can be complex. When multiple utilities are combined in a single data base, the problem becomes even more complex. A GIS that serves multiple users often encounters difficulties when users attempt to agree on acceptable symbology. Although each organization has its unique set of symbols, when the various symbol sets are combined, some symbols may be duplicated, requiring differentiation and redefinition.

The attribute and related nongraphic data of engineered facilities consist of three basic types: (1) The attribute data describing the facilities themselves, recording the characteristics of individual devices, such as the type, size, material, and manufacturer; (2) management data describing installation dates, maintenance dates, and history, and other information useful for managing and maintaining the facilities; and (3) customer data identifying the location (address) of the service and the characteristics of the customer (e.g., residential or commercial, consumption rate, special services, or conditions).

Most existing facility attribute data are automated before the GIS is implemented. The GIS design can include an electronic link with these systems to support data sharing. As the nongraphic data and analytic applications have become more important, techniques have evolved to share or to use graphic and nongraphic facility data jointly. In some cases, the solution is a gateway between different systems; in others, it is to install the GIS software on the same computer with the management and customer data, or to install the GIS on a hardware platform linked directly to the system supporting the nongraphic facility data.

In any case, the nongraphic facility data in a GIS are vitally important to facility management. The nongraphic data must be linked logically in the GIS to the graphic features they describe. This involves development of a logical relationship between the graphic elements and the attribute records. Identifiers also must be defined and entered in each data set to accommodate the linkage. Development of the GIS data base requires entry of these identifiers and careful quality control to support linkage of the data elements.

A subset of engineered facilities is composed of interconnected networks or systems providing access, distribution, and collection functions. Some of the systems are closed networks; others form tree or inverted-tree structures. These characteristics are important to the logic of the data structure and data base definition. The ways in which individual elements are connected and the relationships of that connectivity (e.g., flow in one or both directions) are important characteristics that must be recorded and managed by the GIS. This typically is handled through the explicit encoding of the topological structure of the network and the individual elements. Given topological characteristics, it is possible to model and analyze flows through and disruption to the system.

The level of accuracy required for engineered facilities differs among organizations. Sometimes, the GIS is used only as a reference or index and a high level of positional accuracy is not necessary. The map may, for example, simply relate the number of poles and the correct side for a street segment. In this case, the GIS is used to retrieve, analyze, and display data, but not to locate items precisely in the field. In other organizations, a high level of accuracy is required to support engineering activities and to facilitate field location. Measurement of distances and relative positions must be reliable, and high levels of accuracy are essential. These differences in approach may lead to separate systems or databases among organizations serving the same area, or separate levels of detail within one database.

Data base development for facilities information is always a demanding project

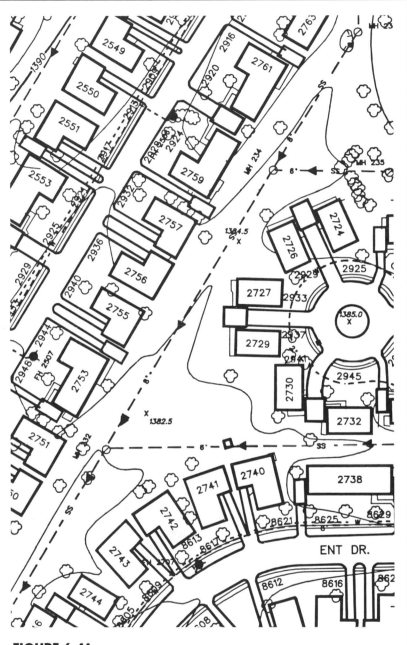

FIGURE 6-11

The location of gravity flow sanitary sewer facilities on a planimetric topographic base map (courtesy of KGIS).

that often involves extensive preparation or scrubbing of source materials. Facilities data are entered from existing maps and drawings of varying quality that may not be produced to a specific or single scale. Before digitizing, these materials must be reviewed, corrected or clarified, and layouts made on maps of the desired scale. The preparation and entry of large volumes of nongraphic data also may represent a major effort. Because of the complexity and magnitude of these tasks, quality control is critical.

Some facility data bases are developed within the user organization, while other organizations choose to contract with firms that specialize in that service. Numerous techniques have been developed to improve the efficiency of the process, including division of the effort into specialized tasks, establishment of mass production procedures, use of low-cost entry devices, and development of special software to facilitate entry and quality control.

Each type of facility has its unique requirements and problems. Entire industries have grown up around individual utilities, providing specialized systems, applications, and data base development services. The following summaries outline each facility's characteristics.

Water systems are composed of mains that carry large volumes of water from treatment plants to general distribution areas and of smaller lines that distribute water locally, terminating in service to customers. The lines are connected through valves and other joints important in the logical construct of the system. Most water systems operate under pressure and their service area is divided into pressure zones controlled by designated valves. Valves and hydrants are particularly important devices in the systems, accessed often for maintenance and emergency response. Some important GIS applications include the ability to locate appropriate valves to isolate problems and perform maintenance, to identify customers tied to specific pipe runs in order to track problems and evaluate the impact of valve closure and water demand, and to analyze pressure regimens so that an adequate water flow for fighting fires as well as domestic and commercial consumption, is ensured.

Water system map features frequently include:

- Mains
- Distribution lines
- Service taps
- Valves
- Fittings
- Hydrants
- Pump stations
- Treatment plants
- Meters
- Reservoirs
- Water quality monitors
- Soil borings
- Aquifers

Attribute data include:

- Type
- Size
- Material
- Source
- Capacity
- Maintenance status
- Hydrant flush date
- "As-built" reference number
- Installation date
- Leak history
- Valve direction turn
- Maintenance history
- Facility number

Sanitary sewers are underground systems that take the form of inverted trees. Unlike water systems, most sewer systems operate by gravity feed, although some pumping facilities are used where terrain makes gravity flow impractical. Also, unlike the water distribution system that radiates *from* a treatment plant, the sewerage system is a collection system leading *to* a treatment plant. The system consists of pipes and mains of gradually increasing sizes. Manholes are important features for maintenance access. Because the flow relies on gravity, elevation and slope are important characteristics of the system.

Customer information is used for service calls, for analyzing flows, and, in some cases, for calculating fees based on physical conditions.

Sanitary sewer map features include:

- Mains
- Manholes

- Pump stations
- Treatment plants

Attribute data include:

- Type
- Size
- Material
- Date installed
- Depth or invert elevations
- Maintenance status

Storm sewers are similar to sanitary sewers, although they may be divided into smaller local systems leading to a stream or river. Elevation and grade are important characteristics. The storm sewer system also is a collection system, but the sources are collectors and culverts rather than customer buildings.

Storm sewer map features include:

- Catch basins
- Culverts
- Lines
- Outfalls

Attribute data include:

- Type
- Size
- Material
- Date installed

Electrical systems feature two major components, transmission and distribution. Transmission systems are major above-ground, high-voltage facilities for the movement of power from generating plants to distribution stations. Often, transmission line rights-of-way are equally important features because of their environmental impact, management requirements, and the value of the real estate.

Distribution systems are networks of lines radiating from distribution stations or substations to customer sites. Poles, switches, and transformers are important map features. The relationships among circuits, information on devices and lines composing a circuit, and the customers served by a particular circuit are important characteristics.

Electric distribution systems may be above or below ground. Above-ground lines and most devices are located on poles. Underground systems include lines, conduits, and vaults, among other features.

Facilities management, operations, and maintenance are typical applications requiring the maps and attribute data of the GIS. An important application is response to outages, wherein the GIS is used to identify problems based on outage reports from customers by relating the customer information to the lines, transformers, and switches of the distribution system. The customer information also can be used to identify special situations, such as persons on life-support systems who may require special assistance during outages.

Many electric organizations are private, investor-owned companies. As such, they pay taxes on their capital assets and are subject to regulation by a public utilities commission. These two conditions create a need for information about facilities included in the rate base and on inventories of facilities within tax districts. Facility inventories or Continuing Property Records (CPR) are a typical GIS application addressing both information requirements.

Electrical system map features include:

- Lines
- Poles
- Transformers
- Switches
- Meters
- Vaults
- Substations
- Rights-of-way

Attribute data include:

- Capacity
- Type
- Manufacturer
- Model
- Date installed

Similar to water, *gas facilities* are closed systems of distribution lines that operate under pressure, primarily underground.

Major gas transmission lines span the continent, and their rights-of-way are of great importance to numerous GIS applications. Local distribution systems are composed of mains of various sizes and connection and control devices, including valves and service connections. The logical relationships of lines and valves are important for control, maintenance, and rapid response to problems. Gas companies often are private, investor-owned utilities, subject to public utility regulation and taxes. They too require maintenance of information to support rate-base justification and tax liability.

Gas facility map features include:

- Transmission mains
- Distribution mains
- Service taps
- Services
- Regulators
- Valves
- Control fittings
- Reducers
- Fittings (tees, crosses, elbows)
- Insulated fittings
- Anodes and other cathodic protection devices
- Leaks
- Abandoned/Retired facilities

Attribute data include:

- Material
- Size
- Diameter
- Design pressure
- Actual pressure
- Depth
- Installation date
- "As-built" reference
- Location
- Number of leaks
- Direction for rotation (for valves)
- Last inspection

NATURAL FEATURES

Numerous layers of natural features may be included in a GIS data base. They may be conditions on the surface of the earth, such as wetlands or vegetation; below the surface, such as geology or groundwater; or above it, such as noise or air quality. Within a layer, natural features most often are defined by irregular patterns of curves as the natural phenomena occur on, above, or below the earth. Frequently, they are defined as closed areas or polygons of a common characteristic, although line and point features also occur.

Natural features include physical and biological conditions. In general, these features are not precisely defined or mapped with high positional accuracy. They approximate the boundaries of changes in environmental conditions. The boundary between soil types or vegetation classes is not a specific line on the earth but often consists of a zone of transition from one condition to another. Many natural features share this characteristic, and it affects the way they are represented in the GIS data base.

Some sets of natural features have relationships that affect their boundaries, tending toward common boundaries among layers. Using the soil and vegetation features again as an example, certain types of soil support specific classes of vegetation. As the soil type changes, the class of vegetation growing in an area also will change along a similar boundary. In some cases, a GIS data base may reduce similar boundaries to identical ones to simplify the data base. This may be done through the definition of integrated terrain units prior to digitizing, or through automated polygon reconciliation after digitizing.

In the early stages of GIS technology, natural features were encoded as values for uniform grid cells. This grid-cell approach allowed primitive systems to reduce irregular patterns of the various features to a common structure for comparison and overlaying. The patterns of most natural features were roughly approximated with a grid structure of appropriate resolution for the application requirements. Often, these grid-cell data bases were built for large areas with cells of 10 to 100 acres. More detailed data bases were also developed for localized analyses with cells representing one to a few acres.

As GIS technology has become more sophisticated, a movement to use the irregular boundaries of the actual feature has arisen. These boundaries are digitized (still typically at relatively small scales, such as 1:24,000) from source maps, and the system's software can overlay and process the combinations of features, often generalizing the data to remove the inevitable slivers and other small insignificant polygons that result from the overlaying.

Natural features are stored in the GIS as points, lines, and areas (polygons). They may or may not have unique identifiers (other than those generated by the system for data management), since there may be no need to differentiate between, for example, soil polygons of the same type. The feature's type or class will be attached, along with other relevant characteristics. Related characteristics also may be stored in separate tables in the data base. For example, interpretive characteristics of a feature (such as the suitability of a soil for various types of development or its susceptibility to erosion) may be stored in a table for each soil type. This approach is efficient, since all polygons of the same soil type will have the same interpretive characteristics.

A set of derived features may be generated and stored in the GIS data base. Two types of functions may be used to derive a feature layer. In one, derivative boundaries may be computed from physical characteristic data. An example of this is generation of slope polygons, wherein areas of common slope gradient are computed from contour or elevation data stored in the GIS. In the other type, two or more sets of features or layers will be overlaid and the resultant combination polygons stored as a data set. An example of this might be the combination of soil, slope, vegetation, and hydrography to define areas with various classes of suitability for development. Another example might be the identification of environmentally critical areas from a combination of sensitivity indicators stored as layers in the data base.

An important aspect of natural features in a GIS is the use of remote sensing and image processing as data sources. Natural resource analysis and environmental assessment were among the first remote-sensing applications. Many natural features can be determined and mapped using various remote-sensing techniques such as infrared photography. These data are collected by satellites or aircraft at various altitudes and recorded in raster format as values for individual pixels in the image. Image-processing technology is used to process, analyze, and display the data in raster format. Recent improvements in GIS technology, both hardware and software, allow the raster image and vector data to be merged and displayed simultaneously. The raster images in the GIS data base are managed in a data structure separate from the vector data and processed with complementary software.

Commercial firms such as the Earth Observation Satellite Company (EOSAT) or the SPOT Image Corporation supply satellite data as standard product. The US Geological Survey EROS Data Center and other federal agencies that have developed data for specific projects and programs are alternate sources.

REFERENCES

Federal Geodetic Control Committee (1980). Horizontal control data. *Input Formats and Specifications of the National Geodetic Survey Data Base*, Vol. 1, Federal Geodetic Control Committee; Rockville, Maryland.

Marks G. W., ed. (1980). *The Planning and Engineering Interface with a Modernized Land Data System*. American Society of Civil Engineers.

North Carolina Department on Natural Resources and Community Development (1987, September). Technical Specifications for Base, Cadastral and Digital Mapping. Raleigh, North Carolina.

SUGGESTED READINGS

Reid W. B. (April, 1990). Landbase, scanning USGS quads or other existing records.'' *Proceedings of AM/FM International Conference XIII.* Baltimore, Maryland; pp. 323–328.

Ripple W. J., ed. (November, 1986). *Geographic Information Systems for Resource Management: A Compendium.* American Society for Photogrammetry and Remote Sensing and American Congress on Surveying and Mapping: Falls Church, VA.

Chapman W. H., (November, 1990). Impace of NAVD 88 on the National Mapping Program. *Proceedings of the 1990 ACSM-ASPRS Fall Convention.* Anaheim, CA., pp. 42–51.

Kevany M. J. (1986). The implementation of geodetic control in the development of a geographic information system. *Geographic Information Systems in Government, Vol. 2 A.* Deepak Publishing, Hampton, VA, pp. 555–566.

Chapter 7

Hardware

Computer hardware that supports geographic analysis and mapping is comparable to that used in business and scientific environments. The increasing popularity of geographic information technology, as well as the technological advances in all types of computer hardware over the last twenty years, have encouraged the development of hardware devices tailored to mapping and other computer graphics applications. This chapter provides an overview of the terms, characteristics, and concepts of the hardware needed to support a geographic information system, and provides the foundation needed to understand computer system configuration concepts, discussed in Chapter 9.

HARDWARE COMPONENTS

The term *computer hardware* refers to any physical device used as part of a computer system. Hardware includes the processing units that execute programs, auxiliary storage units that maintain data and programs, and devices that are used to enter data and to generate maps and reports, such as terminals, plotters, and printers. Computer hardware devices may be classified into two main categories: the processing unit and peripheral devices. The central processing unit directs and executes system operations. Peripheral devices are physically separate from the processing unit. Peripheral devices include auxiliary storage units; devices used to enter, analyze, and display information and generate hard copy; and devices designed to facilitate communications.

The Central Processing Unit

The central processing unit (CPU) directs and supervises all of a computer system's functions. The CPU receives data and commands, performs requested operations, and generates a result or product that is delivered to the user through one of the peripheral devices mentioned above. Functions of the CPU include these:

- Controls user access to the system and monitors security
- Compiles high-level language into machine-readable form
- Controls communications with peripheral devices
- Performs system accounting and diagnostics
- Executes all user program commands in an interactive or batch mode

The concept of a CPU implies centralization of a computer system's "intelligence" and central execution of all in-

When selecting a processing unit for a GIS, a good way to evaluate how much power will be needed is to examine user satisfaction and response time at existing installations with similar applications.

133

structions (Fig. 7-1). Traditionally, most computer systems have used a centralized processing approach, with a "host" processing unit to control all functions of the system. Centralized processing arrangements are still used today, although in recent years the trend has been toward distribution. In distributed processing systems, some processing tasks (such as communications, plot generation, and program execution) are distributed from a main host processing unit and handled by separate processors. Concepts of distributed processing are discussed in more detail in Chapter 9.

Whether centralized or distributed processing is employed, the CPU model can be used to describe the general operation and performance of a computer system. No major distinction is made here between centralized and distributed systems.

In the CPU, the control unit and the arithmetic and logic unit (ALU) contain the electronic circuitry designed to process instructions (Fig. 7-2). The execution of programs in the CPU is directed by the control unit, which accepts computer instructions from the memory, issues instructions to

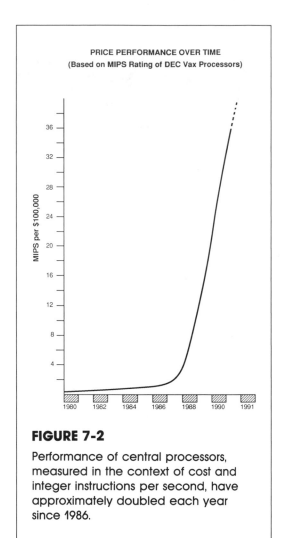

FIGURE 7-2

Performance of central processors, measured in the context of cost and integer instructions per second, have approximately doubled each year since 1986.

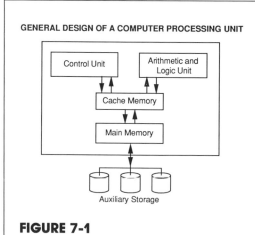

FIGURE 7-1

A simplified view of a computer processing unit includes a four-component core and external or auxiliary storage devices.

the ALU, and controls input/output operations. Information is stored in the ALU and control unit in *registers,* which are temporary storage areas for computer instructions and data. Both computer instructions and data are treated as sequences of binary digits.

The main memory unit provides high-speed access to data and program statements during program execution. The main memory unit is also called *core memory,* or in the microcomputer industry, *random access memory* (RAM). Many computer processors incorporate *cache memory* for the temporary storage of frequently accessed data and programs.

Cache memory is based on the fact that most programs frequently and repetitively access a limited group of instructions and data. When a program is executed by the CPU, the cache is first examined for the required data or instructions and slower auxiliary storage devices are accessed only if the information is not found in cache. Cache memory units and their interactions with other components of the CPU are designed to maintain a high *hit rate,* thereby lowering execution time and increasing performance.

Data is moved between certain components of a computer system on high-speed paths called *buses.* These physical paths connect components within the CPU as well as link it to auxiliary storage units. Data travels over buses in a synchronous manner: A sequence of *bits,* representing one computer *word* or part of a computer word, is transmitted simultaneously. The information-carrying capacity of a bus depends on its *width,* or the number of bits that can be transferred simultaneously. Buses vary in width depending on the design and size of the computer. Mainframe buses usually are 24, 32, 48, or 64 bits wide; minicomputer and microcomputer buses typically are 8, 16, or 32 bits wide.

Auxiliary Storage Devices

The main memory capacity of a computer processor is usually insufficient to store all of the large amounts of data and programs to which users need access. Peripheral devices called *auxiliary storage devices* or *mass storage devices,* are used to store large volumes of data that are accessible by the processor. Auxiliary storage devices are categorized as *off-line* or *direct access.* The most popular device used for off-line storage is the *tape drive,* which archives data sets on magnetic tape for later access. Tape drives are considered an off-line technique because a tape must be mounted physically before data can be read. In contrast, direct-access storage devices (DASD) allow users to store and retrieve data directly without loading it from another source. The most popular direct-access storage device, the *disk drive,* uses magnetic or optical technology to store information. Disk drives are

---

BOX 7-1 How Information Is Represented in Computer Systems

Computers store and manipulate data in the form of binary digits or *bits* that have a value of 0 or 1. A *byte* is represented by a sequence of 8 bits and can be used to represent a character of information (letter, number, or special symbol). Since there are 8 bits available in each byte, and each bit has a value of 1 or 0, a possible 256 values can be represented, each identifying a unique code for a character. Eight bits provide adequate room to uniquely identify all 26 letters of the alphabet (upper and lower case), numbers 0 through 9, punctuation marks, and other special keyboard characters. Two common coding schemes for representing characters that have become standard are ASCII (American Standard Code for Information Interchange), used by most minicomputer and microcomputer vendors, and EBCDIC (Extended Binary Coded Decimal Interchange Code), commonly used by IBM mainframe computers. For example, the capital letter *C* is represented as the binary code 01000001 in ASCII and as 11000001 in EBCDIC.

Storage capacities of main memory and auxiliary storage devices usually are defined in terms of bytes. Commonly used measurement units are *kilobyte* (K), representing 1,024 bytes; *megabyte,* 1 million bytes; and *gigabyte,* 1 billion bytes.

Information in a computer normally is processed as a *word,* or a sequence of bits. A word can hold raw data used by a program, or it can represent an instruction. Typical word sizes of computer processing units used as platforms for automated mapping and geoprocessing are 8, 16, and 32 bits. Traditionally, IBM and IBM-compatible microcomputers (XT and AT) use 8- and 16-bit word sizes. Low-end minicomputers such as the DEC PDP series also use 16-bit word sizes. Most minicomputers and super-microcomputers used for geographic analysis applications—such as the DEC VAX series, Hewlett-Packard 9000 series, Prime 50 series, Sun and Apollo workstations, and the IBM RS6000—use a 32-bit word size.

---

connected to the processing unit through high-speed channels so that data can be retrieved directly by system users.

Tape drives are available most commonly in a reel-to-reel or cartridge format. Data is stored sequentially on the tape as magnetic impulses. Tape drives are useful for the archival storage and transport of large volumes of data or software. In computer mapping systems, tape drives typically are used for the following:

- To archive data or software that is not needed frequently
- To perform routine backups of data and software on disk drives to protect against loss in case of a disk failure
- To transport data or software between different computer systems

Disk drives consist of one or more rotating platters that store information to be read through magnetic or optical techniques. For each platter, one or more *heads* glide in very close proximity to the platter to read or write data, like the turntable of a stereo system. The mass storage devices are connected to the processor by high-speed lines that can transfer data quickly between the drive and the main memory.

Input/Output Peripheral Devices

Input/output peripheral devices are accessed by users to enter and manipulate data and to generate screen displays or hard-copy maps and reports. Those commonly used in geographic information systems include digitizing stations, graphic query stations, alphanumeric terminals, plotters, and other devices used for entry, analysis, and generation of products from a tabular or graphic database.

Computer System Operation and Performance

Most computer systems employ an operating approach called *virtual memory*. As discussed, the ALU executes instructions and uses data that is accessed directly in cache or main memory. Because main and cache memory have only limited storage capacity, complete copies of programs and full data sets on which they operate cannot always be stored at one time in main memory. Virtual storage techniques allow a processor to treat auxiliary storage on system disk drives as an extension of main memory. Using a virtual memory approach, computers can process large programs and data sets from multiple users. Programs are segmented into *pages* that

can be accessed by main memory and retrieved from auxiliary storage in separate pieces. The process of retrieving data and program segments from auxiliary storage, however, is slower than direct access from the CPU's main memory. Up to a point, an increase in main memory will allow larger portions of programs and data sets to be stored in the CPU's main memory, which reduces time spent in disk retrieval and results in faster execution speed and higher performance.

Response time largely depends on an interplay of many factors, including the speed of the processor, the size of main memory, and the speed of data transfers between auxiliary storage and main memory. A breakdown or inefficiency in any of the factors can harm system performance. Some applications depend most directly on processor speed; others are more dependent upon efficient access to auxiliary storage. All of these factors should be considered in evaluating computer systems.

Batch Versus Interactive Processing. Specific functions of a GIS can be performed in one of two main processing modes—batch or interactive. In a *batch* or *off-line* processing application, the user is not in direct communication with the processing unit while a program is being executed. The user submits a *job* to the computer processing unit for execution during an available time slot. The job is placed in a batch queue (a temporary storage area). The job contains information necessary for the execution of various programs, an identification of the user who submitted it, and the destination of products to be generated from the job.

Multiple batch queues can be created, each of which may be assigned different priorities for execution. The priority will determine in what order the computer will execute the jobs in the queues. Batch queues can be set up to execute jobs during times when demand is light. Applications that require significant computer processing time can be sent to batch queues that are activated at night, on weekends, or other times of low demand.

In an interactive environment, a command or program is submitted immediately to the processing unit, and output is returned as soon as the operation is complete. Ideally, interactive operations result in immediate response from the processing unit when a user issues a command. In practice, however, response depends on the performance factors described above and the overall load placed on the processing unit by all system users. Applications such as digitizing, graphic editing, and database queries normally are carried out in interactive sessions.

GIS applications use both batch and interactive processing. Many routine functions such as map updating, data base queries, and graphic display operations depend on a fast response and therefore require interactive processing. Other applications, such as standard report generation, map plotting, and complex geographic analysis (e.g., polygon overlay modeling), require substantial processing time and do not always depend on immediate response. These applications can use batch processing effectively. Both batch and interactive processing usually are required to efficiently balance user demands with available resources.

Measures of Computer Performance. In reference to computer systems, the word *performance* has no specific definition but is generally reflected by overall user satisfaction. One of the most important variables influencing user satisfaction is response time. As noted earlier, many GIS operations depend on quick response. While certain measures provide a basis for comparison of discrete variables in raw system performance, they give only a rough indication of expected response time from the user's perspective (Table 7-1). Million instructions per second (MIPS) or million floating point operations per second (MFLOPS) are routinely used to indicate power or speed of a processing unit. Although these measures often can be used to compare processing units of the same family (similar architectures and operating systems), they are only marginally useful when comparing processing units manufactured by different vendors that run on dissimilar operating systems.

The classic benchmarks, such as the Dhrystone, Whetstone, or Linpack, measure the joint performance of the central processor units (CPU) and cache memory; the latter needed to swap segments of complex and lengthy computer programs in and out of active memory. Still yet other system characteristics that affect overall performance of the workstation go unmeasured by these select benchmarks. Other functions that may be critical to a workstation's performance for a given GIS application or organizational environment include the size and accessibility of the system's main memory, disk input and output (I/O), graphic operations, and operating system, with particular emphasis on those supporting multi-tasking and server or network communication demands. Benchmarks exist to evaluate these other functions, singularly and in combi-

TABLE 7-1 Measures of System Performance

Measure	Explanation
Clock speed	Time interval determining the cycling of data through the internal system bus. Normally measured as megahertz (millions of cycles per second).
Million instructions per second (MIPS)	Speed at which the processor executes instructions in machine language.
Million of floating operations (MFLOPS)	Speed at which the processor executes numeric operations with decimal numbers.
Channel speed	The speed of data transfer between mass storage and main memory. Normally measured as megabits per second.
Disk transfer rate	The speed at which a disk drive can access and read or write data. Normally expressed in bytes per second.

nations, though the results may be expressed in indices different than MIPS and MFLOPS. Other benchmarks mix various combinations of tests called *suites* to reflect better an operational environment or to emphasize a particular performance strength (Antenucci, 1991).

System performance may be an issue during the initial selection of hardware or during the evaluation of an existing system in anticipation of an upgrade. Evaluation of a GIS's performance may focus on user transactions, a simple example of which includes the interactive addition of map features and geographic query of a tabular database. A transaction can also be complex, such as a map overlay operation between multiple layers or a tracing analysis through a linear network.

However, since there is no single definition of "transaction" in a geographic information system, it is impossible to quantitatively rate system performance using transaction as a yardstick. Whereas the quantitative measures described above can be used as general indicators of performance, when evaluating processing unit requirements, it is perhaps best to focus on the qualitative aspects related to anticipated system applications. Vendors often provide good rules of thumb to estimate performance needs. These rules may state the number of devices of a certain type that can be supported by a given processing unit without degrading system performance significantly. When selecting a processing unit for a GIS, a good way to evaluate how much power will be needed is to examine user satisfaction and response time at existing installations with similar applications. By evaluating the type of processing unit, the number and types of peripheral devices, and the applications for which the system is being used, it is possible to gauge relative needs for processing power.

Operating Systems

Every computer processing unit, from small microcomputers to supercomputers, performs functions under the control of an *operating system*. The operating system is the software that interacts directly with the processing unit hardware to control all basic functions of the system. It controls memory space, allocates disk space and user-access security, loads programs for execution, provides error recovery services, supervises input/output operations, and other system-level operations. Operating systems are discussed in more detail in Chapter 8.

CHARACTERISTICS AND TRENDS OF COMPUTER PROCESSING UNITS

The performance-to-cost ratio of computer processing units improved dramatically in the 1980s and early 1990s. Astounding improvement in speed, measured in MIPS, was achieved. Comparing MIPS for a family of processing units manufactured by a major computer vendor illustrates the dramatic rise in the cost/performance ratio from 1980 to 1989 (Fig. 7-3).

Advances in processing technology have blurred the distinction between the once-commonly accepted terms *microcomputer, minicomputer,* and *mainframe.* In the past, processing units could be so categorized on the basis of word architecture (8-bit, 16-bit, 32-bit, 48-bit, 64-bit), processor speed, number of users supported, main memory capacity, mass storage capacity, and cost range. Performance levels for computers in the mainframe category five years ago are well within the minicomputer range today. The increase in the capabilities of desktop microcomputers has further obscured their distinction from minicomputers. Although the use of these terms to characterize computer processing units is still valid, they indicate general ranges of capabilities rather than specific processing unit types (Table 7-2).

In an attempt to better characterize the diversity of processing units available today, other terms are sometimes used for categorization, including super-microcomputer, super-minicomputer, supercomputer, and mini-supercomputer. Although these terms can be useful in communicating concepts, the context of their use must be understood to avoid confusion.

Accompanying the trend of increased processor performance has been a decrease in the size of computer processing units and a decrease in environmental controls needed to operate them. Space requirements for housing computer processing units of a given performance level have dropped nearly ten times since 1980. Environmental and maintenance limitations have diminished greatly since the late 1970s. Today, many powerful minicomputer systems are suitable for office environments using existing 120-volt power. The trend toward smaller, easier-to-operate systems with low maintenance requirements will continue.

Trends in Processing Unit Architecture

Major technological advances in the architecture of processing units contributed to significant performance increases in the last five years. In particular, parallel-processor architectures and reduced instruction set computing (RISC) processors are driving tremendous increases in processor speed for many applications.

FIGURE 7-3

High performance workstations are increasingly common components in GIS configurations where users and applications are distributed throughout the organization. (Courtesy of Digital Equipment Corporation).

Parallel processing involves the use of more than one processor in a single computer to execute tasks more quickly. Computers using parallel-processor architec-

TABLE 7-2 Characteristics of Common Computer Processing Unit Types

Processing Type	Number of Users	Word Architecture	Main Memory	Role in Network	Cost Range
Microcomputer	Single user	16 bit; recent trend to 32 bit	Commonly 1MB to 4MB	Single user mode or server	$3,000–10,000
Super-micro-computer	5 or fewer	32-bit	4MB to 96MB	Single user node or server	$10,000–90,000
Minicomputer	Up to about 256 devices	32-bit	8MB to 256MB	Central data server or distributed computing	$10,000–450,000
Mainframe	More than 256	32-bit or greater	64MB+	Central processing unit	$400,000+

GIS developers have effectively used advances in hardware to support large geographic databases and interactive processes.

ture execute multiple programs or tasks of the same program in a parallel fashion, as opposed to the sequential execution of instructions that occurs in single-processor systems. Parallel processing improves efficiency in executing applications that can be divided into identifiable parts, particularly when processing tasks operate on distinct units of a database or with applications based on programs with many interactive loops. Theoretically, the overall speed of a system increases linearly with the number of processors. In practice, however, a one-to-one gain in speed does not result as processors are added; execution of tasks must be coordinated, and data must be passed between the multiple processors. Both functions entail management overhead for the system. In addition, programs that depend heavily on floating point operations or other complex computational functions are not well suited to parallel processing.

RISC-based computers run operating systems with commands that execute in one processing cycle. This strategy is in contrast to conventional processing unit architectures that require many machine cycles to complete complex instructions. Program compilers on RISC systems are designed to reduce a program's source code into simple one-cycle instructions that execute very rapidly. Although they offer increased performance for many applications, RISC-based systems are not a panacea. They require large main memories and special compilers to translate program source code into machine language.

Multiuser Processing Units

Until recently, GIS software ran almost exclusively on centralized minicomputer systems. Minicomputers are still the primary platform for most operating installations, but use of mainframes, super-microcomputers, and microcomputers is considerable.

The acceptance of geographic information technology, as well as increases in processing unit performance, have recently attracted many new software ven-

dors to the field. In the minicomputer category, the Digital Equipment Corporation VAX series is most frequently used by software developers, although Prime, Data General, and Hewlett-Packard processing units are popular as well.

Microcomputers

Developers of GIS software use microcomputers increasingly as components in their systems in the following ways:

- Nonintelligent graphic workstations (attached to a host processing unit) that emulate graphic terminals for digitizing and graphic query
- Intelligent workstations (attached to a host processing unit) that perform some local processing but still require the host processing unit for most complex GIS functions
- Stand-alone, single-user systems running CAD-oriented mapping or microcomputer GIS packages
- Data capture stations that batch-load data to and from a host processing unit

As discussed previously, the definition of microcomputer is changing. Until very recently, high-end microcomputers such as the IBM PC-AT and compatibles used 16-bit chip architectures and relied on the MS-DOS operating system that can handle directly a maximum of only 640 kilobytes of RAM and 32 megabytes of mass storage. These microcomputer systems have been used successfully for database management applications and computer-aided drafting (CAD), using software such as AutoCAD, VersaCAD, Intergraph's MicroStation, and other microcomputer CAD packages. In addition to microcomputers running the MS-DOS operating system, several computer mapping packages have been developed for the Apple Macintosh microcomputers.

Most current computer mapping system packages running on microcomputers emphasize map entry, editing, and production and have limited attribute handling and geographic analysis capability. Full GIS function, which includes complex overlay analysis, sophisticated geographic queries, and other spatial analysis, is hindered by the performance and storage limitations of current 16-bit microcomputers. To some extent, however, the performance of 16-bit microcomputers has been extended through the addition of various coprocessors to speed up certain processing tasks, to drive input/output devices, and to support high-resolution interactive graphics.

Desktop microcomputers based on 32-bit processors, which became available in 1988, provide powerful, low-cost platforms for GIS development. These new microcomputers, which use chips from Intel, Motorola, and other manufacturers, provide performance levels that were possible previously only with much more expensive minicomputer systems. Because of the increased performance, 32-bit microcomputers are becoming the platform of choice for microcomputer mapping and GIS software. At this time, however, microcomputer operating systems designed for lower-performance 16-bit computers are still in general use.

The future of 32-bit microcomputers as platforms for computer mapping and geographic analysis depends on whether operating system software that will take full advantage of the 32-bit processing environment can be refined. OS/2, an IBM microcomputer standard, and the UNIX operating system, already accepted in the scientific community, are likely to be used extensively on these microcomputer platforms. When application software for computer mapping becomes available for execution in these operating system environments, the use of 32-bit microcomputers for complex geographic analysis will expand significantly. Use of efficient operating systems also will eliminate the now ill-defined distinction between 32-bit microcomputers and super-microcomputers.

High-Performance Workstations

The term *workstation* often causes confusion. Its meaning is not precise but varies according to its context. In a general sense, a workstation is a device or assemblage of devices that interact with a computer to perform specific tasks. All forms of stations and terminals are called workstations, including digitizing and query stations, alphanumeric terminals, microcomputers, and high-performance workstations. Until about 1986, GIS workstations often were large dual-screen devices configured with a digitizing tablet, keyboard, and cursor. Workstations now are characterized by the use of 32 bit processors and have extensive graphics capabilities drawing upon multiple "windows" functionality to replace the second display device. Computer-mapping vendors offer workstations as single-user devices and in network arrangements with other workstations or larger multiuser processing units.

Super-microcomputer workstations are offered by a number of companies, including Sun, Digital Equipment Corporation, Hewlett-Packard, Intergraph, IBM, and others. These devices, such as the Sun 4 and Sparcstation series, Hewlett-Packard 9000 series, Intergraph's Interpro series, IBM's RS 6000 series, and DEC VAXStations, are used by GIS software vendors as integral parts of their systems.

The Future of Computer Processing Units

The advances in computer processor performance are so rapid that one may think they cannot continue at the current pace. It is likely, however, that improvements in processor performance will continue in response to the demand for more processing power (Manning, 1990).

Along with continued refinement of parallel processing and RISC architectures, many experimental developments on the horizon hold promise, including super-parallel architectures consisting of tens, hundreds, or thousands of individual processors in an integrated system. One extension of the concept of highly parallel systems is the *neural network*, wherein individual processors are linked much as neurons in the human brain. In a neural network, processing power and memory are completely distributed but efficiently connected. Other advances in the development of superconductors will, over the next two decades, eliminate obstacles caused by the relatively inefficient electrical conductors that make up computer circuits today.

In the future, GIS users can expect to see increasingly more powerful computer processors. Continual processor improvements demand that software vendors anticipate these advances and develop new products or enhance existing software to take advantage of them.

TECHNOLOGY AND TRENDS IN AUXILIARY STORAGE DEVICES

Tape Drives

Both cartridge-tape and reel-to-reel drives are used in geographic information systems for archiving and transporting large volumes of data and software. Tape drives provide an effective and relatively low-cost method to store data off-line. Tape storage is particularly well suited to archiving data that are not needed frequently by system users.

Cartridge-tape drives are a standard feature included by many manufacturers of small minicomputer systems. The drives accept a tape cartridge that resembles a video cassette and stores up to about 300 megabytes of data. They are useful for making routine backups and for acquiring software from vendors who supply it in that format.

Reel-to-reel tape drives have been the accepted method of off-line data storage almost since digital computers became commercially viable in the 1950s. The standard reel-to-reel tape is $\frac{3}{4}$-inch wide, wound on a reel that is easily mounted and removed from the drive. The drive itself has a take-up reel and mechanisms that allow the tape reels to be rotated in either direction.

Data is stored on the tape in a sequential manner. Most reel-to-reel tapes have nine tracks, each of which forms a sequence of bits along the length of the tape. The 9 bits arranged across the horizontal axis of the tape include 8 bits, representing a byte of data, plus one *parity bit* used for error checking.

Reel-to-reel drives are classified as either *streaming* or *start/stop,* terms that refer to the method used to write data to the tape. In the streaming method, data is copied on the tape without any physical gaps between files or records and without marks on the tape that delimit files or blocks. A streaming drive can quickly write and back up an entire disk of data, but individual files or directories cannot be retrieved. This method is useful for routine system backups.

A drive that reads and writes in the start/stop mode can place and interpret physical delimitations between files and records (Fig. 7-4). A start/stop drive is suitable for operations in which specific files or disk directories need to be identified and retrieved individually.

Drives that read and write in a start/stop mode can operate at speeds of up to 50 inches per second; streaming drives can operate at speeds of more than 100 inches per second. Some drives have the capability to read and write in both start/stop and streaming modes.

A tape for reading or writing in start/stop mode can be divided into blocks, records, and files for efficient organization of data. These format specifications and other tape parameters can be set by the user. Common parameters that must be taken into account when using tapes include these:

Tape label—optional information written at the beginning of a tape that describes its format

Record length—the size (in bytes) of tape records, which are physically separated from each other on the tape with small gaps

Block size or blocking factor—the number of

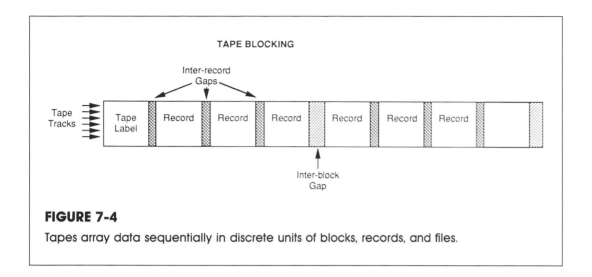

FIGURE 7-4

Tapes array data sequentially in discrete units of blocks, records, and files.

records in one tape block, which is another physical limitation of the tape

Density—the number of bits that are stored per inch (bpi) on the tape. Common densities include 1,600, 3,200, and 6,250 bpi. Many tape drives can be switched to read at different densities

Parity—the convention used in setting the value of the parity bit. It may be set to reflect an odd or even value of the sum of the other 8 bits, or it may be always set to "1" (odd) or "0" (even). An error is flagged when the parity bit deviates from the convention in use.

Disk Drives

Disk drives are the method of choice for the storage and retrieval of large volumes of data for direct access by users. Significant advancements have been made recently in direct-access, mass-storage units using two different technologies: 1) traditional, magnetic disk media; and 2) optical disk media.

Magnetic Disk Media. For magnetic media, disk storage density and data access rates have increased steadily. Since 1980, storage densities of magnetic drives have more than doubled. Removable-pack, magnetic drives have almost been eliminated from

the market in favor of smaller, fixed-media drives that often are designed to be mounted singularly or in multiples within a processor cabinet.

Magnetic disk drives for microcomputers typically have capacities from 20 to 150 megabytes per drive in a 5.25- or 3.5-inch format. In minicomputer systems, magnetic drive capacity ranges from 70 to more than 700 megabytes for drives with a 14-inch disk format. The Digital Equipment Corporation's RA-82 drives, for example, have a capacity of about 620 megabytes and occupy a space of about 3 cubic feet. In larger mainframe computer systems, external disk-drive units generally have capacities of 2 to 3 gigabytes.

The ability of processing units to execute programs at increasingly higher speeds has required faster rates of data transfer between disk drives and the processing unit. To meet this need, hardware developers have developed more efficient read/write mechanisms, higher-capacity data channels or buses, and intelligent disk-controlling devices. GIS developers have used these advances effectively to support large geographic databases and interactive processing.

Optical Disk Media. Using optical disk drives for mass storage is a new technology employed operationally since 1985.

Most optical drives use a technique in which a laser burns small impressions on a disk surface. The drive reads data by sensing variations in the light reflected from this surface. Using the laser technique, optical disks can store more than twenty times more data than magnetic drives. However, with the laser technique, data can be written to a particular portion of the disk only one time, and no data can be erased. This is called *write once–read many* (WORM). WORM technology, which has recently become cost-effective, is used to archive large volumes of data that must be stored and retrieved but rarely changed.

WORM technology is beginning to prove effective for document archive and retrieval of maps and geographically referenced documents. Hard-copy documents can be easily scanned and stored on the drive for quick retrieval. The high capacity of optical drives makes them suitable for storing maps and documents that are used primarily as nonintelligent pictures in raster format. An optical disk drive can store documents (such as engineering drawings and building plans) that are referenced geographically from a vector-based system but do not need to be stored in an intelligent vector format. Several companies (DeLorme, for one) offer optical disk systems that store various maps for reference in document preparation.

Erasable optical drives have begun to appear. Most of these use a *magneto-optical* method, in which a laser directed at the disk surface shifts the polarity of the magnetic field at that point on the disk. These drives are capable of high-storage density and allow reading and writing of data.

The lack of commonly accepted disk formats and interface standards has inhibited the widespread adoption of optical drives in GIS and other data processing systems. Disk access rates for optical drives are still significantly slower than that of magnetic media and, in the case of erasable optical drives, the technology has not been proven completely. Through the next five years optical-drive technology will become increasingly popular in geographic information systems, though magnetic media will continue to predominate.

THE ROLE AND FUNCTION OF INPUT/OUTPUT PERIPHERAL DEVICES

Historical Setting

The types of input/output peripheral devices used in geographic information systems are similar to those in traditional data processing configurations. The distinguishing feature of those in a GIS is the focus on graphics devices that enable the entry, edit, display, and hard-copy generation of maps.

Input/output peripheral devices are hardware components that users access for operations such as digitizing and updating maps, entering tabular data, performing geographic queries, invoking a program for geographic analysis, generating a screen display or hard-copy product, and archiving data to magnetic tape. The types of peripheral devices normally found in geographic information systems are especially suited to these operations.

Vendors of GIS hardware and software offer many models and devices that satisfy the functional requirements of users. Input/output Peripheral devices commonly used in a GIS are grouped here into categories that describe their basic functions:

- Digitizing workstations
- Photogrammetric mapping workstations
- Graphic query/edit workstations
- Screen copy devices
- Alphanumeric terminals
- Pen plotters
- Electrostatic plotters
- System line printers
- High-volume laser printers
- Desktop printers

Advances in processing technology during the last two decades have been equaled by rapid changes in input and output peripheral devices. Improvements in devices for data entry (both graphic and tabular) and for the display and plotting of hard-copy maps have been dramatic. From the late 1960s to the mid-1970s, most computer cartography involved the manual encoding of grid-cell maps and the use of programs like SYMAP, developed at the Harvard Computer Graphics Laboratory, for the production of maps on system line printers that created cartographic patterns by overprinting characters.

Graphic terminals that became generally available in the early 1970s used storage-tube technology, which had some significant limitations, most importantly the lack of color capability. Most graphic monitors today are raster-scan cathode ray tube (CRT) systems that can display a broad range of colors at high levels of resolution. Another trend since the mid-1980s has been decreased reliance on nonintelligent, host-dependent graphic terminals in favor of high-performance workstations with local processing capability.

Throughout the history of computer mapping and graphics systems, pen plotters were commonly used to generate large-format hard-copy maps or drawings. Common frustrations voiced in the 1970s and early 1980s by their users included the slow plotting speed and the pens' tendency to clot during printing, which required users to monitor the plotter at all times during operation. Since the early 1980s, major progress has been made in ameliorating these difficulties with pen plotters.

In the arena of hard-copy map production, electrostatic plotters have made a substantial impact in recent years. Electrostatic plotters produce hard-copy plots from vector files that are first converted to a raster format. Compared to pen plotters, they can produce large-format, hard-copy plots quickly and require relatively little maintenance or attention by users. Traditionally, the disadvantages of electrostatic plotting devices have been their low resolution, lack of color capability, and relatively high cost. The "stair-step" appearance of diagonal lines in the raster format was apparent on the plot. Despite these drawbacks, the popularity of electrostatic plotters encouraged vendors to make improvements that have substantially eliminated these disadvantages.

Many other developments in the 1980s improved peripheral devices, making them flexible tools for input and output operations. Impressive advances have been made in small-format screen-copy devices, particularly for color hard-copy production, and in scanning technology. Video camera systems and film recording devices have been developed for specialized applications. These advances, combined with improved software for processing scanned data, promise to widen the niche for scanners in map-data capture.

Data Entry, Edit, and Query Devices

Devices used to input, process, and display data in a GIS combine graphic and tabular data-display technology with flexible capabilities for data entry and manipulation. The devices included in this category include digitizing workstations, graphic edit and query stations, and alphanumeric terminals. These three types of devices have varying capabilities.

The digitizing workstation and the graphic edit/query workstation consist of graphic display monitors (usually color), keyboards for the entry of alphanumeric data and commands, and various mechanisms for manipulating graphics. In the digitizing and graphic edit/query workstations, screen displays are generated in raster format, which creates the graphic image with a matrix of cells or pixels (called *picture elements*). These devices can process data from a vector-based GIS and display them in raster form.

Graphic monitors are available in various sizes and screen resolutions. Microcomputers typically use 13-inch (diagonal) monitors with a microcomputer graphic "standard" known as the *enhanced graphic adapter* (EGA). EGA allows for a maximum of sixteen colors and a maximum resolution of 640 × 350 pixels. Since 1987, a number of vendors have supplied other graphic adapters for microcomputers to support larger monitors (up to 19 inches) and higher resolutions in the range of 800 × 1,000 pixels. In still-larger systems, graphic monitors of digitizing workstations usually measure 19 inches and have high resolutions (commonly up to 1,280 × 1,024 pixels).

Edit/query workstations consist of an

BOX 7-2 How Color Is Made

Color is used in GIS products to convey information and to capture a viewer's interest. Color adds dimension to maps and drawings and broadens the amount of information the map can convey.

Devices that generate GIS screen displays and hard-copy products use various techniques to create color. In a sense, these devices trick the eye and the brain into perceiving variations of hue, saturation, and intensity.

Graphic screen-display systems and hard-copy devices use fundamentally different techniques to generate color. Graphic screen-display systems use the concept of *additive primary* colors or *colors of transmission,* while hard-copy production devices (except for some film recorders) are based on *subtractive primary* colors or *colors of reflectance.*

In an additive color approach, a system that emits light (e.g., a graphic monitor) mixes various intensities of the three additive primaries–red, green, and blue—to produce a range of hues and shades. An additive color system relies on a light source to transmit colors to the observer's eye. Any color can be produced through combinations of these additive primaries. The colors that can be produced by a given graphic display monitor are limited only by the physical specifications of the devices and the software driving them.

Color hard-copy devices (such as pen plotters, electrostatic plotters, and thermal transfer devices) generate color by using subtractive primaries—magenta, cyan, and yellow—and depositing these on a hard-copy medium in the form of inks, dyes, or pigments. The color that the human sees is reflected from the surface of the medium. Magenta, for example, results when green light is "subtracted" from the visible color spectrum. The additive and subtractive primary colors are complements of one another. Many color hard-copy devices use black ink in addition to the three subtractive primaries, since the combination of yellow, cyan, and magenta does not in practice produce a sufficiently deep black tone. If the white background of the paper is included, a total of eight general hues can be created. Since most color hard-copy devices available today cannot vary the amount of color deposited at any one pixel location, it is not possible to print hues other than the six additive and subtractive primaries (plus black and white). It is theoretically possible to produce any color variation by mixing various amounts of these subtractive primaries.

Most computer mapping systems use raster-scan monitors for color screen display. These devices generate color in a manner similar to a color television set. *Guns* emit a stream of electrons that excite phosphors on the screen's surface (see Fig. 7-5). Each pixel is created by an arrangement of three phosphors (red, green, blue).

Color variations are created by applying different levels of intensity to each gun. The number of colors that are possible is related to the number of intensity values that can be applied to the electron guns, principally determined by the memory capacity of the device.

Most high-resolution hard-copy color devices (above 200 dots per inch) use a technique called *dithering* to simulate additional variations of saturation and hue, thereby extending the range of possible shades. To create apparent variations in saturation (e.g., light blue), not all pixels within an area on the plot are filled in; some white spaces are left between the blue pixels, giving the overall impression of a lighter shade. This works because the pixel resolution is fine enough that the eye and brain do not perceive the individual blue and white pixels. Dithering is also used in some devices to simulate different color hues. For example, if an orange hue is needed, it could be produced by depositing the yellow pigment over the area to be shaded along with smaller number of pixels colored with the magenta pigment. By mixing different relative amounts of subtractive primary colors, a range of hues can be created.

146

integrated set of components similar to that of digitizing stations. Since edit/query workstations are not used for digitizing tasks, they do not require digitizing tablets and often can satisfy user requirements with a lower screen resolution than typically is needed with digitizing stations. Color capabilities vary among vendors.

The raster-scan technology used almost exclusively today differs from older vector storage-tube monitors that dominated the early years of interactive graphics systems. Cathode-ray tube monitors display an image by directing a beam of electrons onto a phosphor-coated surface. In vector-display devices, the vector components of an image are directed to the screen and captured on the phosphor surface. Using an approach similar to color television sets, raster-scan devices create an image that is redrawn or *refreshed* as a series of horizontal scan lines, at a time interval fast enough so that no flicker is perceived by an observer. Common refresh rates used in monitors today are 30 and 60 hertz (cycles per second). An *interlaced* monitor refreshes every other scan line in a refresh period, totally drawing the image every two cycles. Noninterlaced monitors refresh the entire image with each cycle.

Digitizing and graphic edit/query stations use various methods for entering and editing graphics and for entering commands from graphic menus mounted on a digitizing tablet or displayed on the monitor. The most common entry-and-edit mechanisms include cursors and mouses.

Cursors, sometimes called *pucks*, are palm-sized devices consisting of a glass or plastic lens with a cross-hair target and multiple buttons (Fig. 7-5). With these devices, an operator can digitize hard-copy maps mounted to the surface of the tablet. Conductors in the tablet receive electrical signals emitted by the cursor and convert these to relative tablet positions. Buttons on the cursor can control various commands or enter data during a graphic entry or edit session.

A *mouse*, another device sized to fit in the palm, controls a cursor or pointer on the screen of the graphics monitor. A small rotating ball on the mouse's underside or a light-sensing device tracks its movement over a surface. As the mouse is dragged over a surface, the screen pointer moves in relation to its movements. By pushing buttons on the mouse, an operator can invoke commands from a menu and enter-and-edit graphic features (Fig. 7-6).

Other devices for controlling graphic entry-and-edit operations include joysticks and console-mounted track balls. These devices function in a manner similar to the mouse. By rotating or moving them back and forth, the position of the pointer on the graphic monitor can be controlled, allowing the user to invoke menu commands and manipulate graphics. Characteristics of the major types of data entry, edit, and query devices are discussed below.

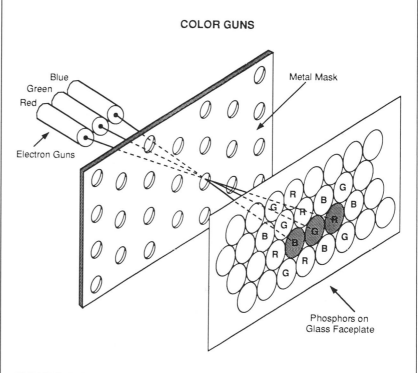

FIGURE 7-5

Electrons excite three-color sensitive phosphors (red, green, and blue) on the terminal screen to produce color images.

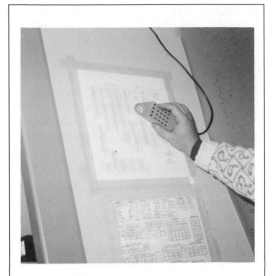

FIGURE 7-6

An operator can establish a menu of commands on the digitizer board's surface and initiate them with the puck. (Courtesy of Kentucky National Resource Information System [KNRIS])

Digitizing workstation. The digitizing station is used to enter, edit, display, and analyze graphic and nongraphic data. The main components in the digitizing workstation include one or two high-resolution monitors with color display capabilities, a digitizing tablet cursor, and a keyboard. Commands are issued and complex graphic features are entered using a cursor and tablet or screen menus.

Digitizing workstations have varying capabilities of local processing power and special system features depending on the vendor. Major types of digitizing workstations include:

- Graphic Terminals

    These digitizing workstations consist of a digitizing tablet configured with a graphic terminal that has little or no local processing capability. Most digitizing and editing functions require direct interactive communication with a host processor. A number of major software vendors support the Tektronix graphic terminals, or those of other manufacturers which are configured with digitizing tablets.

- Microcomputers

    Some digitizing stations use microcomputers (normally IBM PC or compatible) configured with a digitizing tablet and color graphic monitor. Depending on the software vendor, this workstation functions either as a nonintelligent terminal emulating the operation of a graphic terminal (little or no local processing capability), or it may use some of the PC's processing power to perform some functions locally. For vendors that support a microcomputer platform, the IBM XT, AT, or compatible are the most popular. Use of 32-bit personal computers, such as the IBM PS/2 series and the Compaq 386, is increasing.

- Super-microcomputer Workstations

    Some digitizing workstations use 32-bit workstations with or without local mass storage capability that can perform most or all workstation functions locally without using the processing power of a host computer. Configured with high-resolution graphic monitors and a digitizing tablet, they are able to function alone or as part of a network with other workstations or host computers. These workstations support both the functions of the graphic edit and query stations and the alphanumeric terminals.

Photogrammetric Mapping Workstation. These specialized stations capture data from aerial photographs. A stereophotograph pair is mounted on the device, and the resulting three-dimensional image is projected for view by the operator. Using a pointer with hand controls, map features are traced by the operator, who can see and interpret the three-dimensional image. Planimetric features such as roads, building outlines, or topographic contour lines can be mapped in this manner. By identifying known control points and entering their coordinate locations, the resulting map can be tied to a geographic map grid. Most photogrammetric workstations are connected to a dedicated or host processing unit. As they are traced, the map features are coded and entered in a digital file for final editing at a later stage.

Graphic Edit and Query WorkStation. Graphic edit and query workstations are similar in function and specifications to the digitizing stations, except that they are not used for digitizing operations. They perform graphic query and display functions and can be used to edit the graphic or attribute data bases of a GIS, including the use of coordinate geometry and precision entry procedures to capture graphic data. A mouse normally is used to manipulate the graphics and enter commands through screen menus. Depending on the vendor, the graphic edit and query workstation may include hardware components similar to the digitizing station but without a large digitizing table (Figs. 7-7, 7-8).

As with digitizing workstations, edit and query workstations can be configured using graphic terminals, microcomputers, or supermicrocomputers workstations.

FIGURE 7-8

Query stations provide ease of access through screen menus and prompts. (Courtesy of Dames & Moore).

FIGURE 7-7

Query stations requiring less precision for interacting with the screen cursor may use an external "mouse."

Alphanumeric Terminal. Alphanumeric terminals are inexpensive devices used to enter and display textual data. These terminals have a standard keyboard and monitor (commonly 9 to 13 inches diagonal) that is capable of displaying at least 80 columns by 24 lines of text. Some alphanumeric terminals include function keys for invoking special operations driven by an application program. Some alphanumeric terminals display text in multiple colors, and some display crude graphics.

Hard-Copy Output Devices

This category of peripheral devices comprises generators of hard-copy tabular or graphic products. Maps and other products can be produced in large-sheet and page-sized formats. Devices such as pen plotters generate products directly in vector form, while most other devices produce hard-copy graphics using raster technology. Devices best suited for printing text include high-speed line printers and low-end dot matrix printers.

Plotting devices can generate a wide

range of plot sizes. In the computer graphics industry, plot size is usually classified in the following categories recognized by the American National Standards Institute (ANSI):

A-Size: 8.5″ × 11″
B-Size: 11″ × 17″
C-Size: 18″ × 24″
D-Size: 22″ × 36″
E-Size: 34″ × 44″

The role of hard-copy output devices in GIS is determined by the requirements of specific applications and, ultimately, the product's user community. Features and characteristics such as size, speed, resolution, color capability, and hard-copy media should be considered. The most common categories of hard-copy output devices in computer mapping systems are discussed below.

Line Printers. For high-volume printing of reports, tables, and other text documents, the line printer (so named because it prints one line at a time) is favored because of its speed and relatively low cost. They generally use one of two techniques that prints either fully formed or dot matrix characters. The full-character method arranges raised characters on a chain or band that rotates horizontally (parallel to the printed text's orientation) at very high speeds. Hammers strike the characters over a ribbon, and the image is formed on the paper being fed through the printer by a sprocket mechanism. Dot-matrix printers have print heads consisting of multiple wires or pins that form a character with a pattern of dots as the head strikes the paper.

Until recent years, matrix printers suffered from poor print quality because of the low density of the print head's dot matrix. Advancements have produced reliable print heads with a greater density of pins, thereby improving the quality of the characters. Dot-matrix line printers provide the flexibility to change fonts and print styles as directed by the software creating the print file or by switches on the printer. There is no need to physically change a chain or band to alter fonts.

These advantages of the matrix printer have contributed to its recent popularity.

Printing speeds for line printers vary from 300 lines per minute to more than 2,000 lines per minute. Printers in the range of 600 to 1,200 lines per minute are typical in geographic information systems that require volume text output.

High-Volume Laser Printers. Laser printing is a relatively new technology used to produce both hard-copy text and graphics. Laser printers produce images one page at a time by using a laser to apply charges to a rotating drum. The drum is then exposed to a dry toner, causing toner particles to adhere to charged portions of the drum. Paper is pressed against the drum and the toner fuses to the paper. One page can be produced with each rotation of the drum.

Laser printers are used to generate high-volume, high-quality text reports and graphics. High-volume laser printers now on the market deliver speeds of up to thirty pages per minute with a resolution of 300 dots per inch. Laser printers have not yet been used extensively in geographic information systems, but their versatility, as well as expected price decreases, makes their increasing use likely.

Pen Plotters. Pen plotters can plot a vector-format file from a host system using multiple line weights and colors (Fig. C-46). Pen plotters are classified into two main categories: flatbed and drum feed.

Flatbed plotters have a horizontal surface and a gantry overhead that houses the pens. This gantry controls the dropping and raising of the pens and the x/y movement over the plotting medium that lies flat on the bed of the plotter.

In *drum-feed* plotters, the plotting medium is moved back and forth along the y axis by the drum and the pens are moved along the x axis. Drum-feed plotters use either a sprocket mechanism or a grit (i.e., high friction) wheel to feed and stabilize the medium during plotting. Sprocket mechanisms are somewhat more reliable, but they cost more to manufacture and require expensive prepunched plotting media. Many *dual mode* drum-feed plotters available today accept roll media and cut sheets. In geographic information systems, drum-feed plotters are by far more popular than flat bed plotters because of their:

- Ability to plot multiple plots continuously using roll media
- Flexible plot length, not restricted by plotter bed dimensions
- Low space requirements
- Competitive cost

Flat bed plotters, however, still have a niche in mapping operations that require high stability, such as scribing.

Plotters come in sizes ranging from small desktop units to large-format devices that can plot at widths up to 44 inches. In most cases, pen plotters are used in GIS for large-format products (24 to 44 inches wide). Plotters can house multiple pens of various types (ballpoint, drafting quality, scribe points, and others). Most plotters can accept different media, including mylar, drafting paper, and scribing film.

Plotters are connected to a host computer and operate by reading a *plot file* that has been prepared using the host system software. The plot file contains the map data in vector form with a series of plotting commands known as *primitives*. These primitives are part of a library of commands that specific vendors use as a convention on their plotters.

Some plotters available today are designed with *intelligence*—the ability to accept a plot file from the host system and further process it to make plotting more efficient. This is done by arranging the vectors in the file to reduce the distance the pen must travel between plotting functions and to reduce the number of up and down repetitions.

The following major features or specifications should be analyzed when evaluating pen plotters:

Resolution, or *accuracy*, is expressed as the maximum expected error with which a pen can be positioned relative to the absolute coordinates being read from a plot file. For mapping operations, a resolution of 0.002 inches or better is standard.

Repeatability is a measure of the accuracy with which a pen can be repositioned at a particular point on the plotting media. For mapping operations, a repeatability of 0.005 inches is standard.

Speed is a measure of how fast the plotter can move the pen across the medium. Specifications on plotter speed are given as maximum values for straight-line movement along one axis. Obviously, actual speeds in a plotting operation will be lower than this ideal rating. The user must be aware that certain types of media and pens may not be able to operate at top plotter speed. Although speed is a good indicator of plotter performance, this specification must be considered along with other factors to assess its overall performance. For plotting operations in a production environment, speeds of 20 inches per second or greater are desirable. Less-intensive plotting operations may be satisfied with lower-speed plotters in the range of 10 to 20 inches per second.

Acceleration is a measure of the force exerted on the plotter's pen, expressed in units of gravity (G). This measure indicates how quickly the pen reaches its maximum speed from a resting position.

The *number of pens* and the *pen-carriage mechanism* both should be considered. Most large format plotters have a four- to eight-pen capacity, which is adequate for most map plotting operations (Fig. 7-9). Drum-feed plotters use two approaches for holding pens, either in a carousel that moves along the *x* axis during plotting, or in a *stall* fixed at one side of the plotter. In the stall design, the pen carriage must move back to the stall to change pens.

The price range for a typical drum-feed

FIGURE 7-9

Pen plotters, once the backbone of map output, are giving way to high density and speed electrostatic plotters. (Courtesy of KNRIS).

plotter used in geographic information systems that can plot E-size products (see above) at speeds above 20 inches per second is $7,000 to $12,000. Over the past decade, major improvements in the design of pen plotters have introduced the following improvements:

- Increase in maximum plot speeds from about 15 inches per second to more than 30 inches per second
- Use of on-board vector optimizers to increase the efficiency of plotting operations and off-load processing from the host
- Ability to accept a wider variety of plotting media
- Advances in plotter inks combining quick-dry, low-viscosity properties and increased color choices
- Increase in the number of pens supported
- Features for easy maintenance such as automatic pen-capping
- Average 30 percent drop in cost

The major advantages of the pen plotter, compared to other large-format plotting devices, are its high-resolution plotting capability (it does not convert data to a raster format) and its relatively low cost. The pen plotter, however, cannot compete with electrostatic plotters when high speed and production volume are re-

quired. The overall time required to produce a plot with a pen plotter increases proportionally with its complexity. The speed of electrostatic plotters is not affected by plot complexity, although it does affect the time required to transform a vector graphic into a raster image.

Electrostatic Plotters. Electrostatic plotters connected to a host processor read graphic files and produce hard-copy plots (Fig. 7-10). In recent years, electrostatic plotters have become the workhorses in the high-volume production of plots generated by geographic information systems. The electrostatic plotter reads a plot file in raster form, in which the image is composed of a matrix of dots. An image is then produced by the electrostatic plotter using the same technique as a photocopy machine: Electrical charges applied to the surface of the plotting medium with a series of small nibs form a pattern of the image being plotted. After the medium is charged, it passes through a liquid toner solution that adheres to the points of the medium where the electrical charges were applied. Excess toner is then washed off and the plot is discharged.

The production rate of electrostatic

FIGURE 7-10

High speed, improved resolution and declining cost have made the electrostatic plotter the choice for large volume production. (Courtesy of Belleview, Washington).

plotters is impacted substantially if vector information from the GIS must be converted to a raster format for the plotter to read it. The first generation of electrostatic plotters used in geographic information systems usually relied on the host processor to perform this rasterization process and to control the plotter during operation. Rasterization can be time-consuming and slowed further by other host operations. Today it is much more common for electrostatic plotters to be equipped with local controllers and rasterizing units so vector information can be sent from the host to the plotter and rasterized there. Unless specially compressed, raster files require much more storage space than their vector counterparts (commonly from five to 30 times more), so plotter rasterizing units include local storage units or *buffers*.

The trend toward offloading processing and plotting functions reduces the burden on the host processor and creates a much more flexible environment for the use of electrostatic plotters. When rasterization is performed by the host, large volumes of raster data must be transferred from the host to the plotter through a high-speed parallel communications line. Line limits require that the plotter be placed no more than about 50 feet from the host. Electrostatic plotters that perform the rasterizing function locally can make effective use of serial communications lines that allow the plotter to be placed at relatively long distances from the host computer.

Electrostatic plotters can print many page sizes, and some print color. Most units accept multiple plotting media, including opaque paper, translucent bond, sepia, and mylar. Resolution is expressed in dots per inch. In the recent past, plotters with a resolution of 200 dots per inch were common, which leaves obvious jagged lines. Although the 200-dot-per-inch plotters are suitable for some purposes, plotters with 400-dot-per-inch resolution are now more popular (Fig. C-47). Most color electrostatic plotters require four complete passes through the charg-ing–toning cycle, one pass for each of three primary colors plus black. However, several companies now offer single-pass color devices, and it is likely that these plotters will dominate the market within several years.

Plotting speed is a primary feature that differentiates electrostatic from pen plotters. Depending on the complexity of the plot, electrostatic plotters can generate a product from five to more than thirty times faster than pen plotters. Electrostatic plotters also depend less on operator intervention since there are no pens to clog or run out of ink. A dramatic drop in the average cost of electrostatic plotters have made them the device of choice for high-volume, large-format production. E-size, black-and-white plotters with local rasterizers range in cost from about $25,000 to $35,000; E-size, color electrostatic plotters cost between $55,000 and $70,000.

Screen Copy Devices. Screen copy devices differ from other types of hard-copy output devices in their function and in the way they normally are configured in a GIS. Screen copy devices are connected to an output port on a graphics workstation, and they allow the quick generation of an image displayed on the station's graphic monitor. To produce the plot, a workstation operator enters a simple command or strikes a function key.

These devices, producing output as a raster image, usually use one of the following techniques to produce a page-size (A or B), hard-copy image:

- Impact printing
- Laser printing
- Electrostatic plotting
- Ink jet plotting
- Thermal transfer

Models available from a variety of vendors produce color or black-and-white copies. Most GIS vendors can support more than one type of screen copy devices.

Many vendors manufacture devices that can generate prints from GIS screens, ranging in cost from about $3,000 to $6,000. The common output technologies used in screen copy devices are discussed below:

- Impact Printing

  Impact printers use a dot matrix to print graphics. The raster image is produced as a pattern of dots generated by small hammers striking a print head over a ribbon. Multiple color ribbons can be used to generate a small range of colors, a maximum of eight in most cases. Dot-matrix printer resolution is limited by the density of pins on the print head; most offer a grapic resolution of 70 to 150 dots per inch. With greater print density, the head is more likely to be damaged by repeated impact. Another disadvantage is the limitation of ribbons for generating high-quality color products. Dot-matrix printers have a tendency to produce colors with inconsistent hue, brightness, and overall clarity. However, dot-matrix printers offer a low-cost option, especially if color consistency or high resolution is not important.

- Laser Printing

  As discussed previously, laser printers produce images one page at a time by using a laser to apply charges to a rotating drum. Low-volume laser printers that print page-size (8.5″ × 11″) documents at a rate of five to eight pages per minute are used extensively in word processing and electronic publishing applications. This technology is fast compared with other screen copy devices, and offers relatively high resolution (300 dots per inch is common). Page-size laser printers are best suited for black-and-white copies. Although several manufacturers recently have offered color models, the technology is costly; it has not yet been accepted in the geographic information industry. Improved color capabilities, however, will likely make these devices more popular in the future.

- Electrostatic Plotting

  Both black-and-white and color electrostatic plotters have been used extensively in geographic information systems to produce page-size screen copies. Resolution for these units ranges from 80 to 400 dots per inch. Electrostatic plotters produce a large range of colors at comparatively high resolutions; they are very popular as screen copy devices.

- Ink Jet Plotting

  In ink jet plotters, colored inks are forced through small jets onto the print media. Ink jet plotters usually include three or four jets (for the three primary colors plus black), each with separate ink reservoirs. Ink jet plotters use either a *drop-on-demand* or *continuous-stream* method. In the drop-on-demand approach, a pulse is applied to the ink reservoir, causing a restriction that forces the ink through the jet. Continuous-stream devices create a continuous flow of ink that is normally deflected by electrodes into a gutter for recirculation. When the image calls for a color to be printed, the stream is deflected to the medium. Ink jet plotters have a reputation for high image quality and brillant colors with consistent tones delivered in area-shading applications (Fig. C-48). Resolutions from 120 to 250 dots per inch are common, and higher resolutions of more than 300 dots per inch are possible. Features for ease of use are being added, including a self-cleansing mechanism that reduces the potential for jet clogging. Ink jet devices have been available since the mid-1970s, but they have become accepted in the GIS arena only since the mid-1980s due to great improvements in speed, resolution, and reliability.

- Thermal Transfer

  Thermal transfer plotters use a heating element to melt dots of waxy ink from a film substrate to a specially treated paper. The thermal print head extends across the width of the paper. Most thermal transfers require three or four passes to plot a color image. One primary color or black is plotted at each of the passes. Resolutions typically range from about 200 to 300 dots per inch.

Special Purpose Devices in GIS Configurations

Digital Capture Devices. These devices include video camera systems and raster scanners. Each of these technologies offers a way to convert an analog image, such as a hard-copy map or aerial photograph, to digital format. In each case, the tone or color variation of the analog image is "read" optically and converted to a raster form suitable for display and analysis. The major types of digital capture devices are discussed below:

• Video Camera Systems

For certain geographic information applications, video camera systems are useful. These provide an interface to a video camera and allow the quick capture of an image (such as an aerial photograph) for subsequent processing. The video camera, normally mounted vertically, scans an image that is mounted on a horizontal surface below the camera. Using image processing software, the image can be displayed and enhanced to aid in various types of analysis. A video capture system is not well suited for applications requiring high resolution and accuracy because of the resolution limitations of the camera and problems with optical and radial distortion. Certain earth science applications in the geologic, forestry, and other disciplines can use video camera systems effectively.

• Optical Scanners

Optical scanners digitally encode information from hard-copy maps and documents consisting of line work, text, and symbols (Fig. 7-11). The scanner senses variations in reflected light from the surface of the document. Any marks on the hard copy, including stains or wrinkles, that exceed a given threshold of contrast between the mark and the background media will be encoded as "information." Scanners on the market today that accept large-format maps and drawings use one of three types of approaches: flat bed, drum, and pass-through.

In *flat-bed* scanners, the hard-copy document is placed on a flat, horizontal surface, and the scan head moves along both x and y axes over the document. In *drum* scanners, the document is mounted on a cylinder that rotates while the scan head moves horizontally across the cylinder. *Pass-through* scanners accept a document that is fed through a slot and read by the scan head line by line.

The digital file that results from the scanning process is in raster (grid) format, and does not generally differentiate between different types of features, annotation, or symbols appearing on the original hard-copy map. The raster image, a nonintelligent picture, cannot identify individual features uniquely or link them to nongraphic files. Most scanning systems come with special software that converts these raster images to a vector format with line segments individually identified. In this format, nongraphic attributes can be linked in a postscan editing session.

Most scanners currently are not able to differentiate easily between different line types, widths, and colors, or to interpret text and special symbols. Scan-

FIGURE 7-11

Scanners use optical techniques to duplicate in a digital format subtle variations in tone (reflectance) from paper or other hard copy map products. (Courtesy of Intergraph).

ners are very sensitive to the quality of the hard-copy documents, and difficulty arises in cases where maps are created with inconsistent line work or are smudged or scratched. As a result, scanning is not suited to create a layered database if the original hard-copy maps contain a mix of features that must be separated into different GIS layers.

The most important specification to consider in evaluating scanners is the scanning resolution (measured in pixels per inch). The resolution should be matched with the content and complexity of the hard-copy products. A resolution that is too high can result in too much "noise" being encoded and can dramatically increase the scanning time per document. Too low a level of resolution does not reliably capture line work in sufficient detail, nor does it recognize two lines that are spaced more closely than the scanning resolution.

Other features to consider include the type of application for which the data will be used, the ability to read text, and the ability to separate features based on line width, color, or gray tones. Special software, which often comes with the scanning system, is required for raster-to-vector conversion and editing of both raster and vector files.

Film Recording Devices. It is useful for some geographic information applications to generate output in the form of film-based products such as microfilm, 35-mm slides, or high-resolution film plates. The high resolution of film makes this medium important for products containing substantial detail, fine line work, and small type. The computer graphics industry provides several products that have been employed by GIS vendors to fill these needs:

- Slide and Print Production Devices
  Presentations often are enhanced by the use of 35-mm slides, 8½" × 11" color transparencies, or color prints. Systems are available that direct graphic output to a camera system for the generation of a color film image. Inexpensive devices, slide or transparency systems may be *terminal driven*, wherein an image is generated directly from the screen display. The resolution of a terminal-driven device depends on the resolution of the screen image, which may not be acceptable for high-quality presentation materials. Other devices directly read a digital vector image from a microcomputer or host system to generate a high-resolution image. Map publishing applications require high-resolution film recorders that can produce color separation plates for an offset printing process.

- Computer Output Microfilm (COM)
  COM devices produce microfilm images that store maps and other graphic documents. The image can be stored on aperture cards and used with a specially designed microfile printer to generate hard copies quickly. This approach is well suited to GIS installations that must handle large volumes of maps or drawings with frequent hard-copy output requirements. COM devices read an x/y image directly from a computer system. Film is exposed and developed by the unit producing the microfilm image. Most COM devices read vector images directly and are not limited by the resolution of raster images from a display monitor. Special printers are manufactured that produce hard copies from microfilm images.

COMMUNICATION HARDWARE DEVICES

Geographic information systems depend on efficient communication between devices, including the host processor and input/output peripheral devices, and between individual processing units in a distributed network. Communication concerns become more critical when devices are connected on a local area network (LAN) or when they are separated by great distances and must be connected through remote lines. Data communication terms and concepts are discussed in detail in Chapter 9, but the major communication devices used to support local area networks and remote communications are described here.

Modems

Modem is the accepted acronym for *modulator-demodulator,* a device used to convert analog signals from a communication carrier to digital signals that a computer can interpret. The analog signal typical of telephone communications transmits voice data as an electrical signal that continuously varies in frequency and amplitude. Other media, such as radio, microwave, broadband cable transmission (i.e., cable TV), and satellite systems, can be used to transmit data. All media require modems to encode and decide the analog signals.

Although many specialized types of modems exist, they all consist of the same basic components: a power supply, a transmitter, and a receiver. In geographic information systems, modems are most commonly used to connect computers and peripheral devices through voice-grade telephone lines. For example, if a graphic query station located at a field office is to communicate with a host processor several miles away, a modem is placed at each end of the line, allowing the query workstation user to communicate directly with the host computer over telephone lines (Fig. 7-12).

Two categories of analog telephone service typically are used to support computer communications: dial-up lines and dedicated phone lines.

Common private-line telephone service uses *dial-up* facilities, and any modem that supports dial-up service can use these lines to transmit and receive data. A dial-up modem may be *direct connect,* in which the telephone jack is connected directly to the modem, or it may be used as an *acoustic coupler,* in which the telephone receiver is placed in a receptacle to send and re-

ceive audio signals then converted by the modem into electrical signals.

Dedicated telephone lines set up point-to-point service between two locations, often using specially conditioned lines that have lower error rates than dial-up facilities. In a dedicated line, only the two connected points have access to the line. The local or long-distance telephone company must be contacted to set up dedicated telephone service. For dedicated lines, there is an initial installation and conditioning charge plus a monthly lease cost. The lease cost for dedicated service generally increases with the distance between the two points.

Modems are classified primarily by their data transmission speeds and the type of analog signals they support. Speed is usually measured in *bits per second.* Another unit of measurement for speed is *baud,* which describes the modulation rate of a wave, or the average frequency of the signal. In the most sophisticated modems, the data rate (bits per second) is greater than the baud rate. Sophisticated modems use various modulation techniques to increase the information-carrying capacity of a communication line.

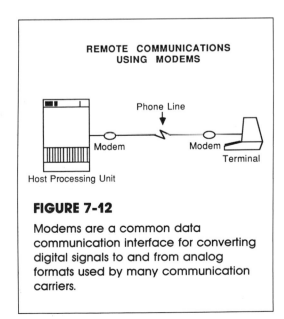

FIGURE 7-12

Modems are a common data communication interface for converting digital signals to and from analog formats used by many communication carriers.

Modems that use analog phone lines can be categorized as follows:

Low-speed modems, which include primarily dial-up modems that can transmit at speeds of up to 1,200 bits per second

Medium-speed modems, which include dial-up or dedicated-line modems transmitting data at speeds from 1,200 to 4,800 bits per second

High-speed modems, which include dial-up or dedicated-line modems transmitting at speeds from 4,800 to 19,200 bits per second

Limited-distance modems (LDMs), designed for transmitting data on dedicated circuits over short distances (usually within two miles, depending on the line quality and speed). Relatively inexpensive, they can transmit data at rates of up to 1 million bits per second. As distance increases, the maximum speed will generally decrease.

Modems adhere to certain standards that define line protocols and error detection schemes. To date, modem manufacturers have complied with standards developed by AT & T, but the international CCITT (Comité Consultatif Internationale de Télégraphique et Telephonique) stan-

dards are gaining greater acceptance. The ability of a modem to transmit data at a given speed depends greatly on the line's quality and the amount of noise or interruption present that may affect the signal. A major feature of sophisticated modems is an ability to detect errors and automatically retransmit data when an error occurs. Also available are modems that can support data communications using radio frequencies, microwaves, fiber-optic lines, and other analog signaling media.

Multiplexors

Multiplexors maximize the efficiency of a communication line by allowing multiple users to communicate on one high-speed link (Fig. 7-13). A multiplexor splits a high-speed line into multiple channels that support multiple devices. Multiplexors thus employed can alleviate the need to set up multiple remote lines, eliminating duplicate lease and modem costs that otherwise may be incurred. The multiplexor is placed in a remote link between the computer hardware component (processor or peripheral device) and the modem.

Multiplexors are of two main types: time division and statistical. *Time-division* multiplexors allocate a specific time segment to each individual channel regardless of whether the channel is being used by the device. This approach is somewhat inefficient, particularly in cases where use is sporadic. *Statistical* multiplexors use techniques to maximize use of the communication line by monitoring and dynamically allocating line capacities to channels based on their respective demand.

Multiplexors can accommodate communication lines ranging in speed from 9,600 to several million bits per second. Typically, maximum capacities of up to 56,000 bits are found in those used to communicate via phone lines. Multiplexors typically support from four to twenty four separate channels.

Digital Service Units

Digital service units can be considered analogous to modems because they con-

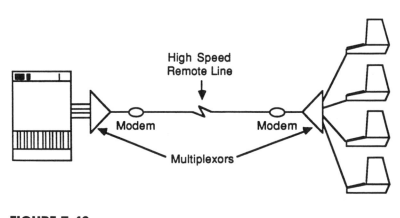

FIGURE 7-13

Multiplexors are used as the data communication interface when high speeds or multiple users are required.

nect computer devices using an outside communication carrier. However, these units connect devices across digital lines when modems are not needed to convert analog signals. In many metropolitan areas, local telephone companies, government organizations, or private companies have instituted digital service to support both voice and data transmission.

Device Servers

Device servers are intelligent hardware components that connect multiple nonintelligent devices, such as dumb terminals, query stations, plotters, and printers, to a local area network. Device servers contain computer processors dedicated to supervising communications between the devices and the network. They translate signals from the devices using a protocol understood by the local area network and route messages from a processor on the network to the appropriate device. Any device that cannot perform communication processing functions on its own must be connected to a local area network through a device server.

Conclusion

The dynamic nature of computer hardware is apparent: Rapid changes and technological advances will bring ever-greater performance at ever-lower costs. This fact sometimes leads to an attitude that "something better will come along tomorrow, so I should wait until next year to start system development." The problem with this philosophy is that something better will always be on the horizon. As long as the demand holds, the ingenuity of hardware designers will continue to produce improved products.

Although it is prudent to anticipate new product announcements when preparing a schedule to procure hardware, it is not wise to delay system development in vague anticipation of a new product.

Hardware configuration design should incorporate the flexibility to upgrade a system as new technology becomes available and as user demand dictates. A logical development plan, taking into account user needs, database requirements, and institutional variables, allows effective use of the many hardware tools now available for mapping and geographic analysis applications.

REFERENCES

Antenucci J. (1991, January), Vol. 1, No. 2, Workstation wars. *Geo InfoSystems*. Aster Publishing Corporation, Oregon.

Manning A. (Fall, 1990). New technology will affect all of us. *Communicator*. EMA, Inc., Minnesota.

SUGGESTED READING

Allison A. (1988, January). "Where there's RISC, there's opportunity. *Mini Micro Systems:* 49–62.

Anderson K (1990, November). Workstation roundup. *Computer Graphics World 13:* 69–76.

Bond J. (1987, June). Parallel processing concepts finally came together in real systems. *Computer Design Vol 26:* 51–74.

Croswell P, Clark S (1988). Trends in geographic information system hardware. *Photogrammetric Engineering and Remote Sensing Vol 54:* 1571–1576.

Dowers S, Gittings B. M., Sloan T. M., Waugh T. C., and Healy R. G. (1990, November). Analysis of GIS performance on parallel architectures and workstation-server systems. *GIS/LIS '90 Proceedings, Vol. 1.* Anaheim, California: American Congress on Surveying and Mapping/American Society for Photogrammetry and Remote Sensing, pp. 551–561.

Falk H. (1988, June). Computing speeds soar with parallel processing. *Computer Design* 27: 49–58.

Ferrari D. (1978). Computer Systems Performance Evaluation, Prentice-Hall, Englewood Cliffs, NJ. p. 554.

Kinnucan P. (1989, July) The technology of cornucopia in color printers. *Computer Graphics Review:* 26–34.

Kinnucan P. (1990, June) The plot thickens. *Computer Graphics Review:* 34–45.

Kurcharvy T. (1988, September/October) Wrestling with the differences between PCs and workstations. *Computer Graphics Review:* 29–38.

Chapter 8

Geographic Information System Software

Software can be described in terms of layers: Low-level software layers drive basic system operations. Upper-level software, or *application* software, consists of the programs used to perform mapping and geographic analysis operations.

The focus of this chapter is on the upper level of GIS software. However, the lower layers of software—the operating system and special system utilities—are discussed briefly to introduce the function of application software.

Stated simply, software is a collection of instructions that tells the computer hardware to execute a task. Software is maintained in a computer as programs stored in the system's memory that provide the computer's basic operating procedures and assignments.

The role and characteristics of software have evolved along with the technological advancements in computer hardware architecture, since computers first became economically viable in the 1950s. Early electronic computers had no software component that could be stored and accessed to drive system applications. In these early systems, the circuitry was set through switches to perform one specific operation (usually complex mathematical functions). It was necessary to reset the switches to initate a different operation.

Since the introduction of those early digital computers, designers have developed more flexible ways to interact with computer systems. Users have become ever further removed from the actual operation of the computer hardware; they now can work on a system with relatively little training. The term *user-friendliness* has been coined to describe the ease of interaction between users and computers.

This trend toward software flexibility or user-friendliness continues today. GIS developers have provided effective software that has made the technology accessible to a wide community of users. Individuals familiar with the details of an application are able to tap into the capabilities of the computer without knowing how the system actually works. The technology is well advanced along the path toward higher levels of interactive flexibility. Exciting advances now on the horizon will increase the effectiveness of geographic information systems.

LAYERS OF SOFTWARE

In modern computer systems, software can be depicted as being composed of layers, as shown in Figure 8-1. This layered model shows the system hardware in the center surrounded by the following three

GIS developers have provided effective software that has made the technology accessible to a wide community of users. Individuals familiar with the details of an application are able to tap into the capabilities of the computer without knowing how the system actually works.

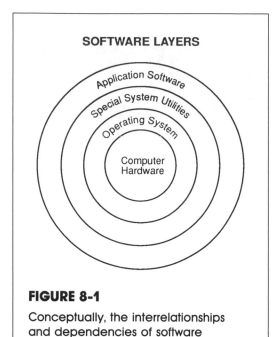

SOFTWARE LAYERS

Application Software

Special System Utilities

Operating System

Computer Hardware

FIGURE 8-1

Conceptually, the interrelationships and dependencies of software supporting GIS and its host processor are layered.

layers of software:

- Operating system
- Special system support programs
- Application software

The outer layers rely on the inner software layers to perform a particular task for a user. The computer system provides the capability for each software layer to interact with the other layers. Although this model vastly simplifies the actual types and relationship between software components, it clarifies the way software interacts within a computer system.

The *operating system* comprises programs that supervise and direct the system operations and that control communications with hardware devices connected to the computer. Special system utilities and application software that are used to perform a task such as the display or plotting of a map access the operating system programs to execute the required functions.

Every computer system includes *special system utilities*. These utilities support programs that perform certain routine functions required frequently by users. They include language compilers, special file management utilities for manipulating files in mass storage, special *drivers* for communication with peripheral devices, such as plotters and tape drives, and a wide range of other programs that help the user and system operator perform day-to-day tasks.

Application software, at the outer layer of our model, consists of the programs directly accessed by a user to generate a particular product. Application software is often described as a *package*, which identifies it as a collection of integrated programs that together perform specific applications. A common example of application software is a word processing package used to enter text and produce documents. Many GIS application packages available today contain multiple programs to accomplish a variety of mapping and geographic analysis applications.

LOWER-LEVEL SOFTWARE

Operating Systems

The operating system provides the interface between the user, the application programs, and the computer system hardware. Although the concept of an operating system continues to evolve in the computer industry and there is no universally accepted set of functions that it should perform, major operating system functions characteristic of all types of computers, from microcomputers to large mainframes, follow.

Memory Management. Operating system programs sometimes called *monitors* or *executives* are used to keep track of the execution of application programs and the scheduling of multiple jobs that have been submitted by users for computation. Memory management programs communicate with the supervisor programs to manage the flow of data and program statements within the central processing unit (CPU) and between the CPU and the mass storage. In multiuser systems, memory management programs control the execution of multiple programs and dynamically allocate portions of main memory to specific programs.

System Access and Accounting. The operating system provides security for data and programs to limit user access. Access rights can usually be assigned at different levels, for example, "read only," "read and modify," and "read, modify, and delete." These levels of access rights are assigned to particular files or groups of files in mass storage according to a user's identification code. When a user logs onto a system and supplies an identification code and a confidential password, the operating system allows access to data and programs according to the user's access rights.

Most operating systems also have programs that track the use of system resources, such as the amount of mass storage space used, the connect time (total time during which a user was in interactive sessions), the time spent processing executed instructions for the user, the time devoted to the use of peripheral devices (i.e., printers and plotters), and other users' accounting tasks.

Communication Control. All computer systems must manage communications between the processing unit and the peripheral devices through which users access the system. Communication programs control this flow of information. Each peripheral device is assigned a device number or *address*, which the operating system uses to track input and output from system users.

Some large computer systems have individual processing units (sometimes called *communication* or *front-end processors*) that are devoted strictly to managing communications. Large computer networks with many peripheral devices and multiple processing units often require special programs that run on these communication processors to coordinate communication tasks with the operating system.

Command Processing. Computer systems offer *command languages*, special sets of instructions through which users can execute operating system functions. To perform a particular task, command processing programs are used to interpret these commands. An example of a command language is the *job control language* (JCL) used to submit batch programs on mainframe and minicomputers. Another example, familiar to personal computer users, is the *DOS* command dialogue, a command language that allows access to operating system functions on IBM PCs.

File and Data Management. This software allows users to store, retrieve, and manipulate files and groups of files maintained in mass storage. It also controls the physical storage of data on the mass storage media. Utilities are provided so that users can access, delete, and copy files, change file names, and perform special functions such as data sorting, file merging, and file comparison.

Until recently, most operating system software was proprietary, designed to run only on a particular family of computers offered by a single vendor. The operating system was acquired as part of the overall hardware package and was installed and activated by the computer vendor. In the computer industry today, the use of such proprietary operating systems is still prevalent. Operating systems predominant in geographic data management systems include VMS (DEC VAX computers), PRIMOS (Prime 50 series computers), VM and MVS (IBM midrange and mainframe computers), and AOS/VS (Data General MV series computers).

Within the past five years, there has been strong interest in the use of *universal* or *portable* operating systems that can be used on different processing units, thereby freeing users and application programs from the limits of a specific computer system. Carried to its full extent, the concept of portability means that any application program (for example, a computer mapping package) could be loaded on another processing unit regardless of the make or model and could operate without program modification. This concept has not yet been fully realized. Currently, most users are confined to particular hardware environments unless they are prepared to perform some often-significant program modification.

One example of a generally accepted operating system "standard" is DOS, used on a majority of 16-bit or 32-bit microcomputers. This operating system, developed by IBM and Microsoft Corporation, has become a de facto standard in the microcomputer industry. The few microcomputer manufacturers that do not promote DOS as a primary operating system (such as the Apple Computer Corporation) often provide an auxiliary ability to run DOS software.

Originally developed by scientists at the University of California at Berkeley and at AT & T Bell Laboratories, the UNIX operating system has seen a dramatic increase in use. Originally developed for scientific applications, it is becoming increasingly popular on multipurpose computers. Many computer system manufacturers have adopted one of a variety of UNIX versions, sometimes offered as an alternative to their proprietary operating system. Its attractiveness as a portable operating system undoubtedly will continue to fuel its rise in popularity.

Special System Utilities and Support Programs

Programs considered special system utilities in one system may fall under the category of operating system or application software in other systems. Despite the somewhat indistinct nature of this type of software, most systems have sets of programs that are accessed by the operating system or application packages to perform routine support functions. Many of these programs are closely associated with the operating system and are provided by the computer vendor as part of the operating system package. Other utility or support programs are delivered with application packages or are acquired from third-party vendors for installation. Major utilities and support software used frequently in geographic information systems are described below.

Language Compilers. Most geographic data management software is written using a programming language such as Assembler, Fortran, COBOL, BASIC, or C. These programming languages use a set of commands based on written English that direct a computer in a structured syntax to perform a certain task. The programs in the form of *source code* must be translated to a more fundamental level that can be understood by the computer. A *compiler* is a program that translates the source code to *machine code,* the binary representation of the program. The machine code can be directly executed by the computer processor. A special compiler is needed for each programming language that will be used on the system.

Device Drivers. Device drivers are programs specifically designed to provide a communication interface between the operating system and application software for support of a specific peripheral device. The device driver translates commands issued through application software into instructions that a device can interpret.

In GISs, device drivers are commonly used to support pen plotters, electrostatic plotters, and other graphic output devices. In these cases, the device driver interprets the specific plot generation commands that describe the appearance and format of the plot, including positioning of the pen on the plot media, line weights and types, shading and colors, annotation, and scale. The driver translates these user commands into instructions or *primitives* that the plotter uses to generate a hard-copy product.

Device drivers are also used in data management systems to support reel-to-reel tape drives or smaller cartridge tape drives found on microcomputers or small multiuser computers. Special devices such as optical scanners may also require device drivers.

Disk Backup Utilities. Most computer systems provide support programs for the efficient tape backup of data and software on a disk drive. These backup procedures protect against potential data loss in the event of a disk failure. Backup utilities may copy entire portions of disk data to the tape or they may allow for *incremental backups,* in which only changes that have occurred since the last backup are written to the tape. Tape backup routines on systems with reel-to-reel tape drives can be used with both streaming and start/stop drives (see Chapter 7).

Subroutine Libraries. Often, subroutine libraries are provided as part of an operating system, sometimes as part of the GIS application software. The subroutines are program modules written in a variety of programming languages, such as Assembler, Fortran, and C, that can allow users to develop sophisticated applications that access system functions.

Special Communication Software. As computer systems become more complex, special communication software is often needed to support the operating system as it controls multiple devices in a network. Typical functions that may be performed by special communication software include the following:

- Clustering of multiple computer processing units and mass storage devices
- Local area network support for multiple computer processing units and peripheral devices
- Support of remote communications using modems or other devices
- Support of gateways for communication with other computer networks
- Device emulation and protocol conversion for access to other systems that do not recognize the same operating system commands or communication protocols

GEOGRAPHIC INFORMATION SYSTEM APPLICATION SOFTWARE

This software is most frequently used to perform tasks in a geographic information system. The application software is provided in the form of software packages, each consisting of multiple programs that are integrated to supply particular capabilities for mapping, management, and analysis of geographic data. Application software developed for geographic information systems can be conceptualized in two parts, as depicted in Figure 8-2:

- A "core" package of basic mapping and data management capabilities
- Separate application packages that are integrated with the core package to perform a specific mapping or geographic analysis application

The specific types of capabilities provided as part of the core package or as special applications will vary among software vendors. In a broad sense, however, the

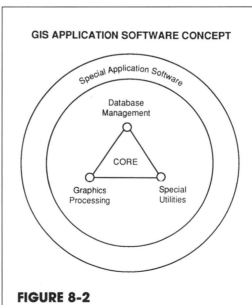

FIGURE 8-2

GIS applications software is built around a set of core functionality.

Choice of GIS software should be based on an evaluation of a product's proven capabilities as demonstrated by its existing user base, as well as the consideration of new software and enhancements that do not have a long history of actual use.

core package will include one or more of the following components:

- Graphics processing functions
- Database management functions
- Basic cartographic and geographic analysis utilities

Software packages that are designed principally for mapping may emphasize graphics processing and have limited database management functions and geographic analysis utilities. Packages offered by software vendors as "full" geographic information systems normally include significant emphasis on both graphic and nongraphic data management with strong links between the map (graphic) database and nongraphic attributes.

Special application packages may be integrated with the core package to address specific user applications, such as network tracing, terrain analysis, special thematic mapping, among others.

Graphic Processing Capabilities

Graphics processing capabilities include functions that allow the user to enter or edit map features and associated annotation, and to generate screen displays or hard-copy maps.

Confusion often arises about the distinction between graphics software developed for computer mapping applications and computer-aided drafting (CAD). There is no absolute definition; it is generally accepted that CAD systems provide special functions for more efficient entry and edit of the graphic features and annotation common in engineering drawings. The engineering drawing may be associated with a mechanical engineering application (e.g., auto body design) in two or three dimensions, or it may be an architectural layout or civil engineering site plan. Graphics capabilities associated with CAD systems include special functions for the efficient entry of line work and geometric forms such as parallel lines, rectangles, arcs, fillets, and ellipses; special geometric functions, such as tangents to a curve and bisecting lines; dimension annotation; and

other graphics capabilities common in the engineering drawing process.

Mapping systems generally are distinguished from CAD systems by their emphasis on processing geographically continuous databases at smaller scales than engineering drawings. Also, systems designed for mapping and geographic analysis traditionally are better able to incorporate tabular data, and depend less upon special geometric entry facilities more suited for engineering structures and features.

It is within the civil engineering disciplines that mapping and engineering applications often overlap. Drafting work in applications such as subdivision design, preparation of engineering plans or as-built drawings for utility lines, and other similar civil engineering applications are closely associated with the mapping process. The general trend has been for vendors of GIS software to incorporate more of the functions traditionally associated with CAD systems into map entry-and-edit packages.

Graphic Entry Capabilities.   Graphic entry capabilities in GIS allow users to input map features and store these as x,y (and sometimes z) coordinates based on a geographic reference grid. Text annotation and feature identifiers also can be entered to define a map feature uniquely and provide a basis for associating the feature with tabular attributes stored in a database.

Major graphic entry capabilities commonly included in the core package of a geographic data management system are described below:

- Interactive Digitizing

  Interactive digitizing is a procedure by which a system user converts a hard-copy map to digital form. The process allows a user to enter the x and y coordinates of map features by freehand tracing from a hard-copy map mounted on the tablet of a graphics station (see Chapter 7). Features are traced using a cursor or puck which consists of a transparent lense and cross hair that the user positions on the line work of the map features to be digitized. Many digitizing programs allow an operator to digitize in manual or stream mode. In the manual mode, users enter coordinate points at their discretion using designated buttons on the cursor. Stream digitizing techniques enter a coordinate at a predetermined x,y movement of the cursor. In stream digitizing, all coordinates are automatically entered as the operator traces the cursor along a map feature boundary.

- Special Feature Entry

Special entry programs allow the user to create lines and geometric shapes (often associated with CAD systems) efficiently in an interactive map entry process. The graphic features are entered by a cursor or mouse using screen or tablet menu commands. The operator is prompted to input basic parameters used by these special programs to create the line work or geometric shapes. The number of these special graphic entry commands provided with GIS software packages varies greatly from vendor to vendor.

Some GIS packages offered by vendors today include graphics components rich in special feature entry routines that have evolved from interactive CAD systems. Other GISs contain fewer special entry functions, emphasizing instead those that are most frequently used in map entry operations.

Some examples of special graphic entry functions are listed below (Fig. 8-3):

- Rectangle
- Three-point arc
- Spline curve

Many other special entry functions are possible, such as angled lines; bisecting lines or tangents, arcs, and circles based on radius and arc length, chamfers, fillets, ellipses, parabolas, and others. Often these special commands can be combined in single drawing operations to generate complex features.

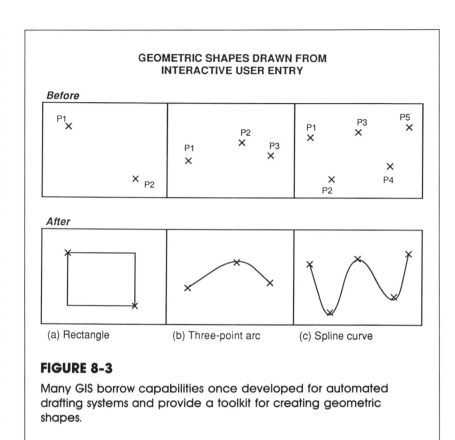

FIGURE 8-3

Many GIS borrow capabilities once developed for automated drafting systems and provide a toolkit for creating geometric shapes.

Annotation Entry. Annotation is text that labels map features in a display and provides other necessary notes in map titles and legends, for example. Geographic information system packages generally create annotation for map features in one of two ways. The first is to position and enter annotation for map features during the map digitizing or entry process. In this case, the operator selects commands to indicate the location, placement, and orientation of the label. The annotation may then be typed on the keyboard at the time of graphic entry.

The other technique, which many systems provide, is to gain access to the nongraphic database and use data elements within it as annotation for specific map features. This capability, illustrated in Figure 8-4, allows the operator to customize a map for display based on elements stored in a nongraphic database. The elements themselves may be entered during the

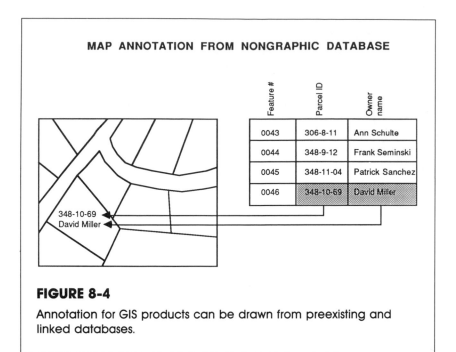

FIGURE 8-4

Annotation for GIS products can be drawn from preexisting and linked databases.

map entry process, in a separate session, or loaded from some preexisting data file. When using this approach, the user must still supply information about the placement and orientation of the annotation or allow the system to automatically align labels based on predetermined parameters.

The decision to store annotation as elements in a database depends on the applications and map display requirements of the user. Some map annotation may not be associated with single map features and therefore may be stored more efficiently in a map layer of the system database as *dumb* annotation. It is generally advisable, for features linked to a nongraphic database, to store all likely annotation as database elements. This provides considerable power and flexibility in map display as well as map query and analysis operations.

Graphic Edit Capabilities. GIS offers special capabilities for the editing of graphic features, including a variety of delete and modification functions. Commonly used graphic edit functions include the following:

- Feature Delete

  Interactive delete capabilities allow the operator to point to a feature at a graphic digitizing or edit station with a mouse or cursor and issue the command to delete the feature from storage. The delete function may be invoked by pushing a designated mouse or cursor button or by entering a command from the keyboard or a menu on the screen or digitizing tablet. Many systems allow the user to define an area or *fence*, in which all enclosed features are deleted when the delete command is issued. In GIS, when a nongraphic database is linked to the graphic feature, the delete function will often automatically delete the appropriate nongraphic elements as well.

- Feature Modification

  Feature modification functions change the shape, orientation, or position of existing features. Typical graphic modification functions that operate on individual features or groups of features include stretching, trimming, rotating, and moving. As in the case of delete commands, some feature modification commands can operate globally on a group of features around which the operator places a boundary or fence.

- Special Edit Commands

  Some software packages have special edit programs that are executed in a batch mode on a digitized map file. These routines analyze the map file and identify (and often automatically correct) apparent cartographic inconsistencies, such as shapes that do not completely close, gaps between lines that should meet, and overshooting lines that should terminate at an intersection with another graphic feature (Figure 8-5). The special edit programs identify these inconsistencies based on user-defined distance tolerances that allow the programs to differentiate between graphic errors and valid cartographic representations. Obviously, the tolerances assigned to these special edit programs must be chosen carefully. If they are too small, inconsistencies will not be identified. If they are too large, there is the chance of altering graphic features that are actually represented correctly.

- Line Generalization and Smoothing

  Some graphics processing software includes programs to generalize or smooth lines stored in vector form. Both of these processes can be considered cosmetic functions that adjust the strings of coordinates defining a line to make it more suitable for display. Line generalization (Fig. 8-6) reduces complexity by eliminating x,y coordinates that define a given line. Coordinates are eliminated based on their spacing along the line or the angular bends in the line. Smoothing programs add additional x,y points to a line's definition to round out sharp angular bends and give the line a more aesthetic appearance.

Graphic Display and Plotting Capabilities. Part of the graphics processing component of a geographic information system includes programs to customize the display and hard-copy production of maps on graphic monitors or plotters. A number of factors are important in display and plotting operations. A GIS has software capabilities that allow an operator to control the appearance and format of the display or plot by setting the following parameters:

- Line patterns and weights (thickness)
- Point symbol types
- Color
- Shading patterns
- Text font types, styles, weight, placement, size, and orientation
- Placement and design of titles, legends, and other special map elements
- Size and/or scale of the plot or display

The software allows an operator to assign parameters to particular map features for display on a graphic monitor or in hard-copy map production. Normally, for each class of map features (e.g., manholes, parcel boundaries, soil polygons), parame-

CARTOGRAPHIC INCONSISTENCIES IDENTIFIED BY SPECIAL GRAPHIC EDIT PROGRAMS

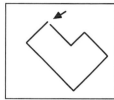 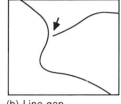 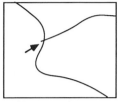

(a) Gap in closed shape   (b) Line gap   (c) Line overshoot

FIGURE 8-5

Correcting cartographic inconsistencies improves the quality of the representation of the data and retains, if appropriate, the topologic integrity of the feature.

ters for the display of lines, areas, point symbols, and annotation are established to govern the way in which the class of features is displayed or plotted.

Most software packages employ the concept of *tables* or *libraries* that store the specifications for the display or plotting of lines, symbols, color, shading patterns, and text (Figs. 8-7, 8-8, 8-9). These tables or libraries store the display specifications that are referenced by unique identifiers.

LINE GENERALIZATION AND SMOOTHING

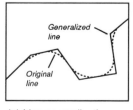 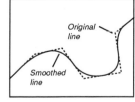

(a) Line generalization   (b) Line smoothing

FIGURE 8-6

Line generalizations and smoothing are facilities that are as important to data compression as they are to graphic representation.

SAMPLE LINE TYPES

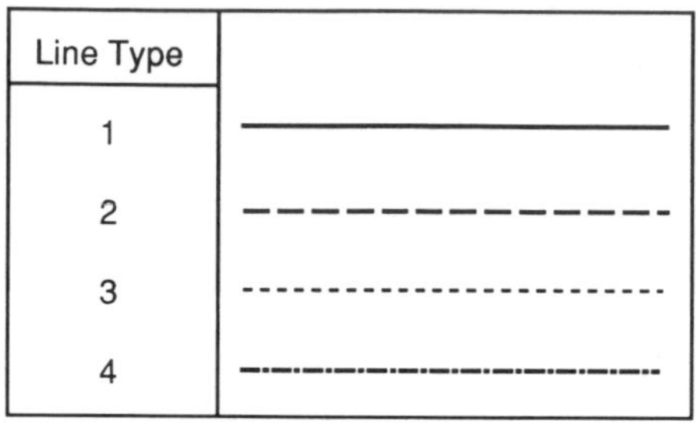

Line Type	
1	————————
2	– – – – – – –
3	··················
4	–·–·–·–·–·–

FIGURE 8-7

Various line types are stored in tables or otherwise (pre)defined and related to features or their attributes.

SAMPLE POINT SYMBOLS

Symbol #	
1	✕
2	○
3	●
4	⦰
5	□
6	▲

FIGURE 8-8

Point symbols are frequently provided by vendors in libraries or constructed using software tools. The symbols are subsequently associated with particular features through tables or linkage to specific attributes.

The desired library identifier is assigned to each map feature and the proper library entry (e.g., specific line type) is accessed during a display or plotting operation. Many GIS packages store the library identifiers as part of a nongraphic database that can be changed easily if a display or plot must be modified (Fig. 8-10).

GIS software packages contain programs to design plots for generation on pen or electrostatic plotters. These plot design programs allow users to create a file of commands that describes the necessary format and plot parameters to a particular plotting device. This plot command file drives a program that generates a plot file from a map file on the system database (Fig. 8-11). The plot file is then generated by the plotter.

Database Management

GIS software packages often include capabilities to store and retrieve nongraphic attribute data associated with map features. For instance, a user may want to store basic ownership data (e.g., parcel number, owner name, site address, appraised value) about land parcels whose boundaries are entered in the graphic database. GIS vendors store nongraphic attributes

SAMPLE TEXT FONT TYPES

Font Type	
1	Helvetica
2	*Helvetica Italic*
3	**Helvetica Bold**
4	Times
5	**Times Bold**
6	Σψμβολ (Symbol)

FIGURE 8-9

Fonts of various types can be associated with annotation; facilities permit changing the fonts prior to printing or displaying graphics.

and link them with their associated map features to support map display and analysis.

These GIS software packages use one of two approaches to manage the nongraphic database:

- Proprietary database management software
- Commercially available third-party database management software

Regardless of the approach, nongraphic database software used in GIS has the following common components:

Data Definition Language (DDL). The DDL allows a user to describe the characteristics or *schema* for files that will contain nongraphic attributes. The schema holds information such as the name of data elements, the size of element fields in bytes or columns, the data element format (e.g., alpha, integer, binary), and other data required by the software to process the attribute data.

Data Entry Procedures. All database management systems offer some degree of interactive nongraphic data entry. The data entry capabilities create a file according to the data definition language and allow users to keypunch data according to system prompts. Error messages appear when a user attempts to enter data that violates rules of format and field length established in the schema.

Most advanced database management systems enable the design of special screen formats for more efficient entry of nongraphic data. These entry format design programs also allow for logical error checking during the entry operation to flag or prevent entries that violate established validity checks. Typical logical error checks include:

- Comparison between the entered value and a list of valid entries
- Comparison against a valid range of values for numeric fields
- Comparison with previously entered fields to automatically enter a value

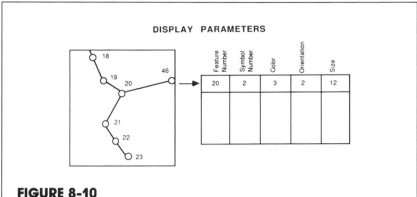

FIGURE 8-10

Display parameters allow the GIS user to customize screen displays or map output.

Data Manipulation Language (DML). The true benefit of a database management system is its ability to extract data based on user-defined criteria. Information that meets these criteria is extracted from the database through a query language that uses high-level, easy-to-understand commands. Data query languages generate searches using commands that operate on elements of the database. These commands can normally be combined in complex sequences using the conjunctions

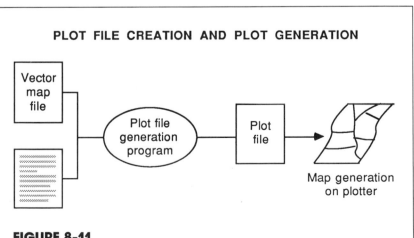

FIGURE 8-11

A plot generation program modifies vector and raster data prior to map production by a particular plotting device.

AND and OR. Typical query commands may include "equal to," "less than," "less than or equal to," "greater than," "greater than or equal to," substring searches for text fields, and others.

For example, a soil scientist may want to identify all soils on slopes of greater than 3 percent that are highly suitable for installation of septic tanks. The query may take the following form with a database of soil types:

IDENTIFY SLOPE GT 3 AND SEPTIC EQ HIGH

This command line may drive a search that would identify all soil types meeting the stated criteria.

Although there is no universally accepted data query language, there has been considerable movement toward standardization with the Structured Query Language (SQL), developed by IBM for use with several of its mainframe database management packages. Many developers of GIS software have adopted SQL. An example of the general syntax of SQL commands follows:

SELECT SOILTYPE FROM SOIL-LAYER
    WHERE (SLOPE > 3)
    AND (SEPTIC = "HIGH")

Linkage of Nongraphic Data with Graphic Map Features. The power of a GIS is its ability to query and analyze both graphic and nongraphic data together. This concept of graphic/nongraphic data linkage provides powerful capabilities for mapping and geographic analysis. For example, using the soil database described above, a planner may want to generate a map of septic tank suitability in the review of building permits. The suitability rankings stored for each soil unit in the nongraphic database would be used to control display or plot criteria to generate the suitability map, with appropriate shading and symbology depicting the different suitability categories (Fig. 8-12).

In addition to the basic characteristics described above, some database management systems include other features

that increase their flexibility. Capabilities for complex report design using high-level programming languages, sometimes called *fourth-generation languages* (4GL), can be used to develop applications.

Basic Cartographic and Geographic Analysis Utilities

Most GIS packages come with a set of special utilities for performing routine mapping and geographic analysis. This tool box of programs is invoked by simple commands that operate on graphic and nongraphic database elements. The utilities can be used alone in simple applications or combined with others to build more complex applications.

The particular set of basic cartographic and analysis utility programs included as part of the core software package will, of course, vary among commercial software vendors. Depending upon the particular vendor package, some of the programs described here are part of a special application package rather than core software. They are, however, standard utilities required by many users.

Map Coordinate and Geometric Transformation. These utility programs are used to adjust and transform the coordinates of an automated map so that it conforms to a desired map reference grid or a set of base map features. These programs transform the coordinates and internal geometry of the map so that it fits a designated coordinate system or map projection.

- Map Coordinate Translation

Most maps are based on a geographic reference grid that determines the horizontal (x,y) coordinates of all features included on the map. The reference grid is the "glue" that holds the map together and is the basis for overlaying multiple map layers. The grid is defined by an origin point, and coordinates are defined as increments from this origin along the x and y axes. The reference grid may be based on an arbitrary local coordinate system or a global reference

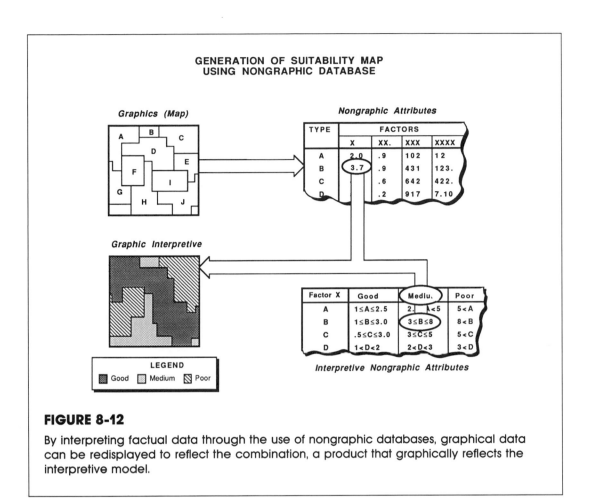

**GENERATION OF SUITABILITY MAP
USING NONGRAPHIC DATABASE**

Graphics (Map)

Nongraphic Attributes

TYPE	FACTORS			
	X	XX.	XXX	XXXX
A	2.0	.9	102	12
B	3.7	.9	431	123.
C		.6	642	422.
D		.2	917	7.10

Graphic Interpretive

Factor X	Good	Medlu.	Poor
A	1≤A≤2.5	2.1≤A<5	5<A
B	1≤B≤3.0	3≤B≤8	8<B
C	.5≤C≤3.0	3≤C≤5	5<C
D	1<D<2	2<D<3	3<D

Interpretive Nongraphic Attributes

LEGEND

Good   Medium   Poor

FIGURE 8-12

By interpreting factual data through the use of nongraphic databases, graphical data can be redisplayed to reflect the combination, a product that graphically reflects the interpretive model.

grid (e.g., latitude/longitude) that accurately references map features to absolute positions on the earth's surface.

Map coordinate translation programs recalculate the coordinates of all control points on the map and all map features. The coordinate translation may be simple in the case of a shift of the origin, requiring only a constant addition or subtraction to all map coordinates. However, coordinate translations are more complex in cases where a map projection transformation is required as well.

- Map Projection Transformation

Map projections have been developed to provide methods to display the curved surface of the earth on a flat surface. This mapping process results in some distortion that may affect the size,

shape, direction, or distance relationships of features on the map. Projections are models that are used to mathematically "project" the features from the curved surface onto a flat map. Many different types of map projections have been developed, each of which minimizes certain types of distortions in the mapping process. Because the map projection establishes a mathematical relationship between the earth's surface and the map, coordinates of map features can be determined reliably.

Some geographic data management operations require maps that conform to a specific projection to ensure a proper fit. If a digitized map must be reliably overlaid with another digital map that was based on a different map projection, a transformation program can be run to geometrically adjust one set of map features to fit those on the other map.

- Rubber Sheeting

   The rubber sheeting process geometrically adjusts map features to "force" a digital map to fit a designated base, so called because the map is stretched to fit the base, given a set of known coordinate values whose locations are defined on both the base map and the map to be rubber sheeted (Fig. 8-13).

   It is important to understand that rubber sheeting actually distorts the position and shape of map features from the original map using mathematical operations. Rubber sheeting should thus be judiciously applied or it can undermine the integrity of the map.

   An example of the reasonable use of rubber sheeting is in the case of a land cover map prepared directly from an unrectified aerial photograph. This map can be rubber sheeted to fit a planimetrically correct base such as a USGS 7.5' topographic map. The coordinates of

control points, such as road intersections, that can be located both on the topographic map and the photograph are entered into the rubber sheeting program. The land cover map is then stretched to fit the topographic map base. Some distortion of land cover boundaries will occur in this process, but the distortion is below an unacceptable level, given the required accuracy and scale of the map.

Basic Cartographic Operations. This category of utility programs performs basic operations to merge, extract, and edgematch digital maps.

- Edgematching

   Edgematch programs join two digital maps that are geographically adjacent to create one spatially continuous map. The edgematching is governed by control point coordinates, i.e., tic marks on the map boundary (Fig. 8-14). In many cases, edgematching programs adjust map features that cross over the original map edges to preserve the connectivity of lines and boundaries across the map edge. This feature adjustment is guided by user-supplied tolerances.

- Map Merging

   This operation merges a portion of a digital map into a preexisting map layer. Merging adds data to a map layer and is one way to update the map layer as new features are captured individually in digital form. A map merge operation may be performed on a map layer containing parcel boundaries and rights-of-way (Fig. 8-15). A new subdivision plat may be automated in an independent step, perhaps using a coordinate geometry procedure, and merged with the existing parcel layer.

   If the subdivision plat refers to the same geographic coordinates as the map layer, the map merge operation can proceed without any special translation of coordinates. If the coordinate systems used by the map layer and the plat are different, the subdivision plat coordinates may first be translated or the plat

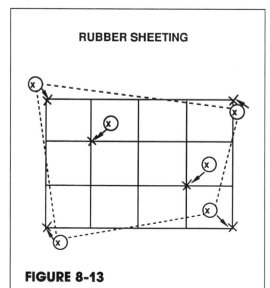

RUBBER SHEETING

FIGURE 8-13

Rubber sheeting force-fits a select number of points to predefined positions, typically defined through coordinate values.

may be fit into the map layer through the identification of common control points.

- Window or Extract

Window or extract programs allow a user to copy the graphic and, often, the nongraphic features of an area encompassed by existing map layers and move them to a separate file or layer for processing independent of the map layer from which they were copied. In a sense, this operation works like a cookie cutter to create a separate digital map database for display, modification, or special analysis (Fig. 8-16).

The window may be a box or other regular shape defined by an interactive or batch input of vertex coordinates or it may be an irregular shape defined by interactive digitizing. In some cases, the window may be a previously defined region (e.g., county planning district) in the graphic database.

Special Mapping Utilities. In some GIS packages, programs are included as part of the core software to perform special map production. These programs use the system's graphic display and plot capabilities and often draw from the nongraphic database to generate special maps in the following ways:

- Graphic Overlay

Graphic overlay is a relatively simple operation that combines features and their associated annotation from multiple map layers. Graphic display or plotting programs are then used to generate a hard-copy map or a screen display. The graphic overlay process will not correct for cosmetic problems resulting from an overplotting of lines, symbols, and annotation when multiple layers are combined. Sometimes the user must perform some interactive graphic editing to prepare a cartographically aesthetic map.

- Thematic mapping

Thematic maps portray some geographic variable or theme by shading or

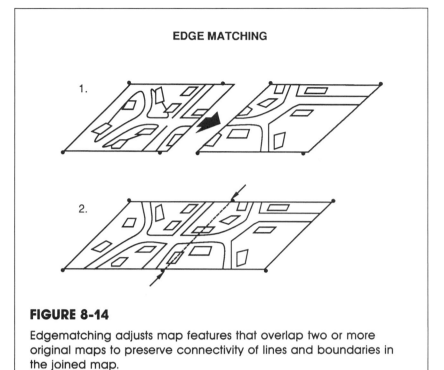

EDGE MATCHING

1.

2.

FIGURE 8-14

Edgematching adjusts map features that overlap two or more original maps to preserve connectivity of lines and boundaries in the joined map.

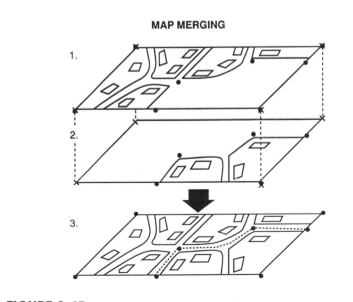

MAP MERGING

1.

2.

3.

FIGURE 8-15

Software combines two or more "layers" of graphic data into a single "layer" or two or more tabular files into a single file.

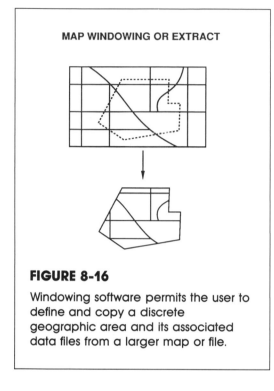

MAP WINDOWING OR EXTRACT

FIGURE 8-16

Windowing software permits the user to define and copy a discrete geographic area and its associated data files from a larger map or file.

county, or they may represent quantitative data, such as population density (Fig. 8-17).

- Address Matching and Mapping

Many cities and counties that implement GIS need to analyze and map geographic data referenced by street address. An address provides the locational key to a variety of information, such as building permits, emergency response incidents (e.g., fires or crimes), and other data that may be stored in existing data files and referenced to site addresses.

The address-matching capability can generate point locations that represent features or events on a map containing a street network or other base features. Point symbols may be defined to denote various types or categories of incidents (Fig. 8-18). These incident maps are usually generated on a street base map in which address ranges (for the right and left sides of the street) are assigned to each block or other defined street segment. The address-matching program reads the address from the incident data file and estimates a map position by in-

symbolizing areas to represent the value or class of that theme. Thematic maps may depict qualitative map information, such as the land-use categories for a

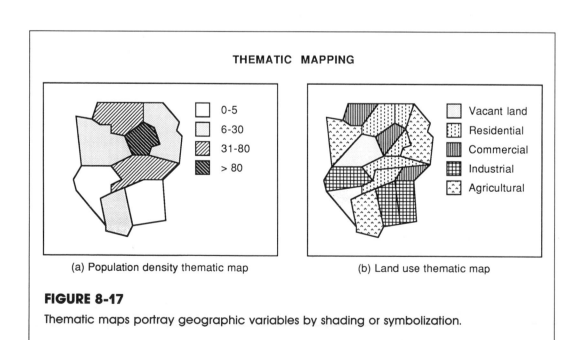

THEMATIC MAPPING

	0-5
	6-30
	31-80
	> 80

(a) Population density thematic map

	Vacant land
	Residential
	Commercial
	Industrial
	Agricultural

(b) Land use thematic map

FIGURE 8-17

Thematic maps portray geographic variables by shading or symbolization.

terpolating between street blocks or other assigned segments.

Although this process does not plot the incidents' exact positions, it normally gives an adequate representation of the spatial distribution of address data. It is based, however, on the assumption that addresses are numerically sequential and that even and odd addresses occur on opposite sides of the street. Although this is generally the case in cities and counties, anomalies often occur that can lead to address-matching problems.

Polygon Overlay Analysis. Many GIS packages allow map features to be stored as *polygons*. The polygon represents a geographic feature that encompasses an area (e.g., soil unit, land parcel, census tract), shown as a closed boundary on a map. Nongraphic attribute data are associated with the interior of the bounded area, though they may be linked to a centroid point located inside the polygon. A polygon structure allows geographic analysis of the entire area it covers. Systems that allow storage of polygons and linkage to nongraphic attribute data files often provide programs for special overlay analysis such as those described below.

- Polygon-on-Polygon Overlay

   This process involves the overlay of two or more polygon maps to produce a new set of polygons formed by the intersection of their boundaries from the original maps. The user may perform this type of operation, for example, to determine the mixture of land use that occurs within planning districts. Such an analysis would be accompanied by a statistical aggregation of land-use categories and areas for each planning district. A polygon-on-polygon overlay may be used in a more sophisticated model that overlays and compares many maps to evaluate the relative suitability for a particular development activity (e.g., land-fill construction).

   Many GIS packages have overlay routines that generate derivative polygon

ADDRESS MATCHING AND INCIDENT MAPPING

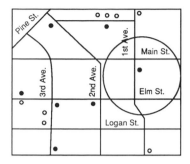 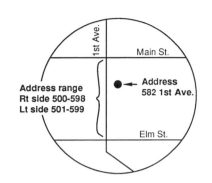

FIGURE 8-18

Street addresses are a common geographic identifier with which people, features, or events are associated with and mapped.

layers based on the union, intersection, or subtraction of polygons in the original layers.

- Point-in-Polygon Overlay

   In the point-in-polygon process, a polygon map is overlaid on a map of point features (e.g., manholes, water wells) to determine which of the point features is located within a particular polygon. For example, a manager in a public works department may want to generate a list of manholes by maintenance district to plan an inspection of a city's sewer system. The point-in-polygon function could aggregate the statistics for this list.

- Line-in-Polygon Overlay

   Line-in-polygon routines are similar to the point-in-polygon processes, except that the polygon map is overlaid on a map of line features. For instance, if a state highway official needs to tabulate the mileage of certain types of roads located within counties or highway districts, this operation would aggregate the statistics for such a tabulation.

Map Analysis Utilities Many GIS software packages include capabilities for special map analysis. These functions provide rudimentary analytical tools that are used independently or are incorporated into more sophisticated analysis routines.

- Distance, Perimeter, and Area Measurements

    These programs calculate distances and areas on a map. Depending on the software package used, these programs may operate in an interactive mode, in which the user traces lines or areas with the digitizing cursor, or they may perform calculations on existing map features in the system's database. These programs may generate results in measurement units on which the map grid is based (often feet) or they may translate the measurement into other units such as meters or miles.

- Radius Search

    A radius or buffer generation program performs a search around a point at a distance specified by the user. This type of operation locates features or aggregates statistics that fall within the search radius.

- Buffer Generation

    This procedure generates new polygons around existing point, line, or polygon features (Fig. 8-19). The user specifies a particular buffer distance and the program computes the boundary of a polygon that includes the specified distance around the feature. The results of the buffer generation could be used in a corridor analysis, wherein a utility planner needs to determine property owners that fall within a specified distance of a planned electric transmission line. The proposed line could be plotted and a buffer zone generated for comparison to a property ownership database. The use of a polygon-on-polygon overlay program could determine property owners within the designated utility corridor.

Other Utility Programs   With increasing frequency, GIS software contains utility programs written to accommodate data of different formats and sources. Two of the

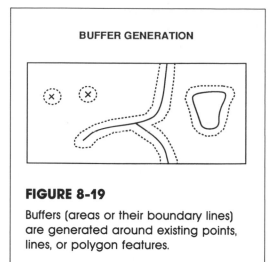

BUFFER GENERATION

FIGURE 8-19

Buffers (areas or their boundary lines) are generated around existing points, lines, or polygon features.

most commonly occurring are those bridging raster and vector data formats and the translation of data from one proprietary system format to another.

- Raster/Vector Conversion

    In some cases, it is desirable to convert raster (grid) data to vector format or vice versa to conduct a particular mapping or geographic analysis. Some GIS software packages have the ability to process data in both vector and raster form and can convert from one data format to the other (Fig. 8-20). One example is the conversion of land cover data generated from satellite data (Landsat or SPOT) from its original raster format to a vector equivalent for use in a map overlay with data from a vector-based mapping system. In the vector-to-raster conversion process, map features in point, line, and area form are translated into single cells or groups of cells in the grid matrix. The conversion program often allows a user to specify the grid-cell size.

- Grid Overlay

    GIS packages that store and process map data in raster form normally have programs that allow users to overlay multiple grid map files to perform some analyses efficiently. A soil scientist may be interested in overlaying soil, slope, and land cover data that has been automated in grid format to analyze soil erosion. A grid overlay program would allow the scientist to assign values to the grid cells and generate a composite picture of soil erosion intensity. The grid-based package was the earliest form of GIS and though vector-based systems have recently been more popular, grid-based analysis capabilities continue to be effective for many applications.

- Vector/Raster Overlay

This is a special program offered as part of some geographic data management packages that displays map data in vector form along with a raster map or image. This graphic overlay program simultaneously displays the vector and raster data, and has important uses in many disciplines wherein users maintain raster data, such as satellite images or scanned photographs, in addition to vector maps. The raster image may be used as a "base map" that provides a convenient backdrop for the display of other map data.

- Format Translation Programs

With many GIS software packages, special programs are available that carry out a data translation (for graphic and/or nongraphic data) for compatibility with another software package. These programs reformat the structure of the data files so that data can be transferred among a variety of systems. In GIS packages, data translation programs perform the reformatting in one of two ways:

- Direct translation
- Translation through an intermediate data format

Direct translation routines are written to reformat data directly from a source data structure to a destination system without an intermediate file format. Other programs reformat the data into a neutral exchange format that can then be transferred into the format of a second system. Although there is no single accepted file exchange format standard, several [such as the Intergraph Standard Interchange Format (ISIF)] are recog-

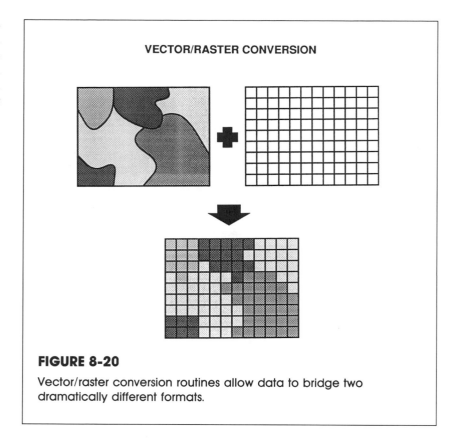

VECTOR/RASTER CONVERSION

FIGURE 8-20

Vector/raster conversion routines allow data to bridge two dramatically different formats.

nized by many software vendors as popular formats for exchanging data between systems. The Spatial Data Transfer Specifications (SDTS) developed under the auspices of the National Committee for Digital Cartographic Data Standards (NCDCDS) is proposed as a federal standard for the exchange of cartographic data and will quickly increase in prominence in the GIS industry.

Programs for translation of data to and from standard nonproprietary formats for spatial analysis are also available. The geographic base file/dual independent map encoding (GBF/DIME) and recently developed topological integrated geographic encoding and referencing (TIGER) formats supported by the United States Census Bureau are standard formats for storing street- and address-related geocoded data. The digital line graph (DLG) format is used by the United States Geological Survey for storing a variety of cartographic data.

SPECIAL APPLICATION SOFTWARE AND APPLICATION DEVELOPMENT TOOLS

Special Application Software

In addition to the core capabilities of the GIS software described above, there are often special programs available that are considered separate from the core package. These programs are designed to fulfill a specific application area and are smoothly integrated with the core package so that users may have easy access to its capabilities.

In the GIS industry today, there are three categories of special application software:

- Software offered by the developer of the core package and fully integrated with the core
- Software developed by a third-party vendor and integrated with the core package
- Software developed by a system user and integrated with the core package

Application software is as diverse as the disciplines and needs of GIS users. See Table 8-1 for a sample of some important applications often addressed by GIS vendors and users.

User Command Interfaces

A command is the user's directive to the system to initiate a system software operation. Commands invoke functions of the operating system, support programs, and run application software. Two basic methods to issue commands are provided by software developers. The first method, traditionally used with most interactive computer systems, is the *command language,* which employs a specific text-based dialogue for the keyboard entry of commands. The other major method is the *menu-driven interface* or graphical user interface (GUI), which uses menus on the computer screen or digitizing tablet. The user points to specific commands without entering a line of text.

Command Languages

Command languages are used at both the operating system and application software levels. Command languages use a specific standard syntax through which a user enters a command at a keyboard, as illustrated below:

prompt>COMMAND OPERAND(s) PARAMETERS

The prompt is a standard symbol or text string generated by the software to tell the user that the system is ready to accept a command. The COMMAND specifies the task or function to be performed by the software. OPERAND and OPERANDS represent the entity or entities (i.e., file, group of files, data element) on which the task or function will operate. In some cases, the user may have the option to include PARAMETERS that provide additional information about how the task will be performed or the format of the result. The example below shows common command syntax.

TABLE 8-1 Sample GIS Applications

Application Category Example	Specific Application
Network analysis	Vehicle routing; water flow analysis
Coordinate geometry	Entry of subdivision and lot lines from plat
Terrain analysis	Slope or drainage analysis
Surface-Subsurface modeling	Mining reserve estimation; analysis of geologic structures
Engineering/Site design	Bill-of-materials estimation; cut-and-fill analysis
Work order processing	Work print generation; automatic map posting

Example: This MS-DOS operating system command requests a screen display of the contents of a file directory.

prompt> DIR SAMPLEDIR /W

The DIR command tells the MS-DOS operating system to display all contents of a file directory (SAMPLEDIR) on the screen and to arrange the display horizontally in rows across the screen.

Command languages provide a structured dialogue that allows users to communicate with a system and request specific tasks. Most of these languages now use standard English commands that reduce the need to learn an entirely new language in order to communicate with the system. However, since these command languages use a structured syntax and specific vocabulary, they do require training and practice for a user to become proficient.

A current trend in software industry research is to develop command and programming languages that are less structured and closer to conversational English. Ideally, these languages (sometimes called *natural languages*) will accept commands in forms that require less-stringent rules for syntax and specific command vocabularies. The underlying software is sophisticated enough to convert the user commands into system instructions. Unstructured languages of this type have not yet had a major impact on geographic data management systems, but as their development proceeds, they likely will have an increased effect.

Menu-Driven Commands

In the last ten years, the menu-driven approach has been used increasingly by software developers as a way to provide a more flexible way of issuing commands. The menu is, simply, an area on a display screen or digitizing tablet that is divided into cells, each of which corresponds to a command. By using various pointing mechanisms such as arrow keys or a mouse (for screen menus), or a digitizing cursor (for tablet menus), the user locates and selects the desired command.

Tablet menus have been in common use since the late 1970s in automated mapping and drafting systems as a flexible way to initiate a specific entry or edit function. The menu is taped to the digitizing tablet and corner points are entered to register it and properly orient the individual command cells. The user points to a particular command cell with the digitizing cursor, pushes a button on the cursor pad, and then proceeds with the function. An example is the entry of an arc determined by three input points. To create this graphic feature, the user selects the "three point arc" command cell on the tablet menu and then enters the three points. The system draws the specified arc through the three points.

Many application software packages come with standard menus that offer a set of commands commonly needed by system users. Very often, these software packages allow users to customize menus to better suit a specific user community or application.

Sophisticated display screen menus are relatively new features in GIS packages. Flexible screen menus have become possible with increases in display screen resolution and processing speeds. Some of the early development of screen menu techniques was carried out by the Xerox Corporation in the early 1980s in their STAR document processing systems. The Apple Computer Corporation further refined this technique in the development of their Macintosh microcomputers.

Since the mid-1980s, most software packages developed for geographic data management employ some level of screen menu interfaces. Users point to areas on the screen using a mouse, digitizing cursor, arrow keys, or other mechanism for issuing the commands.

Menu-driven approaches have become quite popular because, if designed properly, they make it easier for a user to communicate with the system, reducing the time and inefficiency of typed commands.

Many software developers have designed icon-based menus, in which commands are represented by pictures or symbols that serve as mnemonic devices a user can easily relate to a specific function.

In menu-driven systems, there is sometimes a trade-off between user-friendliness and efficiency. Initiating some functions may require the user to make multiple selections from a series of menus as opposed to invoking the operation from one typed command line. Most software packages that use menus also allow the user to bypass the menus and issue text commands for many functions, which can speed up the process.

Macro Programming

The term *macro programming* has different meanings, depending upon the particular software environment. It is generally accepted, however, that a macro program in its simplest sense consists of a sequence of software commands that in combination perform a complex operation. The sequence of commands is stored in a file (the macro program) that is executed by a single user command (e.g., RUN MACRO1). Most GIS software packages provide macro programming features. Often, macro programs can be invoked through selection from a screen or tablet menu.

The building of macro programs for frequent mapping or geographic analysis functions increases efficiency by eliminating the need to remember and enter multiple commands to initiate an operation.

An example of a geographic query application common in many planning agencies that could be made more efficient with a macro program is a search to determine owners of property within a given distance of a certain parcel. To perform this operation, the system would have to carry out the following steps:

- Identify the target search parcel
- Search parcels within a given radius (e.g., 500′)
- Access nongraphic data records for parcels within the search area

- Generate a report with owner names and addresses
- Generate a screen display or plot of a map showing all identified parcels

Depending on the software package being used, one or more commands must be issued to complete each of these steps. These individual commands can be entered and stored in a macro program that through a single command may be executed each time the property search is required. This obviously reduces time and the chance for error.

Some GIS packages offer macro features beyond simple command sequences. Such features may include the following:

- Prompts for interactive data input from users
- Storage of temporary variables for use by the program
- Arithmetic functions
- Conditional (if–then–else) logical branching

COMMERCIALLY AVAILABLE GEOGRAPHIC DATA MANAGEMENT SOFTWARE

Packages for performing mapping and geographic analysis first became available on the market in the last 1960s. Some were based on a grid data structure; others used vector formats for storing graphics. By today's standards, these early packages performed only rudimentary operations.

In the last five years, there has been a rapid increase in both the number of vendors entering the software market and the capabilities offered by GIS software. Software is available for hardware platforms ranging from mainframe computers to microcomputers. The market and research arena for geographic data management software is volatile; new developments and product offerings appear frequently.

Because developments occur so rapidly, selection of GIS software can be a complex decision. The choice should be based on an evaluation of a product's proven capabilities as demonstrated by its existing user base, as well as the consideration of new software and enhancements that do not have a long history of actual use. Software vendors continue to enhance their products, and users should consider these new capabilities in their evaluations.

As in the development of computer hardware, the software industry is thriving. Developers offer ever more creative solutions to the complex problems of geographic data management. As long as the demand for software remains high, users can expect to see continued developments that offer increased flexibility in addressing the needs of mapping and geographic analysis. However, we are not limited to passive observation as the future of software development unfolds. Users have a responsibility to articulate their needs to provide the perspective and focus lacked by the software development industry. It is therefore necessary for the user community to become educated about geographic data management systems and how the technology can best be applied to solve everyday problems.

SUGGESTED READING

Carlyle R. Leaping ahead in software productivity. *Datamation 35:* 22–31.

Croswell P. (1988). Definition of applications as a basis for GIS planning and system procurement. *URISA Proceedings,* Vol. II, pp. 13.

Croswell P. (1990). GIS analysis functions. *Selected Bibliography on GIS Concepts and Techniques.*

1990 GIS Sourcebook, GIS World, Inc. Fort Collins, Colorado.

Sena M. L. Do-it-yourself mapping. *Computer Graphics World 13:* 62–66.

Wadland K. R. (1990, January). Graphical user interfaces: A window to the future—Part II. *Computer Graphics Review:* 36–40. January 1990.

Chapter 9

System Configurations and Data Communications

In the past ten years, a flood of hardware and software advances has created many new opportunities and options for configuring computer systems. The concept of distributed systems, wherein computer processing power and data may be separated from a host computer, has had a profound impact on geographic data management systems.

Chapters 7 and 8 discussed computer hardware and software components of geographic data management systems. This chapter brings those concepts together with a view of the total system, the network that integrates the functions of hardware and software, and the support provided by data communications.

Until the late 1970s, most computer configurations were relatively simple, based upon a centralized scheme in which all processing and data storage occurred at a host mainframe or minicomputer. Peripheral devices such as terminals, graphics workstations, and plotters connected to the host computer were, at best, able to perform very limited functions locally.

In the past ten years, a flood of hardware and software advances has created many new opportunities and options for configuring computer systems. The concept of distributed systems, wherein computer processing power and data may be separated from a host computer, has had a profound impact on geographic data management systems. The growth of high-speed local computer networks, increased capabilities for remote data communications, and the development of techniques and standards for linking different types of computers will continue to influence system configurations.

COMPUTER NETWORK CONCEPTS

Speed and Distance Concepts

A computer network may include devices that occupy a single room or are distributed across a continent. A single-user microcomputer mapping system may consist of the microcomputer, disk drive, color graphics monitor, a pen plotter, and a printer in close proximity, all connected by various types of cables. More complex systems may extend cabling to connect multiple devices and perhaps multiple processing units throughout a building or group of buildings. Remote communication technologies using the telephone system, microwave transmission, or satellites are able to connect devices in global networks.

Communicating between devices in a computer network is as simple as sending bits of data from one device connected to others by various means, as discussed below.

Analog Versus Digital Communication

Two basic modes of electronic communication are used to transmit data in computer networks. The *analog* method transmits a signal that may vary continuously in frequency or intensity within a given interval. *Digital* transmission, on the other hand, transmits data as discrete impulses in two states that represent the binary digits, 0 and 1. These two methods of communication can be explained using the analogy of ceiling light fixtures connected to either a dimmer mechanism or a simple on/off switch. The dimming mechanism allows the light intensity to be adjusted on a continuum (analog) whereas the on/off switch offers only two different values (digital).

The most common example of analog communications is the telephone system. Human speech consists of a continuous range of sound frequencies and volumes. When a person speaks into a telephone receiver, the speech is converted to an electrical signal that also varies in frequency and intensity corresponding to the speech pattern. The telephone on the other end of the line converts the electrical signal back into sound waves that are understood as speech by the listener.

Digital communications, in which the electronic signal has two states (high voltage/low voltage), is used in the internal circuitry of most computers and in communication with peripheral devices directly connected to the computer through various types of cabling.

Both digital and analog communications techniques are used to transmit data in computer networks (Fig. 9-1). Analog communications are used primarily where distances preclude a direct cable connection between computer devices. Since computers require digital signals to store data and execute instructions, *mod*ulator-*dem*odulator (modem) devices must be used in cases where analog media, such as the telephone system or microwave, are employed to carry data. Modems convert the analog signals to a digital form and vice versa.

Digital Transmission Modes

Digital transmission techniques use cables to connect the internal components of a

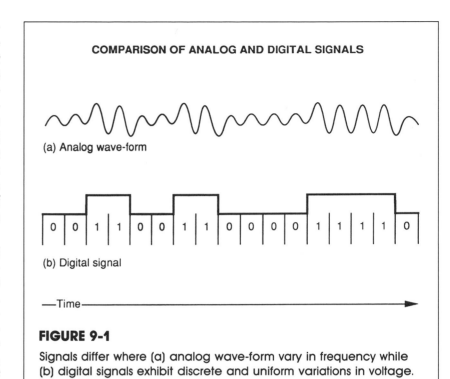

COMPARISON OF ANALOG AND DIGITAL SIGNALS

(a) Analog wave-form

| 0 | 0 | 1 | 1 | 0 | 0 | 1 | 1 | 0 | 0 | 0 | 0 | 1 | 1 | 1 | 1 | 0 |

(b) Digital signal

—Time—

FIGURE 9-1

Signals differ where (a) analog wave-form vary in frequency while (b) digital signals exhibit discrete and uniform variations in voltage.

computer processing unit and peripheral devices in a network. These cables carry digital signals as high and low voltages that can be interpreted as binary digits. Digital communications can be broadly classified as either *parallel* or *serial*.

In parallel communications, cables consisting of multiple wires, each of which carries one data bit at a time, can transmit a sequence of bits simultaneously (Fig. 9-2). Parallel communication lines transmit high volumes of data very quickly. They are used in the internal circuitry of computer processors and to connect devices requiring very high transmission speeds, such as mass storage devices and high-speed line printers. Parallel lines have major distance restrictions and therefore most often connect components and devices in close proximity, usually in the same room.

Serial lines, used commonly to connect peripheral devices in computer networks, transmit data one bit at a time. Within this category of serial communications, a distinction is made between *asynchronous* and *synchronous* (Fig. 9-2). Asynchronous com-

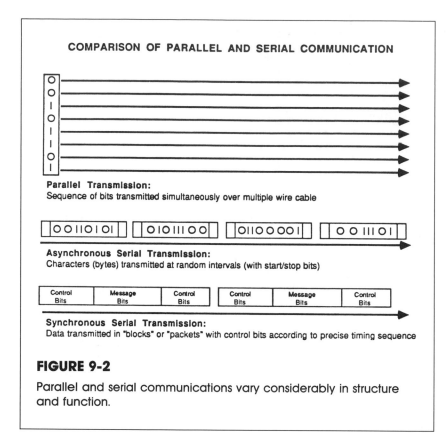

COMPARISON OF PARALLEL AND SERIAL COMMUNICATION

Parallel Transmission:
Sequence of bits transmitted simultaneously over multiple wire cable

Asynchronous Serial Transmission:
Characters (bytes) transmitted at random intervals (with start/stop bits)

Synchronous Serial Transmission:
Data transmitted in "blocks" or "packets" with control bits according to precise timing sequence

FIGURE 9-2

Parallel and serial communications vary considerably in structure and function.

munication lines transmit data without a specific timing sequence. In synchronous communications, blocks or *packets* of bits containing data are sent according to an established timing sequence. Each packet is made up of a sequence of bits representing a message and other sets of bits that supply information to help control the transmission.

Asynchronous transmission is quite common in minicomputer or microcomputer systems that use the popular RS-232 cabling scheme to connect peripheral devices such as terminals, printers, or plotters. Each device has an individual line to the host computer and bytes (8 bits) of information are sent individually across the line. An extra bit called the *start bit* precedes each byte and most asynchronous transmissions also add a *stop bit* at the end of each byte.

Synchronous communication is used in many mainframe computer systems and in computer configurations with high-

speed local area networks (LAN). This method of communication is used, for instance, in IBM configurations using the Systems Network Architecture (SNA) arrangement or in local area networks employing Ethernet standards. The packets or blocks of information in synchronous communication are structured to adhere to a particular protocol understood by devices on the network.

Speed and Distance Considerations

All computer networks have limitations on how fast and how far data can be transferred. These speed and distance limitations are determined in part by the type of transmission media and by environmental conditions that may affect the strength and quality of a signal. RS-232 cables, for instance, send and receive data reliably at distances of up to about 200 feet. Inherent attenuation characteristics of the cable and outside conditions, such as electrical interference within a building, can affect the quality of transmission beyond this distance.

In the world of data communication, speed is normally expressed in bits per second. This measure indicates the volume of data that can be transferred within a given time period. The term *bandwidth*, also in a data communications context, expresses transmission speed. The phrase "bandwidth of 10 megabits" means that the line can transfer data at the rate of 10 million bits per second.

A measure of speed often referred to in discussions of analog transmission is *baud* or one *signal unit* per second. If an analog signal is viewed as an electromagnetic wave, one complete wavelength or *cycle* is equivalent to a signal unit. Baud has often been used synonymously with bits per second. With some analog transmission techniques, the baud rate may be equal to bits per second, but special modulation techniques frequently deliver a higher bits-per-second rate.

Different devices and geographic data management applications have varying speed and distance requirements. For instance, a pen plotter that plots a map relatively slowly has modest speed demands that may be met by connection to a host computer through an RS-232 cable or even a dial-up telephone line using a 1,200 bits-per-second modem. Connections to support interactive data access between multiple computer processing units, on the other hand, might require the higher speeds of a local area network. It is necessary to closely evaluate the specific needs of the system to choose the optimal means of communication.

Local Communications

Local communication implies a direct cable connection between devices in a computer network that allows data to be transmitted in digital form. Local communications may support devices within the same room, an office building, or a collection of buildings within close proximity. Generally, it becomes less technically and/or economically feasible to use local communication techniques when devices are separated by greater distances.

A variety of cabling types support local data communications, each of which has its own speed and distance limitations (Table 9-1). A special category of local communications techniques is the LAN, which supports multiple devices in a high-speed network (see below).

Remote Communications

Remote communication techniques are used when technical or cost restrictions preclude local communication connections. Depending on the particular system, remote devices could be in the same building or distributed around the world. Remote communications often use the facilities of a third-party communication carrier via such analog transmission as a local or long-distance telephone company.

The local or long-distance telephone system is the most frequently used remote communication medium. Where low-speed communications are acceptable, standard voice-grade lines offer an inexpensive alternative. Inexpensive dial-up modems that can transmit data reliably at speeds ranging from 300 bits per second to 4,800 bits per second are required. Dial-up modems are currently available for transmitting data at a maximum rate of 9,600

TABLE 9-1 Comparison of Speed and Distance Limitations of Cabling Schemes for Digital Transmission[1]

Cable Type	Maximum Transmission Speed[2]	Maximum Length between Devices[3]
Twisted pair copper wire	4 million bps[4]	1,500 feet
RS-232 cable	19,200 bps	150 feet
Standard coaxial cable	20 million bps[5]	2,000 feet
Thin-wire coaxial cable	10 million bps	1,000 feet
Multimode fiber optic cable	1 gigabit per second	30 miles

[1] Figures are dependent on site conditions and the specific communication hardware and software used.
[2] Speed in bits per second (bps). Speed depends partly upon total length of cable.
[3] Maximum expected distances without repeaters.
[4] Capacity can be extended to over 10 million bps within limited distances.
[5] Higher speeds up to about 50 million bps can be achieved over short distances (less than 100 feet).

bits per second, but the actual transmittal of data for these high-speed devices depends on the quality of the line and error rates that may severely reduce the real throughput of the line. Telephone companies also offer dedicated leased lines in which a direct point-to-point connection between two computer devices is established. Leased lines offer higher speed, a better-quality signal, and greater security, but their initial installation charges, special modem requirements, and monthly lease fees make them a much more expensive option.

In many metropolitan areas, local telephone companies provide special digital service for both voice and data communications. Different classes of digital service are often available to serve low- and medium-speed requirements (up to 56,000 bits per second), and high-volume needs such as T-1 and T-3 lines (up to 1.5 and 45 million bits per second, respectively). Digital service units, not modems, are required for proper connection to digital telephone facilities. With T-1 and T-3 service, multiplexors are used to divide the high-speed transmission into individual channels to support multiple devices.

Many other media are available for remote data transmission, each with its own advantages and limitations in specific situations (Table 9-2). Some of the more popular remote communications facilities used to transmit data include microwave systems, broadband cable networks (cable TV), satellite systems, radio transceivers, and infrared systems.

Wide Area Networks

Wide area network (WAN) is a general term referring to a computer network where devices are connected by various types of remote communication links. Often describing networks that extend throughout a defined geographical region such as a metropolitan area or service district, WANs use a variety of remote linking techniques (analog telephone lines, digital telephone service, microwave). The types of remote communications used depend on the

speed requirements of users in the networks as well as the availability and cost of various communication facilities.

Network Topologies and Protocols

Network Topology. Network topology refers to the arrangement or layout of the devices and communication lines in a computer network. The topology explicitly defines how devices interconnect and, therefore, how data flows in a network. The devices or interconnection points are called *nodes*. From a user's perspective, a particular network configuration should offer: 1) Sufficient transmission speed to support devices and applications; 2) flexible access to data and the ability to move data to different points in the network; and 3) adequate quality control and protection to ensure error-free, secure transmission. Manufacturers of computer hardware and software offer systems based on a variety of network topologies that address these needs and provide for efficient management of the network.

Three major network topologies, or hybrids thereof, are in common use:

- Hierarchical or Star

  As shown in Figure 9-3, a star or hierarchical network topology consists of a central node to which all other nodes are connected. This arrangement implies some level of centralized control of data and/or applications that are accessed by users at other points in the network. All traditional centralized configurations use this type of network topology to support peripheral devices. This is the case with small minicomputer systems that support terminals, plotters, and other devices through individual RS-232 connections. A variation of the star topology is the tree structure (Fig. 9-3), in which one or several devices control many other subordinates. As a result, the network is vulnerable to interruptions from a failure at one of the controlling nodes.

  A predominant network standard in the mainframe computer environment is

TABLE 9-2   Comparison of Remote Communication Environments[1]

Method	Normal Transmission Speeds[2]	Distance Limitation between Devices
Dial-up telephone line	300 bps to 9,600 bps	Theoretically unlimited where telephone service is in place
Dedicated analog telephone line	300 bps to 19,200 bps	Theoretically unlimited where telephone service is in place
Digital telephone service (e.g., DDS, Synchronet)	1,200 bps to 56,000 bps	Normally within metropolitan areas or specified service regions
T-1 digital lines	Up to 1.54 million bps	Normally within metropolitan areas or specified service regions
Broadband analog transmission on coaxial cable	Up to about 50 million bps	Up to about 20 miles
Microwave	—	Up to about 25 miles[3] depending upon signal frequency (requires intervisibility)
Radio	Up to 9,600 bps	Theoretically up to about 50 miles[4]
Laser or infrared beams	Up to 19,200 bps	Up to about 1 mile depending on atmospheric conditions (requires intervisibility)

[1] Figures are highly dependent upon environmental conditions, and the specific communications hardware and software used.
[2] Speed in bits per second (bps).
[3] Dependent upon line-of-site path between transmitter and receiver.
[4] Normally used with a 5-mile range.

IBM's Systems Network Architecture (SNA), based on a tree structure. Ultimate control of the network is assigned to one or more communication processors, which support all network nodes (Fig. 9-4). A program operating at the communication processor assigns unit numbers to each network node. Physical unit (PU) and logical unit (LU) numbers determine the nodes' characteristics and function. Recently, IBM has expanded the SNA environment to provide direct communication between subordinate nodes on the network without the direct control of the communications processor.

● Ring

In ring topology, devices are oriented along a closed path, so that data are transmitted in one direction around the ring (Fig. 9-5). Each device on the ring has some intelligence and can receive and send messages. When a message is

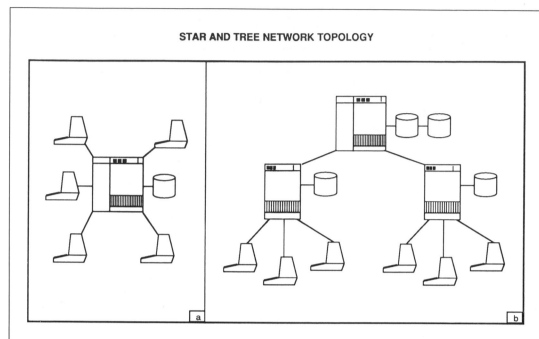

STAR AND TREE NETWORK TOPOLOGY

FIGURE 9-3

The Tree Network (b) is a variation of the Star Network (a), which assigns a hierarchy of network control to branching nodes on the network.

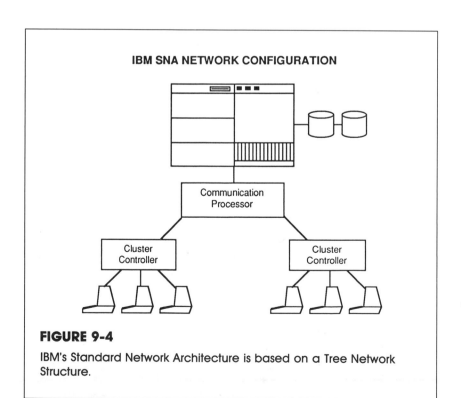

IBM SNA NETWORK CONFIGURATION

Communication Processor

Cluster Controller

Cluster Controller

FIGURE 9-4

IBM's Standard Network Architecture is based on a Tree Network Structure.

passed from one node, it contains an *address* in the form of a sequence of control bits that determines the receiving node(s). Depending on the specific system, the ring topology may be vulnerable to interrupted service if one of the nodes suffers a failure. Many actual implementations, however, provide continued routing of messages through the ring should one node be eliminated from service.

Examples of the ring topology are IBM's Token Ring local area network for personal computers, and the Prime Computer Ringnet for linking multiple minicomputer processing units.

- Linear or Bus

This networking scheme is based on a linear "backbone" to which devices can be attached, as shown in Figure 9-6. In this network topology, intelligent nodes tap into the network to send and receive messages. As in the case of the ring topology, messages are sent with an address for routing to the proper node. Since messages travel in both directions along the linear backbone, there is a possibility for collision between conflicting messages. Implementations of the bus topology use sophisticated software for the detection and correction of errors caused by these collisions. The most prevalent example of the linear topology in GIS configurations is local area networks using the Ethernet standard.

Network Protocols. A network protocol is a fixed set of rules specifying the format of data exchange. A particular protocol defines how the individual bits are to be arranged when transmitting a message so that it can be received and interpreted correctly by another node on the network.

The previous discussion of asynchronous communications described simple point-to-point connections where individual bytes, with start and stop bits, are sent between a host computer and a peripheral device. This type of asynchronous communication can technically be considered a protocol in a simple sense. Protocols become more sophisticated, however, as the complexity of the network increases. The network must support high-speed data transmission, an effective mechanism to get a message from one device to the correct network node, and efficient means must be available to check for errors and to resend a message in case an error has been detected.

Protocols are established to:

- Identify the sender and receiver of a message
- Define the beginning and ending of a transmission
- Specify the controlling device of a network (if any exists)
- Set and coordinate the timing for data transmission
- Establish control information for error detection and correction
- Specify the address and proper routing of a message

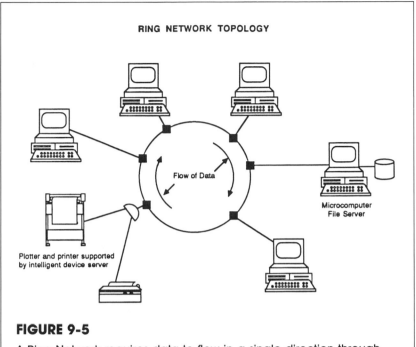

FIGURE 9-5

A Ring Network requires data to flow in a single direction through the network.

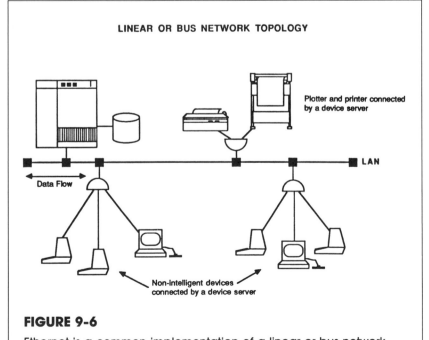

FIGURE 9-6

Ethernet is a common implementation of a linear or bus network scheme where data may flow in both directions.

Up until recent years, one of the most commonly used protocols for synchronous communications has been Binary Synchronous Control (BSC), often referred to as *bisynch*. Originally developed by IBM, this protocol was adopted by many mainframe and minicomputer vendors in the 1970s and early 1980s. Bisynch is classified as a *character-oriented* protocol because a data transmission consists of frames of characters (8-bit strings), each representing a control signal or a portion of the message. Significant disadvantages of this protocol for interactive processing in complex networks led to the development in the late 1970s of *bit-oriented* protocols.

Bit-oriented protocols structure variable-length frames of bits that contain a message component and control bits. Considered more efficient than character-oriented protocols because the message component of each frame can hold more information, bit-oriented protocols in addition offer more efficient error checking, two-way simultaneous communication, and sophisticated routing. A popular bit-oriented protocol is Synchronous Data Link Control (SDLC), developed in the late 1970s by IBM as part of its SNA network concept. Many major computer vendors support the SDLC protocol and often have developed other proprietary bit-oriented protocols.

Although bit-oriented protocols have become common in networks where interactive processing is required, bisynch transfer is still used, particularly in batch-file transfer operations. Other special-purpose protocols have been developed to support communication in special network environments such as public access networks (packet switching), electronic mail systems, and various remote communication environments.

Local Area Networks. The local area network (LAN) is a special type of network designed to support multiple devices within close proximity. Although there is no universally accepted definition of the term, LANs generally have the following characteristics:

- Operated and administered within one organization, not subject to direct governmental regulation
- Integrated into a discrete physical entity through a continuous cabling scheme
- Support high-speed data transmission, allowing mutual access between all network nodes (subject to established security levels)
- Limited geographical distribution of network nodes, normally to a radius of several miles

Many geographic information systems rely on the Ethernet standard, which uses a bus topology to transfer data at a maximum speed of 10 million bits per second. Local area networks based on Ethernet allow multiple devices to tap into the network. The Ethernet-based LAN, which frequently uses coaxial cable, can connect multiple processing units, intelligent workstations, and peripheral devices to allow users to share all resources located on the network (i.e., processing power, data, output devices). Nonintelligent devices such as dumb terminals and plotters are connected to the network through a device server. The device server is a small computer processor that interprets the signals on the LAN and properly routes messages to and from the peripheral devices. Remote devices can gain access to the LAN through modems or digital service units.

The 10-million bit-per-second transfer rate allows all users interactive access to network nodes. Using standard coaxial cable, an Ethernet-based LAN can be extended a total distance of about 1,500 meters. Some system vendors offer Ethernet LANs based on twisted-pair (telephone) wire with significant reductions in maximum distances.

Another popular LAN is the IBM Token Ring. This LAN uses a ring topology to support and share resources among personal computers. Many other computer system vendors have also developed LANs using the token ring design for personal computers and for larger processing units. Token ring networks may use twisted-pair wires, coaxial cable, fiber-optic cable, or may use the facilities of a broadband transmission media to transfer data in analog form. The distances that LAN nodes can be extended range from about 500 feet for twisted-pair to several miles for fiber-optic or broadband cable.

With Ethernet and other LANs, distances are extended through the use of one or more *repeaters* that are placed along the LAN to strengthen the signal. *Bridges* are intelligent devices that connect and allow the transfer of data between two discrete LANs that share the same operating protocols.

Other LAN schemes are in development and promise to deliver greater speed to satisfy the increasing demands of GIS users. Since the quick transfer of graphic information requires high capacity, it is likely that networks supporting speeds of up to 100 megabits per second will become popular over the next decade.

COMPUTER SYSTEM CONFIGURATIONS

Centralized Versus Distributed Systems

An ongoing debate for the past several years has compared the relative merits of centralized and distributed configurations. The intensity of this debate might lead one to believe that there are no options other than a centralized or a distributed system. Configuration approaches possible for GISs actually represent a continuous range of opportunities, from fully centralized to fully distributed.

To understand the concepts of centralization and distribution, it is necessary to focus on the arrangement of both the processing performance of the systems and the data that resides in a computer system. In a purely centralized configuration, a host processing unit stores all data and provides all the processing power to support devices connected to it. Peripheral de-

vices in this traditional centralized model are not capable of local processing: they rely on the host for the execution of all tasks. Today there are few, if any, multiple-user systems that follow this absolute centralized model. In most current systems, there is some distribution of processing power and, less frequently, of data.

Distribution of Processing Power. The idea behind the distribution of processing power is to segregate processing activities among multiple units to relieve one host unit of all the processing tasks, thereby making the system more efficient. Although systems today are still characterized as being centralized or distributed, most have some elements of both.

Distributed processing has been put into practice for many years by large mainframe systems that use a special communication processor, connected to the host processing unit, to manage peripheral devices in the mainframe network. Another common practice is to place specialized processors in the peripheral devices themselves so they can perform some operations independently of the host. This is true of alphanumeric terminals with some local text manipulation capability; electrostatic plotters with special processors to "rasterize" vector data from a map plot file; pen plotters capable of locally sorting vectors in a plot file for efficient map plotting; and graphic query stations with local intelligence for manipulating the graphic display.

In addition to the use of specialized processors, major developments have been made in the networking of two or more multiuser processing units. The distribution of processing power by networking multiuser computers is a method for placing the processing power at the work-group level. The work-group is usually a discrete organizational unit within a larger agency, such as a division within a department, or a department within a larger agency, field, or branch office.

The distribution of processing power in this manner may or may not include the

Configuration approaches possible for GISs actually represent a continuous range of opportunities, from fully centralized to fully distributed.

distribution of data. On a high-speed network, it is possible to configure multiple processing units with no mass storage capability. These units can access data interactively from another node, such as a centralized database or database server (Fig. 9-7). In a network of this type, devices may be connected directly to ports on one of the multiuser computers, or directly to the network through device servers to gain access to computer resources at any node on the network.

Another tier in the distribution of processing power is the microcomputer or super-microcomputer workstation level. When configured as part of a network, these devices may be considered "peripherals," but they have a substantial amount of processing power and often mass storage capability. In this scenario, microcomputers or workstations may be configured as single-user devices in a network with a host processing unit. The microcomputers or super-microcomputer workstations may interactively access data stored at the host, or data files may be temporarily downloaded for local processing at the microcomputer or super-microcomputer.

Mapping and geographic analysis applications are executed at the microcomputer or super-microcomputer workstation. Maps or reports are generated on peripheral devices attached directly to the microcomputer or workstation or on peripheral devices connected to the network through device servers that allow them to communicate with a processing unit. As discussed in Chapter 7, microcomputers or super-microcomputer workstations may also function independently as stand-alone systems or in networks with other microcomputers and super-microcomputer.

In the computer industry, distribution of processing capability has been promoted within a network computing approach referred to as the "client-server architecture." This model includes the server, or front end, that stores a corporate database. "Clients," which are workstations on the network, submit requests to the server, which extracts necessary data from the corporate database and transmits it to the client for further processing. GIS software vendors aer only beginning to apply client-server concepts, but it is likely that this approach, allowing data access in a transparent, high-speed environment, will become popular in GIS environments.

Distribution of Data. A database is a collection of automated files forming a logical set of information that the user can apply to perform various tasks. A geographic database may consist of map features and associate nongraphic attribute elements structured in such a way to facilitate mapping and geographic analysis. A database is considered distributed if portions of it are stored and maintained at more than one node in the network. Two levels of data distribution are possible:

- Distributed data with duplicate copies
- Fully distributed database without redundant storage

It is often efficient to distribute data with duplicate copies at nodes when the network cannot support high-speed inter-

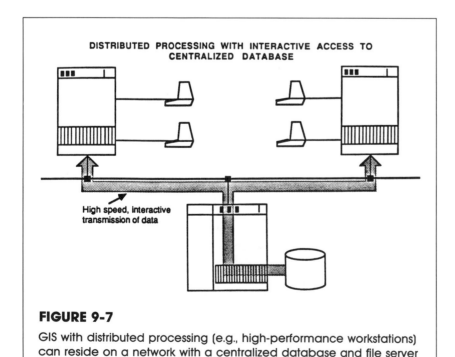

DISTRIBUTED PROCESSING WITH INTERACTIVE ACCESS TO CENTRALIZED DATABASE

High speed, interactive transmission of data

FIGURE 9-7

GIS with distributed processing (e.g., high-performance workstations) can reside on a network with a centralized database and file server (processor).

active access to all its points. For example, a GIS may be implemented in a local government where major user organizations include the property assessment office, water and sewer department, and planning department. Each of these agencies maintains different layers in the geographic database (e.g., land parcels, water and sewer systems, zoning, district boundaries) and each is located in a different area of a large city, making direct high-speed communications impossible. Each of the groups, however, frequently needs data maintained by the others.

In this case, it would be reasonable to distribute duplicate data at the three nodes. In other words, each node redundantly stores all commonly needed database layers (Fig. 9-8), ensuring that each user will have quick on-site access to the data without waiting for slow transmission on remote communication lines from the node where the data is updated. An inherent problem with this arrangement is that a given user, if not at the node where data is updated, could access an out-of-date copy. There is also the potential for conflicting updates of the data layers at different locations.

Such problems can be minimized with effective system management and maintenance procedures. Updates of a particular layer at the appointed node must be made to each of the other nodes storing that layer. An appropriate time interval for transferring updates to the other nodes must be established, i.e., weekly, daily, or more frequently, depending on the application. Sound system management policies along with network and data management software controls can be used to direct this process systematically.

A network configuration that allows interactive, real-time access to data stored at other nodes can eliminate the limitations of the duplicate storage scenario. This type of configuration (Fig. 9-7) relies on a high-speed flow of data that allows programs executed at one processing unit to access data interactively from another node as needed by the program. Duplicate storage of data is not necessary and all us-

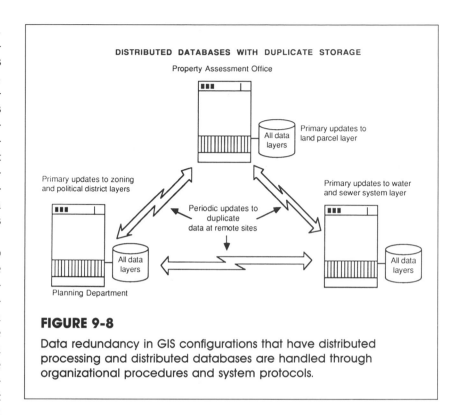

FIGURE 9-8

Data redundancy in GIS configurations that have distributed processing and distributed databases are handled through organizational procedures and system protocols.

ers can access data stored at the primary site—theoretically, completely up to date. This approach depends upon high-speed communications that are best suited to a local area network environment.

Advantages and Limitations of Centralized and Distributed Approaches

As hardware and software technology has advanced to allow more distributed computing, the approach has become popular among many user groups who want more direct control of and access to their data and the system's processing resources. Distributed computing reduces response-time problems sometimes associated with centralized mainframe systems. User groups may develop their own applications rather than depend upon a central data processing staff to carry out programming and data management tasks.

The distribution of computer systems provides many attractive opportunities, both technical and administrative. Distributed systems are more complex and therefore require more complex management and control. Generally, centralized computer systems are easier to manage but are less responsive to users and offer fewer options for incremental upgrades. Distributed systems alleviate these problems but require more sophisticated management and often some duplication of staff resources to maintain the system and support its users.

In designing a computer configuration, many options are available. System design is not simply a question of centralization versus distribution; the most effective approach may be a hybrid configuration containing centralized and distributed elements. Users should consider the following factors in evaluating network configurations:

- Physical distribution of sites
- Existing communication facilities
- Local processing requirements of users
- Database content
- Frequency and volume of data update and usage at different sites
- Access security requirements
- Anticipated hardware upgrade needs in the future
- Institutional factors relating to system management and "ownership" of data

Managing a Computer Network

All computer systems, from small microcomputer systems to large mainframe configurations, require efficient management to establish and enforce administrative procedures and to oversee the technical operation of the system. The technical management requirements of computer network operation are extensive and, obviously, increase as the system becomes more complex. To be effective, network and data management software tools, along with proper management procedures, should be used cooperatively. Some of the major concerns important in routine system management are listed below:

- Establishing and maintaining network access security

Software tools are used to assign access rights to portions of the database in the network by user password, physical device, or other types of control. Monitoring tools can be used to detect unauthorized access.

- Maintaining data directories

Networks with distributed systems require global directories, or a "database about the system database," containing information about what data is stored where and what access procedures have been established. Data directories simplify access to the network and free users from the need to know where data is stored and its local naming conventions.

- System accounting

Usage of system resources, including the processing unit and peripheral devices, is tracked by a user or project account. Statistics on usage collected by software can help plan allocation and upgrades of those resources and tabulate costs by user.

- File and record locking

Locking is a data management procedure employed by multiuser systems to prevent one user from modifying some portion of the database (map feature, layer, data element, or entire file) while another user is making changes. Most locking procedures allow view-only access while the data is being updated.

- Database update control

In a network configuration that relies on duplicate storage of data (Fig. 9-8), procedures must be established to transmit updates from one node to create duplicate databases at the other node(s). Much of this process can be automated by writing update procedure programs and invoking those programs through the system clock on an established update schedule.

- Diagnosing and correcting network problems

Complex networks are subject to occasional problems resulting from equipment failures, software problems, damage to or outage of communication lines, and other mechanical problems. Management can monitor the network automatically to identify and isolate problems and sometimes provide fixes to keep operating.

RESOURCE SHARING AND DATA EXCHANGE BETWEEN COMPUTER SYSTEMS

Geographic Analysis in a Multivendor System Environment

Geographic information systems often are implemented in organizations where geographically referenced data and special analysis programs reside on a variety of computer systems. When systems serve multiple departments, geographic data usually is maintained on systems including microcomputers, minicomputers, or mainframes, even though these computers may not be the primary platforms for carrying out mapping and geographic analysis.

The potential need to exchange data between systems to support geographic data management applications is extensive. Several possible scenarios are presented below:

Scenario 1. A property accessor's office requires parcel-level land ownership data from a mainframe computer to support standard parcel mapping or special tax district analysis with the geographic data management software on a minicomputer.

Scenario 2. The local planning department requires the same ownership data to conduct property searches or to prepare special thematic mapping to support subdivision reviews.

Scenario 3. A gas or electric distribution utility company uses a GIS to identify customers affected by interruptions at points on the network or during planned shutoffs in portions of the network. Account number and customer data stored on a separate customer information system must be transmitted to another computer.

Scenario 4. A police department routinely needs to analyze crime history data to determine geographic and temporal trends. A GIS running on a microcomputer is used to display on a city street map and symbolize by type the location of crime incidents keyed to street addresses. To do this, data must be downloaded from a police dispatch system operating on a minicomputer.

Scenario 5. A large retail department store chain considering a major expansion program is using a GIS to evaluate site suitability across the nation. Various automated demographic data tied to map locations must be acquired from the United States Census Bureau and loaded into the system for the geographic analysis.

Scenario 6. A state natural resources regulatory agency uses a GIS to assist in evaluating the environmental impact of proposed programs. This analysis requires a proximity search and evaluation of existing permitted sites in the vicinity. A permit tracking system is maintained on the agency's mainframe computer and includes locational coordinates for existing permits. This data must be transferred to the GIS to conduct the analysis.

Each of these scenarios has its own requirements for the type and amount of data that must be exchanged and the frequency with which it must occur. These requirements are specific to the users' applications. A number of techniques for data exchange between disparate systems, both simple and complex, are available. Selecting a particular approach depends on the specific application and, ultimately, the cost of implementation.

Types of Data Exchange

Methods used to transfer data between computer systems vary in complexity and effectiveness. When exchanging data between computer systems made by different vendors, it is usually necessary to convert data coding and protocols so that the data can be recognized by each system. Data exchange techniques can be grouped into four main categories:

- Device emulation
- Media transfer
- Direct batch transfer
- Interactive link

Device Emulation. In a pure sense is not really a data exchange at all, but describes the use of software or hardware to enable communication between a peripheral device (often a terminal) and a foreign computer system. For instance, a user may wish to access an IBM mainframe from a minicomputer terminal that recognizes ASCII character coding and a specific format and protocol for displaying text or graphics. A software package or hardware card placed in the terminal would accomplish all necessary code conversion and protocol translation to make the terminal seem like an IBM 3270 terminal to the IBM mainframe computer. This would allow the user to have direct access to applications and data on both systems through one minicomputer terminal.

Media Transfer. Is an efficient and cost-effective way to transfer large amounts of data between systems. This method is sometimes referred to casually as a *sneaker link,* since tapes, disks, or other transportable storage media are hand-carried between the systems. Data is written on the source system and copied to the destination system. If operators at the destination know the format specifications of the storage media and have the equipment and software to read them, the process is straightforward. Of course, the primary limitations are the time it takes to transfer the media between locations and the potential for delays or human mistakes. The simplicity of this approach makes it very attractive when the geographic data management applications do not require frequent transfer of data and when delays will not create major problems.

Direct Batch Transfer. Implies that a communication link is set up between the computers to exchange the data. The link could use either local or remote communications, depending on the particular conditions and geographic distance between the sites. In direct batch transfer, a data file is prepared on one system and transmitted directly to the other system for storage. The transfer itself uses specific programs to initiate the transfer and communication protocols that both systems understand. In this batch environment, a job is submitted to a batch queue and the transfer executed according to its priority in the queue. For many batch transfers, it is convenient to set up a transfer during non-peak hours of operation.

In the media transfer and direct batch transfer methods described above, a data file is actually transported and stored redundantly on another system. When this process is used to transfer data in a multivendor computer environment, some data files must be prepared before the transfer can take place.

First, the data elements required for the geographic application must be extracted from the database of the source system and placed in a data file to be transported. Next, the data file is transported and stored on the destination computer. Following this transfer, the file may need some additional reformatting to match the specifications of the geographic data management software on the destination system. The last step is to load the data file into the geographic database so it can be used by the software. This process is fairly straightforward. In the past few years, the widespread use of relational database systems and high-level programming languages has allowed users to manipulate data without complex programming ability.

Interactive Communication Links. Allow users or programs on two or more computer systems to communicate with each other in a real-time environment. Any system can initiate a data query or run a geographic analysis program and access data directly from another system. In this type of communication, no data files are actually transferred. A program on one system uses the resources of the other in a dynamic relationship (see above).

Interactive links generally require high-speed communications delivered by local area networks or other high-capacity networking schemes. Many hardware vendors have developed effective interactive communications between their own computers. Interactive communication links between computers from different vendors are currently more difficult to implement. Establishing such a network generally requires special hardware and customized software.

There are many hardware and software products on the market today designed to assist communications between computers and devices that recognize different communication standards. A large array of products for microcomputers enable data files to be downloaded from a host mainframe computer. These micro-to-mainframe link products may be in the form of interface cards installed in the microcomputer, software loaded on the microcomputer and/or the host computer, or a combination of hardware and software.

Multiple computer networks can be linked through gateways that combine hardware and software. The gateway converts communication protocols and can offer device emulation, batch-file transfer, or, in some cases, interactive links. The gateway link essentially allows one computer system or network to function logically as a node on the other computer network.

THE GROWTH OF COMMUNICATION STANDARDS

There has been an encouraging trend in the computer industry since the mid-1980s toward the establishment of communication standards. In response to strong user demands, it is becoming easier to share resources and create networks with computers obtained from multiple vendors. A number of national and international organizations today actively promote communication standards, the best illustration of these efforts is the work of the International Standards Organization (ISO) which comprises government and industry representatives from many countries.

The ISO has been developing the Open Systems Interconnect (OSI) model to provide a framework for the development and adoption of standard interfaces and communication protocols for computer devices. These standards include basic requirements for physical and electrical interconnections; specific protocols for transmitting a message in a network; more sophisticated protocols to facilitate complex routing and error detection; and higher-level protocols that describe how data will be formatted and presented. The OSI model does not describe particular protocols but a layered approach, in which higher layers use the standards of the next lower layer to send and receive data (Table 9-3).

The progress made by the ISO and other organizations is impressive. Standard protocols have been recommended for many layers of the OSI model and some have been adopted for the lower layers. The greater demand for open-system computing environments will continue to fuel the development of standards at all levels. The trend toward open standards, however, is somewhat inhibited by the interests of vendors who anticipate a negative impact on the market for their products that now use proprietary protocols and operating systems. Most vendors, however, are actively, but cautiously, pursuing the development of products that comply with open-system standards. It is certain that the future holds great promise for more flexible intervendor networking.

DESIGNING A SYSTEM CONFIGURATION

The Conceptual Design

The first logical step in the design of most hardware/software configurations is a con-

TABLE 9-3   International Standards Organization Open Systems Interconnect Model

Layer	Purpose
1. Physical	Governs the physical coupling of cabling and the electrical standards for data transmission (e.g., RS-232).
2. Data link	Establishes basic protocols for data transmission across the physical link. Governs the blocking of bits and the timing of transmission (e.g., SDLC).
3. Network	Provides services for routing and switching of messages in complex networks (e.g., $\times.25$ packet switching protocol).
4. Transport	Provides for transparent communications between source and termination points and provides services for transmission error connection.
5. Session	Sets up basic procedures for communication between programs on different systems. Initiates and terminates program-to-program sessions.
6. Presentation	Performs functions for standard formatting of data for use by different systems (e.g., data compression, encryption, and deencryption for data security, other types of data translation and reformatting).
7. Application	Includes high-level services for the transparent use of application programs on a network such as network performance monitoring, system accounting, and special record of file transactions.

ceptual design that focuses on the functional requirements of users rather than actual models of hardware devices and software packages. The conceptual design should be based on: 1) Sound familiarity with the intended applications of system users; 2) knowledge of the fundamental concepts of computer hardware, software, and data communications; 3) familiarity with site conditions, existing systems, and communication facilities already in place that may influence the design; and 4) knowledge of the current status of computer technology and new developments that may affect the system.

The conceptual design should describe the function and purpose of hardware and software and the interrelationship between devices and applications in a configuration. It provides an initial basis to review system requirements, make refinements, estimate costs, and lay a foundation for the more detailed specification definition that precedes system procurement. The conceptual design should describe the expansion of the system in phases reflecting expected growth with new applications, new users, and an expanded database.

SYSTEM CONFIGURATION PROFILES

The choices today are almost limitless for configuring GIS systems. Various levels of centralization, distribution, and processing power must be matched with the specific site conditions and application needs of users. Figures 9-9 through 9-16 illustrate examples of characteristic conceptual configurations that have met the needs of various user environments. Each configuration is accompanied by a description of the type of organization, the level of distribution, and the system's upgrade possibilities.

**System Profile A
Super-microcomputer Workstation with
Peripheral Devices**

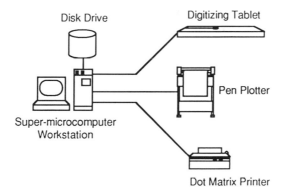

FIGURE 9-9

Description of System Profile A—**Type of System and Organization:** Suitable for a small number of users in single department of local government, small engineering firm, etc.; involved in data capture, map production. **Distribution of Sites:** Single site. **Data Transfer Operations:** No major requirements for transferring data with other sites exist. Data can be exchanged via tape or direct batch transfer. **Upgrade Possibilities:** Expansion into host-based network or network with other intelligent workstations is possible. **Institutional Issues:** No professional DP staffing; no direct supervision by central DP agency.

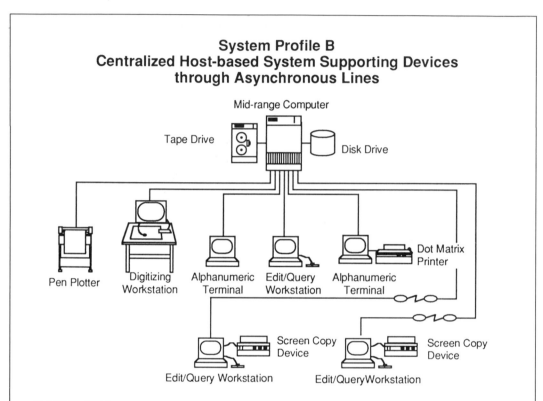

**System Profile B
Centralized Host-based System Supporting Devices
through Asynchronous Lines**

FIGURE 9-10

Description of System Profile B—**Type of System and Organization:** Suitable as a small multiuser system supporting data capture, map production, and database and graphic query applications serving one department or several related departments in a government agency or private firm. **Distribution of Sites:** Central processing unit with most peripheral devices are distributed on one floor or several adjacent floors of a building. Remote users on dedicated phone lines may be connected to the system. **Data Transfer Operations:** Data from other sources is exchanged via magnetic tape or direct batch-file transfer. **Upgrade Possibilities:** Some additional devices may be added on available asynchronous ports; may be expanded into a local area network to support additional users. **Institutional Issues:** A trained system manager is required to oversee system operation; no direct supervision by DP staff normally exists. Additional full or part-time technical support may be needed.

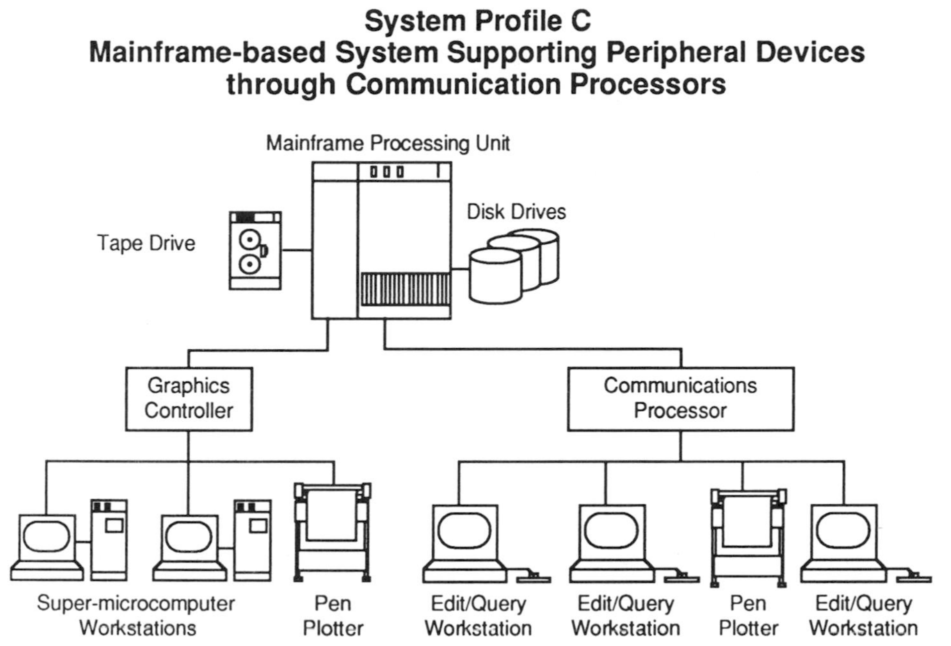

System Profile C
Mainframe-based System Supporting Peripheral Devices through Communication Processors

Mainframe Processing Unit

Disk Drives

Tape Drive

Graphics Controller

Communications Processor

Super-microcomputer Workstations   Pen Plotter   Edit/Query Workstation   Edit/Query Workstation   Pen Plotter   Edit/Query Workstation

FIGURE 9-11

Description of System Profile C—**Type of System and Organization.** Suitable for a medium to large government or private organization. Users require routine access to mainframe database for geographic queries, special analysis, and mapping. **Distribution of Sites.** Multiple locations are directly connected inside single buildings. Additional workstations and other devices connected by remote link (digital or analog line) to graphics controllers or communications processor. **Data Transfer Operations:** Periodic exchange of data with other systems occurs via tape. Possibility for direct gateways to other computer networks. Microcomputer data transfer is easily established. **Upgrade Possibilities:** Additional devices may be added flexibly up to the computing capacity of the mainframe processor. Potential network with another mainframe processor (in close proximity) to share resources. **Institutional Issues:** A system manager and one or more technical support staff are required. The system may be operated under auspices of the organization's main DP section.

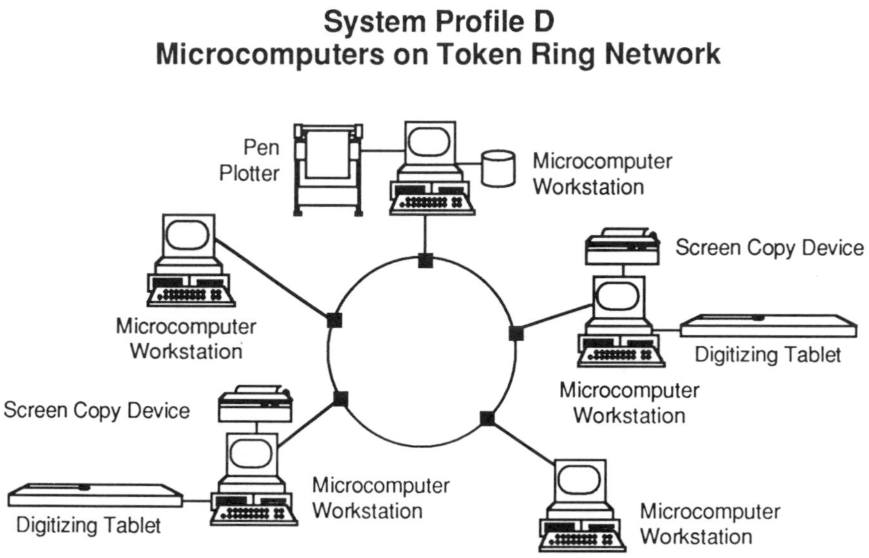

System Profile D
Microcomputers on Token Ring Network

FIGURE 9-12

Description of System Profile D—**Type of System and Organization:** Suitable for a small to medium-sized government department, group of departments, or private firm (engineering, landscape architecture, consulting). **Distribution of Sites:** Devices are distributed within one building. **Data Transfer Operations:** Data may be exchanged with other sites by microcomputer disk transfer or dial-up phone links. **Upgrade Possibilities:** Additional microcomputers and servers can be easily added to ring network up to the physical limits of the network. The potential exists for gateways to other token ring or nontoken ring networks or mainframe computers. **Institutional Issues:** The network is managed and maintained by the user organization without direct involvement by a DP organization. One or more technical specialists within the user organization are delegated time to oversee network operations and provide user support.

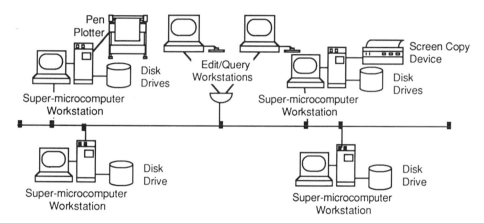

**System Profile E
Distributed Processing and Database with
Super-microcomputer Workstations on LAN**

FIGURE 9-13

Description of System Profile E—**Type of System and Organization:** Suitable for multiple divisions or departments in a medium to large government, utility, or private organization requiring a wide range of mapping and geographic analysis capability. **Distribution of Sites:** User sites are located on a LAN within one building or cluster of buildings in close proximity. Other user sites may link to the LAN through a remote communication line. **Data Transfer Operations:** Data is exchanged interactively between workstations on the network. Remote sites may exchange data through direct batch file transfers or tape transfer. Direct gateways with other networks are possible. **Upgrade Possibilities:** Additional workstations and peripheral devices may be easily added up to the physical limits of the LAN. The LAN can be extended over a larger area through use of line repeaters. Multiple LANs can be linked together through the network bridges. **Institutional Issues:** Each individual user group operates semi-independently with no direct control by a DP organization. A coordinating body and an overall manager are assigned to set and enforce database standards, access security, and operating procedures. Individual user sites develop in-house technical expertise.

**System Profile F
Local Area Network with Main Processing Unit,
Super-microcomputer Workstations, and
Peripheral Devices**

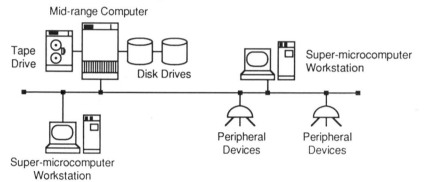

FIGURE 9-14

Description of System Profile F—**Type of System and Organization:** Configured on a local area network supporting multiple users with a host processing unit. Suitable for a multidepartmental setting (e.g., multiple departments in mid-size city or county government or utility company). **Distribution of Sites:** User sites are located within one building or several adjacent buildings. Other users may connect through remote communication lines. **Data Transfer Operations:** Main database is maintained at host processor with temporary storage for local processing at intelligent workstations. Tape transfer and direct batch-file transfer with other computer systems are possible. **Upgrade Possibilities:** New peripheral devices, workstations, or processing units may be added up to the physical limits of the LAN and processing power available on the network. The LAN may be extended to other buildings using line repeaters. LAN bridges may be used to link multiple LANs and gateways may be added to allow direct access to other types of networks. **Institutional Issues:** Trained system manager and support staff are required to oversee system operation and provide technical support to users. The system may or may not be operated under the supervision of a DP organization.

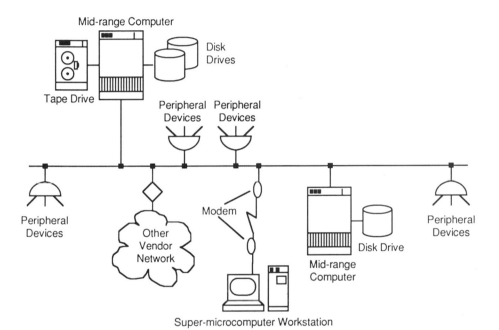

System Profile G
Local Area Network with Multiple
Processing Units and Central Data Storage

Mid-range Computer

Disk Drives

Tape Drive

Peripheral Devices

Peripheral Devices

Peripheral Devices

Other Vendor Network

Modem

Mid-range Computer

Disk Drive

Peripheral Devices

Super-microcomputer Workstation

FIGURE 9-15

Description of System Profile G—**Type of System and Organization:** Multiple processing units and peripheral devices are configured on a local area network supporting multiple functionally related organizations (e.g., county government agencies with public water utility organization). **Distribution of Sites:** Processing units and most peripheral devices are located at multiple buildings within the central business district of a city. Some devices are connected to the LAN through remote lines. **Data Transfer Operations:** Users access a distributed database on the LAN; batch and real-time access to data on other networks may be provided by network gateways. **Upgrade Possibilities:** New peripheral devices, workstations, and processing units may be directly connected to the network up to the physical and logical limits of the LAN. Extension of the LAN or connections to other LANs may be accomplished with other buildings with line repeaters or LAN bridges. **Institutional Issues:** A trained system manager with technical support staff is required. A multiorganization committee may be formed to direct system activities. Supervision or support from the DP organization is likely.

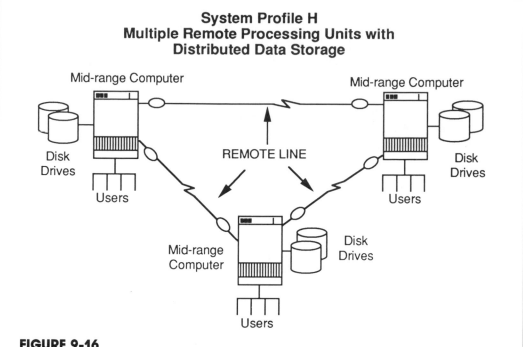

System Profile H
Multiple Remote Processing Units with
Distributed Data Storage

FIGURE 9-16

Description of System Profile H—**Type of System and Organization:** Suitable for multiple organizations in a system network to share common data (e.g., consortium of city government, county government, utility company) or a single organization with geographically distributed offices requiring substantial processing power. **Distribution of Sites:** Multiple processing units are remotely configured with each supporting a database and multiple system users. **Data Transfer Operations:** Data updates at any site are made in batch-file transfer to redundant databases at other sites. Interactive access between multiple sites may be possible depending on speed of remote lines. Gateways with other networks may be added. Common data required routinely are stored redundantly at each site. **Upgrade Possibilities:** Full range of expansion is possible at each site. Additional processing units could be added to the remote network. **Institutional Issues:** A system manager and technical support staff are required at each processor site. A multiorganization committee with an overall chairman or manager is needed to formulate policies and direct system activities.

SUGGESTED READINGS

Croswell P. (1986). Developments in data transfer between Geographic Information Systems and mainframe computer database. *URISA Proceedings*, Vol. II. Urban and Regional Information Association, pp. 47.

Datapro Corporation (1990). An overview of data communication standards. *Datapro Reports on Data Communications*, Section C07-010-103. McGraw-Hill Inc., Delran, New Jersey.

Love K. (1988). Distributed Processing/Distributed Databases for GIS Applications—Basic Concepts and Issues. *URISA Proceedings*, Vol. III. Urban and Regional Information Association, p. 228.

McQuillan J. (1990, June). Broadband networks: The end of distance? *Data Communications:* 76–86.

McQuillan, J. (1990, March 19). The unraveling of interoperability. *Network World:* 31.

Naughton P., Schechter G. (1990, June). Developing trends in high-speed graphics data transmission. *Computer Graphics Review:* 46–51.

Part III

MANAGING CHANGE

Chapter 10

Implementation

Plunging into new technology can invoke exhilaration, fright, frustration, and more. Exhilaration comes from applying technological advancements to routine and exotic problems. Fear is provoked by the constant rapid advancement of technology and the resource commitments it demands. Frustration swells from the tedious pace of implementation and the inertia that generally greets any change in organizations, some even becoming paralyzed by politics and associated turf battles over information.

After fifteen years of maturation, geographic information technology has reached puberty. As with most adolescents, what it needs most now is refinement and self-confidence to fulfill the promise of a prosperous future.

The line between needs and embellishments is not always clear. Human nature, abetted by the industry's aggressive marketing, tempts both the experienced and inexperienced computer user to take on too much too quickly. Attracted by glitter and promises, practicalities are often ignored, along with consideration of budgets, human resources, and time. The lessons of others' experience—both successful and unsuccessful—are instructive as one approaches the challenge of implementing a GIS.

Over the past decade, a model for the design and implementation of a GIS has emerged. The process discussed in this chapter can be divided broadly into five stages:

Concept—defining user needs and GIS requirements
Design—developing a workable solution and implementation plan
Development—acquiring GIS components and developing user applications
Operation—phasing from manual to automated activities
Audit—evaluating operations and planning for the future

GIS DESIGN PHILOSOPHY

Organizational inertia often poses a challenge to the introduction of GIS, and its successful implementation depends on sufficient institutional support. The best way to generate that support is to formulate and present a clear vision of the role of the GIS within the organization. A consistent philosophy will lead to a clear vision of GIS utility. That philosophy may be simply stated: Applications must drive system design.

The term *system* entails more than hardware and software components, al-

The lessons of experience—both successful and unsuccessful—are instructive as one approaches the challenge of implementing a GIS. It is important not to let the glitter and promise distract one from considerations of budget, human resources, and time.

211

though it certainly incorporates those elements. A system also embodies data base content and structure, personnel, organizational structure, and financial resources (Fig. 10-1). System function, data base content, institutional structure, and procedures all are based on a clear statement of the system's requirements. Too often, hardware and software become the dominant or exclusive focus of GIS decisions.

Requirements of a GIS should be defined in terms of the applications the system is expected to support. As noted above, the technology offers a diversity of

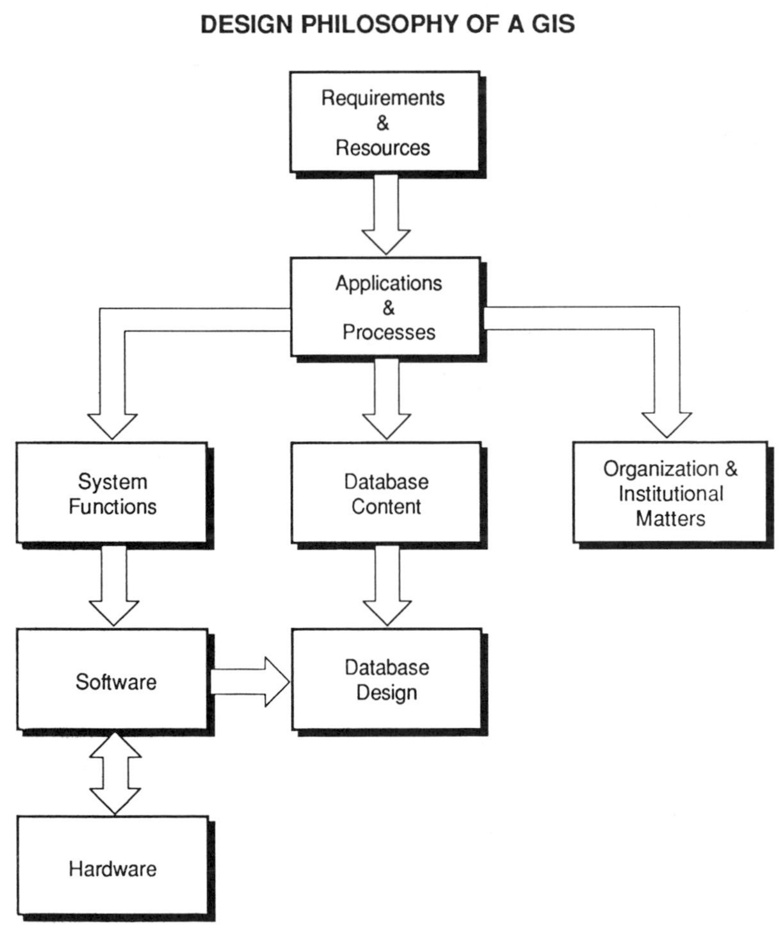

FIGURE 10-1

The philosophy that "Applications must drive system design" will lead to a clear vision of GIS functionality.

potential, not all of which may be purchased from a single vendor. By thoroughly investigating the applications that the user organization(s) must support, and by establishing the relative priorities of each, the definition of a system uniquely appropriate to the organization will emerge.

The design process is iterative, moving from the abstract, to a concept, to a preliminary design, to detailed design, and then to specification. A data base is the foundation needed to perform any application. Whether the data base is built transactionally, over a period of time, or, as is more often the case, is the result of a substantial preliminary effort, its content and accuracy are essential to the success of GIS. The cost of data base development and maintenance will exceed the costs of all other GIS components by several fold. Defining data base content and accuracy, while maintaining applications priorities, is one of the important first steps in system design.

Later in the process, a detailed data base design becomes critically important to system implementation but is not essential early on. Furthermore, the detailed data base design depends on software considerations usually addressed later in the process, and on the institutional procedures that support data maintenance.

The functions of the system can be determined without focusing on specific software and hardware products. Graphic and data base manipulations, analytical functions, and the system's configuration are examples of system functions. By defining these elements, the character of the system takes form and cost estimates become possible. Within the context of the chosen applications, the relative importance of various system functions must be determined:

- Should the system be optimized for the entry of drawings or maps?
- Will the manipulation of graphic data more likely involve the combination of various themes best represented by polygons or the movement and placement of lines and dimensions?

- Will the linear features be schematic representations or accurately depict placement?
- Will linear features be used in a manner that makes their connectivity essential and desirable, or unimportant?
- Should more than one individual have immediate access to graphic representations of the data base?
- Will a centralized or distributed processing equipment configuration best serve the user community?
- Will the equipment support other processing functions that are nonspatial?

The answers to these and other questions form the basis of conceptual system configurations accommodated by multiple vendors. Specific decisions on software and hardware are premature at this stage. In fact, early commitments to vendors can shift the design focus away from solutions that best respond to system requirements.

Computer hardware should be viewed as subordinate to software. Once it becomes clear that one or more software packages support the application requirements, compatible computer processors and operating systems can be chosen. With the increased availability of UNIX and associated technology, software is more easily transferred between computer processors of various makes. Processors and other computer hardware must measure up well to price considerations in addition to their ability to meet specific user needs.

In some situations, the processor and operating system influence, even control, the selection of GIS software. An organization that has accepted a particular vendor as its supplier sometimes limits its software choices. In other instances, the availability of existing processing resources and the need to avoid additional costs governs the choice of a hardware platform.

To realize the full benefits of GIS, institutions must adapt to the technology. Although it is possible to introduce GIS technology as a tool to replace or enhance what is done manually or not at all, its integration into decision making and work flow is likely to require changes in procedures and, perhaps, in the structure of the organization. Identifying potential changes in structure and procedures is another preliminary task.

Structural changes are threatening to any organization. When a system is implemented to increase the productivity or capability of an individual operating unit, structural changes are modest or nonexistent. Perhaps the drafting work force is divided into an automated group and a manual drafting group, resulting in two lead drafters. When GIS is introduced to support an entire agency or company, or multiple government agencies and jurisdictions, the situation usually demands new formal and informal agreements, interactions, user committees, policy or management committees, cost allocation, joint budget procedures, and so on. New lines of communication and assigned responsibilities require structural changes, both subtle and dramatic.

The way a task is accomplished can change. The availability of a single and consistent data base eliminates steps previously taken to process and validate zoning requests, work orders, and information queries. The physical movement of hard-copy data is rendered obsolete. Storage and archival functions change or are eliminated, and a range of simple and complex procedures are modified. The extent to which an organization succeeds in altering its procedures and work flow has a direct bearing on whether the technology will be integrated and the organization will reap its benefits.

The extent to which an organization succeeds in altering its procedures and work flow has a direct bearing on whether the technology will be integrated and the organization will reap its benefits.

GAINING AND MAINTAINING SUPPORT

The decision to proceed with the implementation of a GIS hinges on technical, institutional, and financial factors. Making that decision represents a major commitment for most organizations and requires support at all levels of the organization. Further, that support must be sustained

Organizational inertia often poses a challenge to the introduction of GIS, and its successful implementation depends on sufficient institutional support.

through the tedious and sometimes lengthy process of system implementation.

Three general approaches—technological, political, and educational—can build and maintain organizational support.

The *technological* approach calls for a specific and logical sequence of activities. It requires the definition of user requirements, addresses fundamental hardware and software issues, and defines databases and applications in detail. The results are framed in a cost–benefit analysis. Typically, the technological approach is launched by operational units, such as the comprehensive planning division within the planning and development department in a municipal government or the drafting section within the distribution engineering branch of a utility organization. The technological approach generally is used by individuals and organizational units who are promoting the concept to higher authority levels in the organization.

Senior and mid-level managers usually garner support for GIS with a *political* approach. The aim of this top-down approach is to inspire enthusiasm about the technology and to surmount insecurities associated with change or risk of failure. Senior managers also are in a position to overcome institutional and personal barriers. The political approach employed by senior management typically lacks technological definition. Their interest in the concept more often relates to a desire to enhance efficiency, effectiveness, profit, and state-of-the-art profile. Mid-level managers may find themselves straddling the fence; they might rely on a political approach to win sponsorship from senior management and on a technological approach to gain acceptance of GIS by operational staff.

An *educational* approach most frequently plays a supporting role and varies somewhat given the target group. Senior management needs an overview of the technology that addresses who is using it, with what success, and its costs and benefits. Senior management cares about how

the bottom line performs and the essential factors that influence it, such as cost, schedules, labor, and impact on other priorities.

Mid-level managers require a higher level of understanding of the technology; they are less likely to be persuaded by broad-brush overviews. Labor requirements—expressed in terms of quantity, experience, and availability within the organization—and other logistical issues are likely to be of interest. Financial considerations usually are expressed in terms of fiscal-year budgets and cost per parcel, utility service, or other relevant unit of measure.

At the operational level, the educational requirements—diverse, technical, and interdisciplinary—may add to the marketability of individuals within and beyond the organization. These trained individuals are more likely to seek out additional educational opportunities, such as conferences, workshops, and professional journals. Moreover, they frequently are asked to survey industry offerings and to host demonstrations and sales discussions led by vendors of GIS products and services. The education that operational staff amasses, however, may be fragmented, requiring substantial thought and effort to establish a context relevant to the operational requirements of the organization as a whole.

IMPLEMENTATION METHODOLOGIES

Organizations quickly become enamored with color screen displays, computer processors, and potential applications. The temptation is to buy a system, built a data base on the fly, and get started. Many systems are launched with little planning, little design, and great expectations.

Such an implementation strategy may succeed. However, more thoughtful ways of approaching implementation minimize the risks and match the direction and pace with available resources.

Implementation of a full-scale GIS is a

long, complex process. It generally involves multiple organizations and often costs hundreds of thousands, if not millions, of dollars. The full process, from the decision to implement a GIS to its completion, requires subdivision into logically related steps to aid understanding. Seventeen such steps are described here; for an actual implementation project these steps can be further subdivided into many work units. The sequence of the specific actions will vary from project to project, depending on strategy, priorities, and the local environment. Some steps will be accomplished in sequence, others will occur in parallel, while still others will overlap.

The five broad stages and seventeen steps to successful implementation are as follows (Fig. 10-2):

- *Concept*

Step 1: Requirements analysis
Step 2: Feasibility evaluation

- *Design*

Step 3: Implementation plan
Step 4: System design
Step 5: Data base design

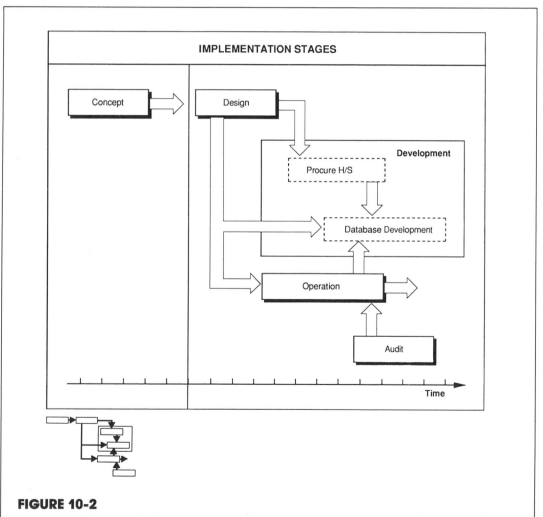

FIGURE 10-2

Five broad stages of GIS implementation provide the framework for 17 steps to success. Figures 10-3 through 10-6 further elaborate the implementation process.

- *Development*

Step 6: System acquisition
Step 7: Data base acquisition
Step 8: Organization, staffing, and training
Step 9: Operating procedure preparation
Step 10: Site preparation

- *Operation*

Step 11: System installation
Step 12: Pilot project
Step 13: Data conversion
Step 14: Applications development
Step 15: Conversion to automated operations

- *Audit*

Step 16: System review
Step 17: System expansion

Concept

During the concept stage of implementation, the project is defined based on an evaluation of user needs and available resources, and feasibility is assessed (Fig. 10-3).

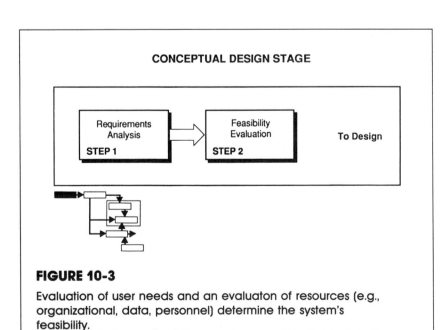

FIGURE 10-3

Evaluation of user needs and an evaluaton of resources (e.g., organizational, data, personnel) determine the system's feasibility.

Step 1—Requirements Analysis. Requirements analysis is the foundation of a successful GIS implementation. The analysis includes the identification of activities relying on map and geographic information throughout the organization, with attention to both graphic elements and nongraphic attributes of the graphic information. In many cases, it is the attribute or geographically related data that is of central interest to the professional and administrative work force. Furthermore, an inventory of current map data files leads to an assessment of redundancy in the collection and handling of data, the identification of information that is desired or needed but not available, and an appreciation of potential data base maintenance procedures. This final component of the analysis stems from the characterization of how information flows through the organization. In many cases, benefits can be derived by simplifying this information flow and eliminating inefficiencies in handling, storage, and manipulation of geographic data among various organizations or units.

Important GIS design considerations related to function, data content, and capacity depend on a clear understanding of what the various units that will use the GIS need. The organization's requirements are identified from a structured analysis of its operations and projections of future conditions. The evaluation of needs should be done rigorously and methodically.

At least the following seven types of requirements are assessed in the analysis:

- Processing functions
- Data contents
- Data standards and characteristics
- System applications and products
- Software functionality
- Hardware devices and capacities
- Communications facilities

The requirements analysis may use several techniques, including interviews, workshops, questionnaires, and modeling. The purpose of the analysis is to describe the present and future requirements of the GIS based on the perceptions and information supplied by current staff. However, since those persons typically are not familiar with GIS technology, it may be necessary to ensure a common basic level of understanding through training sessions, workshops, conference attendance, publications, or other educational techniques.

The requirements-definition process involves identifying those organizations or groups within an organization that may benefit from the GIS and dividing them into functional areas and levels of significance. Individual representatives with extensive knowledge of the organizations and their operations should participate in the analysis.

Interviews are a primary method to obtain required information. They may be structured with specific questions or unstructured, allowing the unique conditions of the organization and the interviewer to guide the discussion. One or more analysts with a thorough understanding of GIS technology should carry out the interviews. If practical, the analyst(s) also should have some knowledge of the organization and its operational function. The persons being interviewed should understand the purposes, management issues, and technical aspects of each function discussed. Conducted with individuals or small groups, interviews cover descriptions of the organization's mission, current operations, and problem areas. During the course of the interviews, sample maps and other supporting materials are collected and reviewed, and the organization's resources and operating environment are defined.

Interviews are relatively time-consuming and labor-intensive. The time is well spent with organizations that will play a key role in the GIS development and use. With other organizations, the effort can be minimized by conducting group interviews or workshops. With this approach, several representatives of multiple organizations meet with an interviewer or moderator to describe and record the needs of the various organizations.

To gather information on maps, data, and application needs, questionnaires may be used. The map features, current map descriptions, scales, data volume, and other information are easily recorded by the participants on properly designed questionnaires. Similarly, information about current costs and resource commitments often can be acquired through questionnaires.

Models or checklists of standard functions and requirements can verify and analyze the information collected. The models are used to analyze data and map features, required relationships among data, processing and display functions, and hardware devices and capacities.

The requirements information is documented for use in subsequent steps. This initial step supplies virtually all of the other steps with information for design, application development, and operation. The documentation describes the scope of participation; the objectives of the system; the data entry, processing, retrieval, and display functions required; the graphic map features and nongraphic data required; and the functional requirements and capacities of hardware devices. The findings are then reviewed with the participants to identify shortcomings, confirm requirements, and clarify ambiguities. Following participant review, comments and corrections are incorporated into the final requirements document.

Internal staff, outside GIS consulting experts, or a combination of both perform the requirements analysis. As mentioned earlier, typically there is a knowledge gap between the GIS experts and the potential users are not fully knowledgeable of GIS. The persons conducting the requirements analysis must bridge that gap by understanding both the GIS technology and its potential for application in the organization, and the functions and operations of the users. The analyst must draw needed information from participants, who because of unfamiliarity may have difficulty stating their requirements specifically in terms of GIS. The number of analysts required depends on the size of the organization, the number of organizations involved, and the complexity of the potential GIS.

A requirements analysis conducted by a skilled team also serves an important educational role if representatives from the participating organizations are sufficiently involved in the process. Sometimes it is difficult to involve users in the analysis and to elicit the specific information needed due to their lack of time, interest, or understanding of the technology's potential.

Four to six months normally is adequate for the requirements analysis.

Step 2—Feasibility Evaluation. Most organizations must evaluate the feasibility of implementing a GIS before making a major investment. The feasibility evaluation generally involves identifying the extent to which the GIS is practical and estimating its costs and benefits. The scope of the feasibility study depends on the size and complexity of the organization and the diversity of potential GIS applications. A large, multiorganizational GIS often requires an extensive analysis, while a single organization, in which a PC-based GIS is the logical solution, may require only a few hours of investigation to prepare a memo on its value.

The feasibility study is based on the requirements analysis. The information on current operations is used to project a baseline condition, assuming that current information management methods will continue. The requirements are used to propose one or more alternative GIS configurations and databases for the organization. These conceptual designs provide the basis for estimating costs and benefits, and may address options such as central and distributed configurations of processing power and databases, extensive or minimal configurations, or other variations.

The costs of acquiring hardware and software, developing the data base, and maintaining and operating the system over a reasonable period are estimated for each alternative. The cost estimates should present a range based on estimates from a number of vendors who can deliver sys-

tems and databases that meet the organization's requirements. Operating costs include a core staff to manage and support the system and its data base, and to develop and maintain application programs.

Using the techniques described in Chapter 4, a benefit-cost analysis serves as the focal point for the feasibility analysis. First, the operation of the GIS and its impacts on the organization are delineated, so that the benefits from implementation can be identified. Where practical, benefits are quantified. Efficiency gained in various functions represents a direct benefit. Most organizations also realize indirect savings in numerous areas. The estimates of these cost savings include both actual labor saved and savings resulting from improved decisions.

The costs and benefits of the alternative GIS configurations are calculated and then compared to current operations costs. The comparison should extend over time to full implementation of the GIS and its data base, and to full operation. Typically, a major GIS requires two to five years for full implementation before all nonrecurring costs and benefits are realized. The feasibility analysis, therefore, should include a seven- to ten-year life-cycle period.

Objectivity and credibility are significant issues in the feasibility analysis. In some cases, a feasibility study is conducted by a person or organization wishing merely to justify acquisition of a system. In other cases, the study is conducted by one who wishes to block acquisition. Neither of these approaches is particularly effective. To ensure credible results, an objective party who will not benefit from a system internally or through subsequent sale of a system or data base services should conduct the study.

Conceptualization of an appropriate GIS design and estimation of costs and benefits requires a high level of expertise and access to information on system and data base costs that may not be available within the organization. Consequently, a consulting expert is often necessary. The consultant also can lend credibility to the findings.

As noted above, the scope of the feasi-

Most organizations must evaluate the feasibility of implementing a GIS before making a major investment. The feasibility evaluation generally involves identifying the extent to which the GIS is practical and estimating its costs and benefits.

bility study may vary widely depending on the organization and its requirements. A thorough, in-depth analysis of a GIS for a large or multiorganizational implementation may take six to nine months or longer. Allowing time for participant review and decision making may extend this period further.

The feasibility evaluation typically plays a persuasive role with senior management, boards, and elected officials in reaching a decision on whether to proceed with what level of effort.

Design

If the feasibility study supports a decision to acquire and implement a GIS, the process moves to the design stage (Fig. 10-4). In this stage, detailed plans for implementation, as well as the system and data base designs, are prepared.

Step 3—Implementation Plan. Preparing an implementation and management plan is the next step. GIS implementation usually involves numerous tasks undertaken by multiple organizations, including system vendors and contractors. Careful planning and coordination is essential. The implementation plan provides all participants with an understanding of individual roles and responsibilities and the relationships among tasks, and it defines and controls all subsequent steps in the process as follows:

- Identifies and describes individual tasks
- Assigns responsibilities for each task
- Indicates the resources committed
- Defines relationships among tasks
- Identifies products and milestones
- Establishes a schedule

The plan documents each factor in an easily comprehensible manner. A computer-based project management program may be used to update the plan and to monitor and report on its status. The plan is updated regularly by the project manager as tasks are accomplished or changes occur.

In addition, the plan contains a network diagram or description of the relationships among tasks and a schedule with

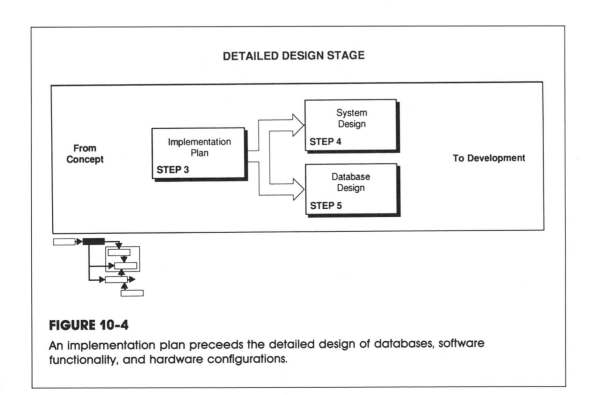

FIGURE 10-4

An implementation plan preceeds the detailed design of databases, software functionality, and hardware configurations.

key milestones noted. The plan defines a "critical path" through institutional and financial considerations, as well as technical efforts. The material is described at a sufficient level of detail to support management control of all tasks and resources. As described in subsequent steps, specific data base development, installation, and plans for phasing from manual to automated data management are incorporated later. A summary of the plan usually is prepared for senior managers and others with a general interest in the project.

Preparing the plan begins with identification of all tasks involved in the implementation project. A general model derived from other successful GIS projects is useful in identifying the tasks. The information from the feasibility study on system configuration, data base contents, cost estimates, and participants is used as the basis for implementation planning.

A standard set of information is recorded for each task, including a description of the task, an identifying name and/or number, the responsible organization and person, the funds, personnel, and other resources allocated, planned and actual beginning and ending dates, products or deliverable items, and the relationships to other tasks. Developing this information requires coordination and agreement among participants, and decisions by management to commit resources and assign responsibilities. Major decisions generally are made at the organization's highest level of management or by a policy body in a multiorganizational project. A draft of the implementation plan should be reviewed by representatives of all participating organizations to ensure its feasibility and to confirm commitments.

The plan should be prepared by someone with prior GIS implementation experience who is familiar with all tasks, and who has the skills needed to estimate the time and resource requirements for each task. A thorough implementation plan for a relatively complex project can be prepared in one to four months by an experienced planner.

Problems may occur in several areas.

Adequate cooperation, coordination, and commitment among all participants may be elusive for a multiorganizational GIS. The participating organizations may have different perceptions of the tasks or effort involved, and rivalries over responsibilities may emerge. Estimates of time and resource requirements also may cause disagreements.

Step 4—System Design. Two parallel tasks are required to prepare the detailed design of the hardware/software system and the data base. GIS is implemented typically by acquiring a commercially available hardware and software system. Numerous GIS systems are available in the marketplace, each with individual strengths and weaknesses that should be assessed in light of the organization's requirements.

Selection of the system may be a formal or informal process. Most government organizations and many utilities use a formal process based on a request for proposal or similar procurement procedure. A detailed design and preparation of specifications is necessary for the acquisition procedure. Because several candidate systems typically are available, the specifications at this stage are defined as functional requirements rather than as specific devices, model numbers, or software packages.

The detailed design builds on the conceptual design of the selected alternative configuration defined in Step 2, and it is guided by the implementation plan from Step 3. The concept is augmented with detailed descriptions of hardware and software function, capabilities, and capacities.

Sometimes it is necessary to collect more detailed information on maps, data, and processing requirements than was needed for the feasibility study. More detailed analyses produce the design specifications that are used subsequently for acquisition of the system. Refinement of the conceptual design is especially important with regard to the specific workstations, plotter, and application software requirements. With the movement to distributed systems and networks, the communications requirements also should receive special attention in the design.

Preparing the detailed design requires special skills and expertise in hardware, software, networks, and communications. The designer must be familiar with the available systems and the methods of system design. The time required depends on the size and complexity of the system and the involvement of the users in the design. A relatively complex system can be designed in six to nine months with time for participant review of drafts and resolution of conflicts.

Problems are sometimes encountered in documenting generic design specifications to solicit proposals from several vendors. An organization may have a problem choosing between components that can be practically justified and those that are requested by various participants. Rapid changes in GIS technology also challenge the designer.

Step 5—Data Base Design. A detailed design of the GIS data base is produced in parallel with the system design. The detailed data base design develops specifications to create and maintain the data base. The specifications guide in the acquisition of data base development services and to encode the data base schema or other definitions necessary for the GIS.

The data base design addresses the contents, specifications, relationships, and sources of data to be incorporated into the GIS data base. A detailed design is prepared for each of the data "themes" to be included in the data base, such as planimetry, topography, ownership, jurisdictional boundaries, utility facilities, and so on.

The detailed design is a refinement and augmentation of the conceptual design developed in the feasibility study. It begins with a review of the feasibility study and conceptual design. In most cases, additional detailed information on the requirements and source materials is gathered. Each source is evaluated thoroughly to verify the availability of the required features and the accuracy, completeness, and ease of use in digitizing. The design document identifies and defines all map features and nongraphic data and their characteristics.

The relationships among the various sets of features or other data are important, especially those of connectivity and adjacency, as well as positional relationships that allow the various overlaid features to be related to each other and positionally adjusted to the control framework. The design specifies the degree of positional accuracy necessary and the hierarchy of layers and features that will achieve it.

Source materials are evaluated to identify one or more source documents or acquisition methods for each required map feature and nongraphic attribute. If no existing source can be identified, a decision is made either to delete the feature or layer from the data base or to obtain it from new aerial photographs or field measurement. Copies of the source materials are compiled by the project manager for use as a sample reference throughout the project.

The source materials are evaluated to verify the area covered by each map feature. Data entry procedures for the identified source materials are designed and evaluated to verify the practical usefulness of the source materials.

The design document is then reviewed by the potential supplies and users of GIS data to verify the sources completeness and practicality. The design is produced by a person (or a team of two or three) skilled in GIS data base methods, cartography, and map compilation procedures. The detailed design of a complex GIS data base takes about six months to complete.

Acquiring and evaluating the source materials are the most significant potential problems. In most organizations, existing maps and data sources are scattered among multiple units and are neither complete nor uniformly standard. Significant effort is required to locate and thoroughly evaluate sources. In the event of conflicting source materials, their precedence must be resolved.

Another challenge is to determine the

Preparing the detailed design requires special skills and expertise in hardware, software, networks, and communications. The designer must be familiar with the available systems and the methods of system design.

relationship of the data base design and the software when the software has not yet been selected. This relationship is particularly critical when the design will incorporate large volumes of nongraphic attribute data whose format must fit a particular structure while retaining their logical relationships to graphic data elements.

Development

In the development stage, an organization acquires GIS software, hardware, and data conversion services, and develops procedures to operate the system (Fig. 10-5).

Step 6—System Acquisition. Acquisition of the GIS hardware and software often is the primary focus of the GIS project, even though several other steps are equally important. The purpose of this step is to acquire the most effective GIS hardware, software, and support at the most advantageous market price.

Systems can be acquired in various ways. An informal review and simple selection may be used, especially by private organizations. Among public and government organizations, however, a formal procurement process generally is required. The formal procurement involves specification of the system requirements, request for proposals or bids, an evaluation of proposals, and selection of the optimal system offering. The level of formality and the complexity of the selection process can vary dramatically, depending on the rules governing procurement within the organization.

Organizations sometimes choose a specific GIS early in the project for a variety of reasons, such as to establish compatibility with a related system, because a key decision maker had prior experience with the system, or because of an effective sales approach. If no specific system has been preselected by this point, a selection process is carried out.

The system acquisition process uses the detailed design specifications produced in Step 4 to prepare the proposal request. Because numerous systems are available, with several viable candidates for any specific acquisition, the selection process should be designed to allow a wide range of practical offerings. First, prepare specifications that describe functional requirements rather than specific devices or software packages. The hardware specifications should describe the types of devices (e.g., workstation, processor, plotter), their functional capabilities (e.g., resolution, speed, capacity), the number of items, and their capacities (e.g., data storage, number of devices supported). The software specifications should describe the functions required for various categories of software (e.g., operating system, graphics processing, data management, applications). Specifications for communications devices and software also are included for systems that will link separate devices.

The specifications are documented in a request format that conforms to the organization's procurement rules. The request for proposals solicits details on items such as documentation, training, maintenance and other support services, and the relationship between the vendors of software and hardware if the sources are not the same. The request for proposals also describes the terms and conditions for the purchase, including insurance requirements, installation dates, payment schedule, and the organization's standard contract language. The procedures and criteria for the evaluation and selection are described, at least generally, in the request for proposals in order to clarify priorities for the vendors.

GIS vendors are invited to submit proposals offering hardware and software systems that meet or exceed the specified requirements. The response to the request for proposals becomes the core of a contract between the vendor and purchaser.

Procedures and criteria for selection are essential to an efficient and fair review and selection process. The procedures used to evaluate proposals include benchmark tests, interviews, reference verification, and others as appropriate. The scor-

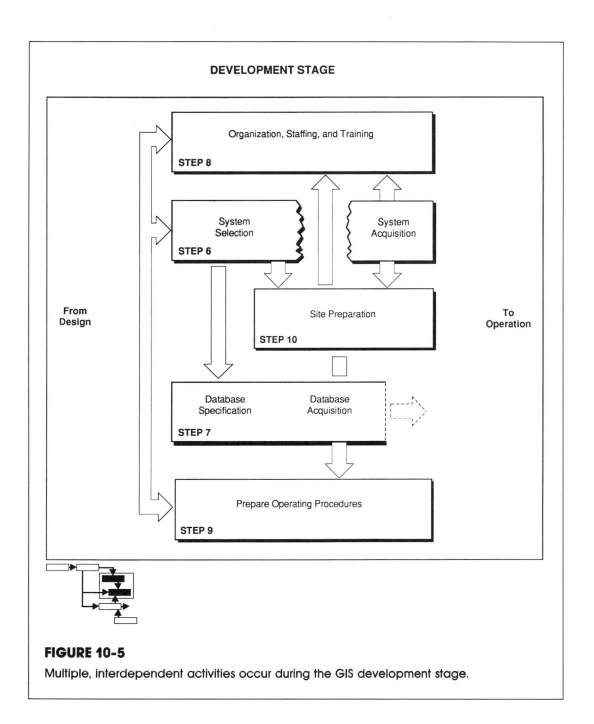

FIGURE 10-5

Multiple, interdependent activities occur during the GIS development stage.

ing and weighting criteria for each step should be defined before proposals are evaluated.

Organizations typically use a team of up to five or six evaluators with expertise in areas such as hardware, software, communications, applications, maps, operations management, and data issues. The team meets prior to procurement for orientation and any necessary training. The team reviews draft specifications and evaluation procedures before distributing the request for proposals. Once proposals are received, the team reviews and scores them, participates in interviews, verifies references, and conducts the benchmark tests. The team should meet at each stage for status reports and discussion of any problems.

Typically, the team selects a "short list" of the three or four highest scoring candidates, interviews each of the short-list vendors, conducts benchmark tests or site visits of the short-list firms, and, finally, selects the most favorable proposal. An evaluation of the technical specifications separate from the cost proposals ensures proper consideration of the technical requirements.

Proposals are ranked by assigning specific scores for each of its components. The scoring may use points on a scale (e.g., 1 to 10) or points for classes of desirability (e.g., "barely satisfactory" to "exceptionally high quality"). Groups of scores may also be assigned weights to balance the importance of each group of factors relative to the others.

Verification of references is an important selection criteria. The experiences of prior clients are particularly important in the areas of vendor support, responsiveness to problems, and the operational functioning of the system.

Interviews with selected vendors offer an open exchange of information and an opportunity to clarify at once any issues that arise. The vendor usually is given the opportunity to discuss the strengths of its proposal and the commitment of the corporation.

Two approaches—a benchmark test or a site test—are used to observe the system in operation. In a benchmark test, a specific test of system capabilities is developed and presented to the vendors. The vendor is given a period of time to prepare for the test, perhaps one to two weeks. The evaluation team then visits the vendor and observes specific tests. Data from a pilot test or another site with similar data bases may be provided to the vendor to populate the test data base. The evaluation team observes and records functionality scores, timing, and other information about the system operations. The test is used to verify the operation of critical functions, ease of use, and other criteria. The evaluation team should develop a test that

is proportionate to the magnitude of the system to be acquired. Benchmark tests are costly, so the purchaser must be reasonable. If an extensive benchmark is necessary, the purchaser should consider partial reimbursement of the vendors for their costs.

Benchmarks traditionally have played a formal role in the selection of computer hardware and software. In the past, they have been used primarily to test basic functions of candidate systems. More recently, as basic GIS capabilities have become more reliable, the focus for benchmarks has shifted to testing a product's utility for selected applications.

The other approach to product verification is a visit to a currently operating customer site. In this case, a specific test usually is not performed. Instead, the evaluation team views actual customer operations and has the opportunity to ask questions of counterparts in a comparable organization. This approach is not as rigorous as a benchmark test, but it provides exposure to the practical realities of a candidate system and its supplier.

After the results of all evaluations are compiled, the evaluation team selects or recommends the best offering. The selected vendor then is invited to negotiate a contract for procurement of the system. The organization prepares for negotiations by identifying modifications needed in the terms of the proposal terms and conditions of the contract. Especially important are agreements for continuing support, maintenance and response to any problems that should arise, including penalty clauses for system failure. The contract should include mechanisms for dealing with any future disagreements.

The length of time for system acquisition may be as short as one month for a simple, informal procedure. Procurements with complex, rigorous procedures that involve multiple organizations may take six months to one year. A procurement process expected to take more than several months must be designed to allow incorporation of emerging changes in technology prior to contract execution.

Problems with system acquisition of-

The experiences of prior clients are particularly important in the areas of vendor support, responsiveness to problems, and the operational functioning of the system.

ten arise in finding a team of persons with adequate knowledge of GIS who also represent the appropriate units of the acquiring organization. It is sometimes difficult to write specifications that are sufficiently detailed, yet still provide equitable opportunities for all appropriate candidate system vendors. Recognizing the subtle differences between systems, evaluating proposed systems, and reaching agreement on a selection may be a problem if the selection team does not have a thorough knowledge of GIS technology.

Step 7—Data Base Acquisition. Data base acquisition is a similar step to system acquisition, and occurs at the same time. Development of a GIS data base—conversion of hard-copy (analog) graphic and nongraphic data and reformatting of existing digital data—requires substantial investment of personnel, time, and financial resources. In most cases, it is impractical for the organization to acquire photogrammetric equipment and workstations and to assign sufficient staff to create the digital data base in a reasonable time. Most organizations developing a GIS today contract with a mapping or data conversion firm for the initial data base development. Conversion vendors can provide a large, skilled labor force, complemented by equipment and software tools optimized for data entry and editing.

GIS data base development generally includes four types of data entry:

- Photogrammetric compilation and digitizing using an analytical stereoplotter
- Automation of existing maps using a digitizer or scanner
- Key entry of data
- Transfer of existing digital data

Acquisition of a GIS data base involves contracting with one or more firms to acquire existing digital data, to digitize map data, or to enter nongraphic data. The acquisition process may be a single contract or may be divided into multiple contracts. To help keep this major effort manageable, data base development sometimes is divided into these logical components: aerial photography and analytical triangulation;

compilation and digitizing of planimetric and topographic features; compilation and digitizing of parcel maps; and digitizing existing maps of other features.

The procurement process for the data base is similar to that used to purchase hardware and software described above in Step 6.

The specifications for data conversion describe all map features and nongraphic data to be entered in the data base, and the standards of accuracy to be achieved. The specifications also describe the source materials, products and their formats, and the relevant characteristics. The request for proposals includes these specifications, contract terms and conditions, evaluation procedures, and directions for submission of the proposal. In addition to commercial information the proposals request solicits information such as conversion procedures and techniques, quality-control procedures and standards, management and staff's experience and availability, equipment and software resources and availability, existing work load, outstanding proposals and potential work load increases, and delivery schedules for preliminary and final products, both hard copy and digital.

As mentioned, there may be a single request, or the overall data base development may be divided among two or more contracts. A set of procedures and criteria is established for evaluating proposals and selecting the most advantageous offer(s). The procedures address the evaluation and scoring of the various components of the proposals received. They also describe the method for conducting and scoring interviews with candidate firms, site visits to review the contractor's facilities and equipment, and the verification of references.

Proposals are reviewed to verify compliance with specifications and to evaluate the capability and capacity of the equipment and personnel that are available to do the work. Each item in the request is scored on its quality. Weighting then is applied to establish the relative importance of various factors.

Quality of service to prior clients is an important factor in selecting a mapping

Probably the most underrated step in GIS implementation is that of organizing for the system.

contractor. This can be verified by contacting references provided in the proposals. Compliance with specifications and timely delivery of products are key elements in the reference evaluations.

Once the proposals are evaluated, the number of candidates are reduced to the best qualified. Site visits to the short list of firms may be appropriate to assess the equipment, staff, and capacity of the firm to handle production in the contract time span.

Cost proposals must consider the relationship between quality and cost. The purchaser must ensure that the vendor is offering services and products of satisfactory quality before considering costs.

In some cases, procurement rules require the award of contracts to the lowest bidder. In a low-bid procurement, only the opening of bids and verification of compliance with specifications is necessary. This is not the preferred approach since it does not allow recognition of differences in a firm's relative capabilities, past performance, and technical approach. In some cases, the higher-quality firms will not bid on such a request, leaving an organization with limited or lower-quality offers. Even those who bid may offer only minimal, lowest-cost services that are not as advantageous.

The specification and solicitation of services should be managed by persons experienced in GIS data base development and knowledgeable about map compilation, map digitizing, and data base development. The selection team also should include persons with an understanding of the GIS map and data requirements and the available source materials.

Solicitation of proposals, selection of the data base contractor, and negotiation of the contract usually takes five to seven months for a relatively large GIS project. Compilation of maps and conversion to digital form may take several years for large areas or detailed databases. The main time limitation is the organization's capacity to prepare source materials and to control the quality of delivered digital products.

Step 8—Organization, Staffing, and Training.   Probably the most underrated step in GIS implementation is that of organizing to implement and operate the system. Postmortems on GIS implementation universally indicate that a group's personnel and how they are organized are key factors in success. When a GIS is managed effectively and staffed properly, its chance for success is high. One of the most serious staffing problems likely to be encountered is inadequacy, either in numbers or in skills.

While the most appropriate approach to GIS organization varies from project to project, certain issues must be addressed in almost every case. Key determinants of the GIS organization are the policies and structure of the overall organization in which it is implemented. Will the GIS be operated within a single organization or shared? If multiple organizations are involved, will the GIS require an interagency structure of some type? Will overall management be assigned to a single organization operating on behalf of all participants, or to a policy board of representatives from the participating organizations? In either case, an executive manager is needed to coordinate the activities of the participants.

The organizational structure may be centralized, with one body providing GIS services to the participants, or it may be distributed, with each participant operating independently of the others. If the participants share data or resources, standards for a compatible definition and structure of the data base and compatibility among the hardware devices and software components of the system must be established by the GIS organization.

The GIS organizational structure defines the authority, responsibility, chain of command, staffing level, and other management characteristics. Defining these clearly is sometimes difficult in a multiorganizational environment, and written agreements between the participants usually are needed to clarify roles and responsibilities. Planning and establishing the organizational structure usually involves the highest levels of management in the participating organizations.

In a decentralized, multiorganizational project, the individual participants develop internal GIS organizations under the overall structure.

Within an individual organization, the GIS responsibility may be assigned to an existing unit, a new GIS unit may be created and placed in an existing department, or an independent GIS department or major unit may be created.

Personnel are designated to perform several specific jobs or functions, including implementation project manager, system manager, and data base administrator. In addition, a GIS organization needs system analysts, programmers, and operators.

Whether shared or operated by a single organization, the GIS requires an implementation project manager responsible for all aspects of the implementation. This person manages the acquisition of the system, development of the data base, assignment and training of staff, development of application programs, and preparation for use of the system.

The system manager is responsible for day-to-day management of all system components and operations, and needs experience and knowledge in all aspects of GIS technology and operation. The system manager supervises the core GIS staff, assists with and coordinates use of the system, and serves as the liaison with the system vendor. The tasks of system manager and implementation project manager may be divided between one or two persons, depending on the scope and magnitude of the project and the operating environment.

The data base administrator manages the resources, quality, maintenance, and use of the data base. This person defines the data base and its component elements, oversees its creation, establishes and monitors quality standards, authorizes and monitors access rights to the data base, and develops procedures for continual updates.

The skills of a system analyst and pro-grammer are required for development and maintenance of application programs, system support, and problem resolution. The complexity of the system, number of users, and sophistication of the applications dictate the number and level of skills of analysts and programmers needed for a specific GIS. The person(s) must have knowledge of the appropriate programming skills, including the data base manager, macro or procedure language, and FORTRAN or another higher-order language provided with the GIS. Experience with the specific GIS software is desirable, but experienced persons are difficult to recruit. It may be necessary to train a person who has the basic skills.

Operators of the system—those who digitize and produce maps, retrieve data, and perform analyses—are spread throughout the user organizations, in most cases. System operators are drafters, engineers, planners, and other employees trained to use the system as it relates to their jobs.

GIS training takes place at multiple levels. The GIS manager and key support staff receive the most intense training, but a wide range of representatives from user organizations also are trained in the system's use and operation.

The most basic training is provided by the system vendor to those who operate workstations. Since the GIS requires special skills typically not available in an organization, additional training for selected persons is necessary. The types of training generally include GIS system management, system software programming, and general GIS concepts. Managers throughout the organization receive training in the applications of GIS technology and its management effects and opportunities. This training may be provided by the system vendor or others, such as a GIS consulting organization, or a combination of both. In addition, the staff should be encouraged to attend GIS workshops, courses, conferences, and other events sponsored by professional societies, universities, or commercial firms.

One of the most common problems is

In the frantic pace of system implementation, small changes in direction, the necessity of addressing immediate and short-term issues, and the press of external demands frequently result in cumulatively large departures from plan.

lack of adequate staffing. Too often, organizations focus on the potential staff efficiencies that will follow implementation and fail to recognize the need for continued staffing of current manual operations and the additional requirements of GIS implementation. In most cases, additional staff are required for a period of time to support the implementation. System operation requires assigned staff resources.

Step 9—Operating Procedure Preparation. The usefulness of GIS to an organization depends on effective operating procedures that are carefully designed and embedded in the activities of the organization. Procedures cover the operation and management of the system, maintenance, update, and administration of the data base, application and use in each participating organization, and production of the standard and special products of the system.

Agreement of all user organizations on an overall structure for system operations is necessary. This structure ensures compatibility among the procedures of the individual units and minimizes redundant activities. The structure guides the flow of initially captured data to the GIS and its data base. It also governs the basic methods by which units use the system to retrieve and manipulate data and produce displays, maps, and reports. It defines the relationships between the system and the end users of its products and capabilities.

Based on this overall structure, the individual operating units design their respective procedures. These procedures describe how to acquire or route data, access and operate the GIS, display or generate maps and reports, deliver products to end users, and use the products in specific operations.

Procedures for use of GIS in functional activities focus on the effectiveness of those activities rather than on the GIS itself. To be effective, the GIS should be an integral part of the operating activities of each unit rather than an appendage that must be dealt with separately.

Significant major operations may need to be designed and documented; other activities will require less formality. For data updates, the procedures elucidate interaction with clients, citizens, or field measurement, data to be acquired, format and characteristics of the data, verification techniques, recording methods, movement of source data to the GIS, entry and quality control, final posting, modification of existing data, and archiving.

Before installation and operation (Step 11) of the system, the organization designs and establishes system management procedures. These procedures govern all aspects of the system's operation. The system management procedures cover activities such as:

- Day-to-day operation of the system components
- Maintenance of the hardware devices and software
- Monitoring of system utilization
- Problem resolution
- Backup of data base and software
- Allocating and authorizing use of system devices and system access
- Management of hardware and software maintenance contracts
- Support for system users
- Authorization of application development and installation
- Upgrading of hardware and software

A set of records describing the system components and procedures for their maintenance is established during installation and before initiation of operations. The records identify components, describe characteristics, and list all problems and maintenance activities over the life of each device.

Procedures address both the GIS operations and their accompanying functional activities. Major procedures are designed through careful analysis (again by a group with adequate expertise) of the function's requirements, the operating environment, and the system capabilities.

GIS application programs are developed as described in Step 14 to support the operating procedures. These two design activities are closely coordinated to ensure integration of the applications into the procedures.

Step 10—Site Preparation. Although electronic equipment requires fewer environmental controls than ever before, it is still necessary to prepare for the installation of GIS hardware devices and the communications network. This step includes selection of optimal locations for the individual devices, allocation of adequate space for all components, preparation of detailed plans for the space, construction of necessary facilities, and installation of an adequate power supply, environmental controls, and communications lines and devices.

The activities of this step are changing as GIS hardware systems change. In the past, virtually all GIS configurations required the construction of an environmentally controlled facility for certain hardware devices. This is still the case with some configurations. Large mainframe and some minicomputer systems require (or operate more efficiently with) an air-conditioned, humidity-controlled, raised-floor environment. Plotters, printers, and disk units often are placed in such a special facility to minimize their impact on the office environment and to control operating conditions.

The typical workstation requires space comparable to what an office desk or drafting table occupies, but other necessary devices require larger accommodations. During the transition from manual to automated operations, space is required for the new GIS equipment in addition to the manual operations that have not yet been phased out.

Another consideration is optimal placement of the system's hardware devices. Some organizations require multiple workstations spread throughout the organization, while others concentrate stations at a single location or share them among multiple users.

While most hardware devices today operate on standard office power, the availability of adequate electricity must be determined prior to installation. Organizations that overlook that seemingly simple detail encounter serious problems. A power upgrade for a building can be time-consuming and costly, but it may be required if the current usage approaches capacity. Site preparation can be complex, involving separate contractors, permits, permissions from building managers, and other logistical arrangements.

A plan is prepared to determine the optimal placement of all devices and necessary support facilities, including wiring and communication devices. The plan is based on the detailed system design configuration, the requirements of each system user, and the physical characteristics of the various devices. Office layouts with GIS devices and other required furniture and equipment, wiring diagrams for both power and communications, construction drawings, and technical descriptions of any materials and equipment to be acquired are included.

Site preparation may include contracts with separate firms for tasks such as moving walls, constructing environmentally controlled areas, and installing heating, ventilation, air conditioning, wiring, or fixtures. This may require planning and scheduling a separate procurement. If construction, wiring, or other significant changes are necessary, permits and permissions from the local building code organization and the building's owner or manager may be required.

Operation

Step 11—System Installation. This step includes the delivery, installation, and initial operation of the GIS hardware and software configuration (Fig. 10-6). An installation plan based on information from the design, acquisition specifications, and contract documents guides all activities.

Installation is conducted under the direction of the organization's system manager. Other participants in installation often include the vendor's project manager and staff, the GIS staff and user organization staff, a communications firm, and technical consultants.

An installation team is designated and tasks assigned as appropriate. The plan for installation is prepared in coordination with the vendor during or immediately following negotiations for the acquisition

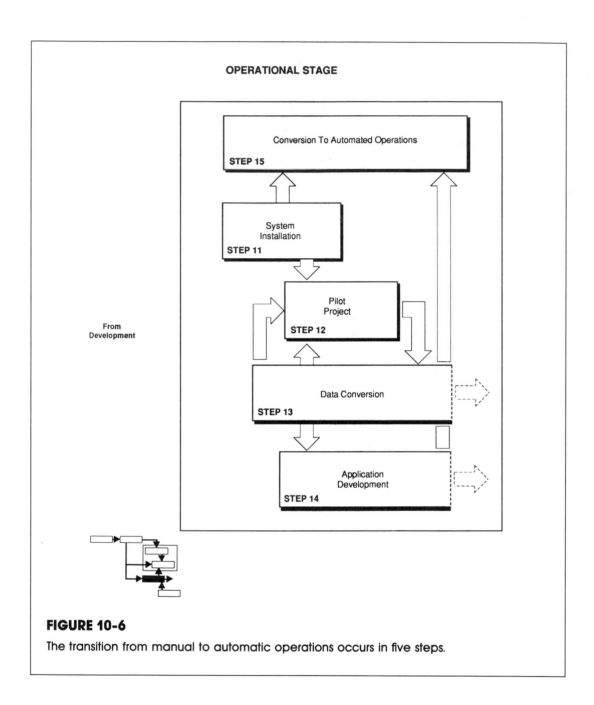

FIGURE 10-6

The transition from manual to automatic operations occurs in five steps.

contract. Preparations for installation are made in the period between contract execution and delivery.

The vendor typically tests all system components when delivered to the customer site. Installation includes cabling of components and the communication network if one is involved. In addition, the software components are loaded into the appropriate processors.

When all components have been installed, a series of tests is conducted. Initially, a site test is performed to verify successful installation and operation of all hardware and software components. These tests are performed by the system vendor and are observed and verified by the purchaser. The purchaser also may have the functionality of specific system components and their ability to meet contract requirements tested by either the vendor or in-house staff. If these tests are completed successfully, the system is accepted and payment is authorized. In many GIS installations, not all hardware components may be available or operating properly, or software may not be fully operable at the time of the tests. In these cases, it is necessary to reach an agreement on a revised operation date, and, if appropriate, penalties that apply. Training for staff in system operation often is offered by the system vendor during the installation period.

Technical problems are often encountered during installation and testing of a system, but most are resolved with minimal difficulty by the installation team. Problems also may develop between vendor and purchaser if tests are not completed successfully. To minimize disagreements over test results, it is important to clearly define all aspects of system capability and capacity in contract specifications and to agree to the tests and criteria for verification of acceptance prior to system delivery.

Step 12—Pilot Project. Because of the magnitude of a GIS project, many organizations begin operations with a pilot project. Several approaches may be taken, depending on the requirements and resources of the organization. Objectives of a pilot vary and are undertaken to do the following:

- Verify the operability and usefulness of GIS technology
- Verify the costs and benefits estimated in the feasibility study
- Create a data base for system selection benchmark testing
- Build prototypical databases that assist the organization in identifying the format and structure requirements of the selected GIS software
- Verify procedures for data base development
- Provide hands-on training for organization staff

Numerous other objectives may be identified. In addition, a pilot study can be useful at several stages of system implementation.

In some cases, a contractor is retained to conduct the pilot or to provide a pilot system and data base before the organization purchases a system. In most cases, a pilot is conducted by the organization itself following acquisition, sometimes with the assistance of vendors or consultants.

Undertaking pilots was once an obligatory step to establish the credibility and performance of software and hardware products. Pilots are now employed most frequently as a vehicle to assess the technical, procedural, and cost implications of data base development, and, secondarily, to establish that critical applications can be accommodated.

A plan for the pilot project defines its objectives, identifies tasks, assigns responsibilities, defines the pilot area and data base contents, and specifies the tests and evaluation criteria.

A small geographic area representative of typical or critical conditions is defined. If an objective is verification and refinement of data base procedures, all or most of the data is entered for the pilot area. If this is not an objective, a subset of the data may provide an adequate sample for the evaluation.

A series of tests or exercises is planned for the pilot; these may include digitizing and entry of data, map production, simple applications, and data retrieval and analysis functions. The tests are selected to verify crucial capabilities and to give prospective users an opportunity to test and observe the system in operation. Some applications may be fully implemented for the pilot; others may be implemented only partially or simulated, depending on the time and resources available.

If a pilot is conducted before system acquisition and a contractor provides the hardware and software, the organization should decide whether the contractor must provide a specific vendor's system. This decision may influence the selection of the contractor. If a specific system is required for the pilot, the qualifications of the pilot contractor and the proposed system must be carefully evaluated. Alternatively, a decision could keep the pilot separate from final software or hardware selection. In either case, the GIS used in the pilot must be able to provide a valid test.

Development of the data base for the pilot includes:

- Digitizing
- Quality control and spatial registration of multiple sets of map features
- Development of the selected application programs
- Preparation and execution of the planned tests
- Evaluation of the experiences, procedures, costs, and other aspects of the pilot

Following the evaluation, system and data base designs, data base development procedures, and decisions on implementation may be revised to respond to the pilot findings.

The pilot requires staff sufficiently knowledgeable to develop applications, operate the system, and conduct the tests. Since the pilot frequently takes place in the early stage of the GIS implementation before the in-house staff has obtained adequate experience, organizations may retain a consultant to conduct some aspects of the pilot. If the pilot is conducted by a contractor or with consultant assistance, the purchaser should participate actively to ensure the necessary organizational involvement and to provide practical, realistic tests of the system capabilities. Also, the pilot serves as an important educational experience for the organization's staff. Its outcome typically leads to design refinements rather than drastic changes in course, particularly if the pilot is a component of a thoughtful approach to system implementation.

The duration of a pilot depends on the magnitude of the GIS project and the extent of the testing. The pilot should, however, be completed within a reasonable time to allow adjustments in the implementation project and to avoid loss of momentum. Three to six months is reasonable for many GIS pilot projects.

Technical problems can be anticipated as the pilot is conducted; their identification is one of the major purposes of the pilot. Problems with staffing and organizational coordination also may arise. If contractors are involved, their contracts should be sufficiently flexible to allow adaptation to changing conditions and preliminary findings.

Step 13—Data Conversion. The creation of the data base is the most time-consuming and costly aspect of the GIS implementation. A detailed plan for the data base development was prepared as part of Step 5. The plan, in addition to the specifications for the data, describes the order of data entry in terms of both geographic areas and map features or layers, the allocation of assignments between contractor(s) and internal staff, and the mechanisms for combining various sets into the integrated data base.

The GIS data base usually is developed through the combined efforts of in-house staff and a contractor. The data base specifications are defined in terms of the system's data management software. As digital data become available, they are loaded into the GIS.

The loading begins with a rigorous quality-control effort. The quality-control procedure evaluates the contents, completeness, positional accuracy, logical consistency, proper definition (e.g., separation into correct layers or types), and other quality characteristics of the data.

As data quality is verified, data are entered into the permanent data base. Subsequently the data base directory and control mechanisms are updated with the characteristics of the loaded data. Any required logical relationships between data base components are established in the loading procedures. A verification ensures that all data have been properly loaded. Upon successful loading, the data base manager certifies the data base for operational use, and data base maintenance procedures are initiated to provide updated map features and data as changes occur.

The initial loading often involves only a portion of the full GIS data base. The

> The creation of the data base is the most time-consuming and costly aspect of the GIS implementation. The GIS data base usually is developed through the combined efforts of in-house staff and a contractor. The data base specifications are defined in terms of the system's data management software.

initial data base may include a part of the geographic extent that will eventually be covered or a subset of the features that will eventually be included. The remainder of the data is loaded in phases as it becomes available from compilation and digitizing activities.

Data base loading is supervised by the data base administrator, who is responsible for the quality of the data and for loading the data in its proper structure. The data base administrator also maintains records on the data base's contents and quality characteristics. In the operational phase, the data base administrator manages the loading of various phases and components until the data base is complete, and then coordinates its maintenance.

Step 14—Applications Development. The GIS extends its efficiencies and benefits to the organization through its application. The designs of specific application programs are a logical outgrowth of their earlier definitions during the requirements analysis.

Generally, the applications support standard, often-used functions of the GIS. They may link together a series of tasks, retrieve or organize data, perform calculations or logical tests, among numerous other operations. These programs are produced using software and macro language tools provided with the GIS, or with a programming language such as FORTRAN or C.

Application programs are developed to support and enhance the efficiencies of numerous activities. Some are very simple programs that automate a few steps in one procedure. Others are complex, incorporating dialogue with the operator and performing data retrieval, computation, quality verification, or other processes. The application programs define parameters for data entry, data to be operated upon, processes to be performed, products (e.g., displays, plots, reports), interactions with the operator, and other functions.

The design of applications may be formal or informal, depending on the complexity of the application. In the most so-phisticated cases, a specific application design task is performed, including detailed requirements analysis, design, programming, testing, debugging, and documentation. In the simplest cases, an operator trained in a macro language prepares a program with little planning and no formal process.

Very sophisticated application programs may be acquired from the GIS vendor or third-party firms specializing in such services. Application programs also may be acquired from other organizations with the same type of GIS and similar functions. These may be obtained free, at low cost, or in exchange for locally prepared programs. User groups formed among those who have acquired a specific vendor's GIS also exchange application programs.

The system manager establishes procedures for the development, maintenance, and control of application programs. The manager controls application development to ensure that applications do not adversely affect operation of the system or data base, that they receive the widest practical use, and that they are maintained properly. Complex applications require detailed documentation for maintenance and may require separate documentation for the operators and users. Training of operators also may be necessary for certain applications.

A set of application programs often is produced in the period between selection of the GIS and its installation so that they are available for initial operations. GIS vendors may also provide applications with the delivery of the system. Following installation, additional applications are developed formally and informally by the support team and system users.

Step 15—Conversion to Automated Operations. In most cases, the GIS is implemented in phases over an extended period of time. Conversion of the current manual operations and activities and initiation of newly available automated operations will require months, if not years. While data base development is underway, manual maintenance of existing maps and day-today operations continues. Resources must be allocated between the data base development tasks, preparation of source materials, data conversion, quality control, and the continuation of current manual operations.

As components of the data base are completed, loaded, and verified, a transition is made to automated maintenance and operations. Because time will have passed since the source materials were committed to conversion, it will be necessary to bring all records up to date. At that point the maintenance will shift to the automated GIS, although a parallel manual operation may continue for some time until automated operations are proven satisfactory. Access controls to system resources and data base components are then initiated and allocated to authorized persons and organizations.

The transition to automated operations must ensure that no gaps or lapses occur in updating records, that no data enter the data base without proper verification, and that resources are not wasted during the transition effort. The transition may involve multiple organizations and persons, including the GIS manager and data base administrator, the new GIS maintenance and operations staff, the manual map maintenance and production staff, the data conversion contractor, and the GIS users and operators.

Audit

In the audit stage, an organization evaluates all aspects of its GIS and plans improvements and expansion.

Step 16—System Review. Things change. Even the best designs and implementations of GIS must remain flexible to shifting organizational mandates and structures, advancements in technology, introduction of new techniques, and personnel changes and additions. Moreover, in the frantic pace of system implementation, small changes in direction, the necessity of addressing immediate and short-term issues, and the press of external demands frequently result in cumulatively large departures from plan.

Periodic reviews, or audits, play an invaluable role in keeping a project on track (or bringing it back onto track). The reviews can be targeted to particular aspects of the GIS implementation, such as personnel, hardware, software, or application development. Or, a more comprehensive view can be taken, comparing each of the diverse range of components to the original plan and the current institutional and technical setting. The audit can take either a high-level, strategic perspective, or one that is more focused on operational issues. Periodic reviews of both types are essential to an audit program.

The methodological foundation of the audits is not unlike the steps taken to develop the initial detailed plans for system implementation. Typically, interviews with technical and managerial staff of the involved organizations are accompanied by a review of prior planning and implementation documents, goals and objectives, and budget and expenditure histories. These efforts are followed by an objective analysis of the project's current status relative to prior plans and the nature of the current institutional and technical setting. Problems, opportunities, and corresponding actions are identified.

The audit culminates with a revised statement of goals, policies, procedures, and actions to be taken within specific time frames and budgets. In those instances where the focus is heavily influenced by operational issues, a revised operational plan often results.

Care must be exercised to insure that the analysis will be objectively accomplished and viewed by others as such. In situations where organizational issues, both internal and external, are influential or where technical considerations require an "arm's-length" evaluation, the use of a third party to conduct the audit is frequently warranted.

Step 17—System Expansion. Audits to support system expansion definitions are a special case of the previously described technical and management audits. Software components of a GIS are ever-changing through new releases of a particular product and the introduction of new products by the industry. As the cost of equipment continues to decrease relative to increased capabilities, review of hardware requirements should take place with some regularity and formality. By doing so, a coherent path can be maintained irrespective of the momentum and hyperbole of new releases, trade shows, advertisements, and sales calls. Similarly, the release of new software from a primary or third-party vendor and the development of application software can best be guided by a regularized review.

In these instances, the methodology is limited to an assessment of contemporary and pending application requirements by organizational unit, identifying software and hardware functions, and proceeding to assess them in a manner comparable to those detailed earlier. The most significant variance is that the software and hardware platforms are likely to be more constrained by the installed system.

ELEMENTS OF SUCCESS AND FAILURE

SUCCESS		FAILURE
Rigorous	PLANNING	Run and Gun
Focused	REQUIREMENTS	Diffused
Realistic	APPRAISAL OF EFFORT	Unrealistic
Dedicated, Motivated Continuity	STAFFING	Turnover
Adequate Finance Plan	FUNDING	Inadequate, Conjecture
Thoughtful	TIME	Rushed or Prolonged
Balanced	EXPECTATIONS	Exaggerated

FIGURE 10-7

The successful and not so successful efforts of GIS implementation durng the last 20 years have contributed valuable insights to assist others.

ELEMENTS OF SUCCESS

Efforts to implement GIS over the past two decades, both successful and otherwise, have extended contemporary understanding of what contributes to success or failure (Fig. 10-7).

Expectations are important. If an organization's expectations of the level of effort, the need for planning, and systems costs are realistic, appropriate levels of resources are more likely to be available in a timely and predictable manner. Success is most likely when an organization's expectations of the system's utility are practical, the system is directed to priority applications, the time allotted for implementation is adequate, and sufficient numbers of experienced staff are hired early in the process.

Instilling realistic expectations within both senior managers and users requires understanding the balance between practical benefits and the resources needed to achieve them. Should unrealistic attitudes develop, even reasonable accomplishments will be measured against unrealizable expectations and the system will be perceived as unsuccessful.

SUGGESTED READINGS

Nyerges T. L. (1989, Fall). Information integration for multipurpose land information systems. *Journal of the Urban and Regional Information Systems Association 1:* 27–38.

Friddle C. M., Walker A. W. (1989, August). Involving nontechnical professionals in the GIS procurement and developments process: The Wake County perspective. *Geographic Information Systems,* Proceedings of the 1989 Annual Conference of the Urban and Regional Information Systems Association II: pp. 66–75.

Antenucci J. C. (1991, February). *Acquiring Technology.* Geo Info Systems, Eugene, Oregon, 44–45.

Foley M. E. (1990, April). Pacing change: Economical AM/FM/Base map maintenance. *Proceedings of AM/FM International Conference XIII.* Baltimore, MD, 121–132.

McFarland R. (1990, November). GIS implementation strategies for the small municipal environment. *Proceedings of the GIS/LIS '90.* Anaheim, California, 167–173.

Antenucci J. C. (1990, October). 20/20 hindsight or if I had to do it again. *Proceedings of the Association for Geographic Information.* Brighton, U.K.; 2.1.1–2.1.3.

Gittings B. (1989). Education and training—the missing link? *The Association for Geographic Information Yearbook 1989.* Taylor & Francis, London; 317–322.

Chapter 11

Legal Issues

Geographic information technology provides one of the most valuable decision support tools ever available, especially for government. The "recombinant" data opportunities provided by spatial analysis and relational database management systems have given GIS managers a resource of extraordinary economic value.

Resolving the issues of GIS data ownership, value, control, and access presents a formidable challenge within current legal and institutional frameworks.

Government agencies and public utilities—custodians of most of the country's maturing GISs—are "creatures of statute" or of constitutions written before the electronic age. The application of old rules to new technology results in an open field for creative interpretations, but the law lacks guidance for uniform public policy regarding the ownership and treatment of intangible information assets.

Analysis of the legal setting in which a GIS is established has become a necessary, practical step in the GIS planning process. Significant investment decisions often depend on resolution of conflicts over who owns and controls information. Such conflicts demand resolution to dissipate tensions among employees fearful of losing control of their information, to eliminate delays in implementation, realize substan-

tial cost savings, and win public support.

Traditional "islands of information" must be broken down to realize the benefits of GIS technology. One of the most important consequences of this technology is the level of interagency cooperation it requires to be useful and cost-effective. Information dissemination forces executives to require unprecedented levels of cooperation and communication between agencies that have historically remained independent.

The legal setting is closely akin to the complexity of the institutional setting. The more institutions that share information and information management resources, the more "distributed" a GIS is said to be. Expanded distribution improves the benefits-to-costs ratio of a GIS, but the institutional and legal issues increase in complexity in direct proportion to the degree of distribution as more laws, charters, preexisting procedures and standards, interagency agreements, goals, purposes, and personalities come into play.

Conflicts over access to and control of government information also are instigated by the private sector "value-added" information industry, which considers government information a public resource. The increased commercial utility of GIS is upsetting the public policy balance that

Resolving the issues of GIS data ownership, value, control, and access presents a formidable challenge within current legal and institutional frameworks.

237

Information wants to be free because it has become so cheap to distribute, copy, and recombine—too cheap to meter. It wants to be expensive because it can be immeasurably valuable to the recipient. That tension will not go away. It leads to endless, wrenching debate about price, copyright, "intellectual property," and the moral rightness of casual distribution, because each round of new devices makes the tension worse, not better.

Stewart Brand in *The Media Lab: Inventing the Future at MIT* (Viking, 1987, p. 202)

historically has sufficed to deal with public access to government information. The costs for the entire community of decision makers, public and private, may be less when data and information are gathered, analyzed, and made compatible in anticipation of need, rather than left to chance at the time decisions must be made. This "public good," combined with the fact that the GIS is built with tax dollars, supports a belief that all access should be free. However, free access to a GIS does not necessarily result in maximum public use and public benefits.

The technical and institutional issues imposed on the GIS custodial organizations are complex, increasingly so for public agencies. Statutes and regulations specify access procedures applicable to custodians of government records, but the law typically does not address the expanded roles of these agencies as information integrators and disseminators, made possible by GIS.

The evolution of GIS into an *information utility* that provides information products and services for standard rates (like a utility company) raises significant policy issues concerning the economics of information and expanded government authority to provide information services. How those legal issues affect GIS is discussed here, along with some of the policy directions lawmakers could take in dealing with the value of information and public access to government-owned information resources. This chapter proposes a progression in the law toward utilitarian use of GIS within public agencies, and reviews legal gaps that must be filled to allow reasonable control of government-owned systems while preserving the public's right to know. In addition to the open-records law, issues examined include the limits of implied proprietary authority and action by government, local government's use of copyright law protection, the role of sovereign immunity and potential exposure to liability for negligence and products liability, equal protection issues, and antitrust laws.

THE VALUE OF INFORMATION

Information is being transformed into a commodity—the result of technological advances in microelectronics, telecommunications, and satellite technology, as well as the development of huge corporate enterprises in these spheres.

The formative and abstract theory concerning the economics of information in the electronic age has not solidified sufficiently to provide a clear guide to GIS. Investigation and refinement of these ideas is critical to establishing the information utility that GIS technology makes possible. Information utilitarianism depends on an evolution in laws, policies, and theories of information economics (see below).

Although it is apparent that information assembled and managed by computer technology has significant value, it is difficult to apply this concept to specific cases. Defining the value of any particular information assemblage as a commodity or intangible asset depends on an acceptable accounting technique. Traditional economic theory has dealt poorly with intellectual property (e.g., patents, copyrights, trademarks, and trade secrets), and it does not begin to deal with intangible assets that have not had the benefit of legal definition.

The Information Industry

The information sector of the economy is outpacing manufacturing, agriculture, and even the service industry. More than half of the American labor force works in the information industry, under a broad definition.

Much of the growth is related to the low capital cost of taking relatively free government information and creating and marketing "value-added" information products. While companies that market repackaged or enhanced information benefit from access to electronic formats, some of these firms are concerned about possible adverse effects from government competition. Industry representatives have realized that the best recourse for them is to support government dissemination of information in raw electronic form without software enhancements or searching aids, but to oppose government dissemination of enhanced or value-added information.

Peter Marx, a director of the Information Industry Association (IIA), testified to Congress in 1985 that more than 400 "information service organizations" in the private sector repackage raw government information. The IIA promotes an open-access policy for raw data, and advocates that all direct beneficiaries of access should bear their fair share of operating and fixed costs. The IIA suggests a mix of filing fees, appropriations, and user charges for funding the agencies, while issuing a warning to agencies that get too independent when their own funding sources allow them to avoid or diminish the traditional purse-string control of their governing legislative body.

More than 850 companies—mostly publishers and database vendors—are members of the IIA, which has lobbied to retain and expand the role of the private sector in providing government information. "The role of the private sector information enterprises is to complement and extend the official distribution system by responding to the public need for sophisticated tools for acquisition, storage, retrieval, analysis, and processing of government information," according to IIA member Paul P. Massa, president of Congressional Information Services, of Bethesda, MD. Robert Simons, DIALOG's general counsel agreed. "It's not a pleasant thing to compete with the federal government," Simons said.

Totally free access or access to the software to analyze and use the raw data would remove the large corporations' advantage in their bid to control the value of information. "Actual-value" cost recovery schemes might cut too deeply into their expected profit margins. They are even concerned that nominal fees established for access not be used to cross-subsidize free or below-cost services to the academic or "needy" users that might be established in the name of equality of access rights.

What was once free for public consumption is increasingly available only to those who can pay. Public institutions are allowing themselves to be restricted in the amount and nature of the information they disseminate as the industry pushes to consolidate its present hold on the information commodity, with a view towards protecting its future expansion.

In spite of federal lobbying efforts, an information marketplace at the state and local levels of government is also coming into existence independently through cost-recovery programs from the sale of information products and services now available from government-owned and -maintained data bases. The evolution toward distributed geographic information systems for cities, counties, and state agencies is forcing renewed consideration of the limits of public access for commercial purposes.

Information Economics

Information management technology has progressed so that creation of wealth increasingly has more to do with information than with the material manipulations in manufacturing. A corollary observation might be that the ability to access taxpayer-financed automated information, particularly land information systems and geographic information systems, is becoming ever more attractive. Should the taxpayer subsidize commercial use of GIS? Should the custodial agency be left without choices between performing its primary statutory mandates and providing commercial vendors with access? Will state open-records policies permit reasonable control by the custodial agency over the financing, timing, and policy effects of meeting the information requests that a mature GIS will generate?

Public goods (in the sense of "goods and services") to an economist have essentially no marginal cost associated with

I think it is likely that the United States will see a vigorous and sustained conflict between governments and private enterprise concerning who owns, controls, distributes, and establishes the value of information. Information leads to knowledge, which has always meant power, wealth, and ultimately, control.

Jack Simpson, President
Mead Data Central

adding distribution. Once the fixed costs of building a GIS have been incurred and the defined applications are utilized by the sponsoring agency, there is only a relatively small additional expense to perform the application for other users as well. The cost–benefit ratio improves with distributed systems, where use of the system and maintenance of the data base are widely shared.

The real costs of providing a GIS product or service are found in the building and maintenance of the database that supports the particular system application or capability, not in the time and materials needed to respond to an access request. If the value of information can be defined and the capital costs of GIS recovered by the system, then the highest cost–benefit ratio can be realized without bankrupting the custodial agency. Without cost recovery for the actual costs of providing the service, the custodial agency cannot afford the progression towards the more efficient distributed structure serving broad segments of the community.

How do economics change when the substance is symbols rather than physical objects? Space and time mean very different things when moving a graphic image and an attribute file rather than tons of steel. There is no definitive work on the economics of information, although it is often stated that there is a Nobel Prize waiting for the first person to fill the gap.

Among traditional concepts, the treatment of intangible assets versus intellectual property is critical.

Intangible Assets. An intangible asset—such as an assembled and trained work force—is created by a business enterprise in order to carry out daily activities. It exists because of the recruiting, hiring, training, and retention activities carried on by the enterprise. Other intangible assets include advertising programs, computer software, licenses, certifications, distributor networks, training materials, parts annuities, customer relationships, advertiser base, and favorable contracts.

Intellectual Property. Intellectual property such as a patent or trademark acquires its essential characteristics, from which value emanates, from the legal system. The property may have resulted from a long and expensive research program, or by fortuitous discovery. Intellectual property includes patents, copyrights, trademarks, proprietary technology, and trade secrets (Smith and Parr 1989).

Good Will. The search for a rule on how to treat information as a commodity with value leads us to the treatment of "good will" as an asset by the business community and the tax laws. Generally Accepted Accounting Principles (GAAP) have compromised the treatment of intangible good will for different situations. Accountants bifurcate the handling of good will, recognizing it in the case of a business acquisition, but leaving it unrecognized in the case of ongoing operations of a business.

The GAAP considers four characteristics important to the recognition of intangible assets on a balance sheet: identifiability, manner of acquisition, determinate life, and transferability. Under tax statutes, good will and any other nonidentifiable intangibles cannot be amortized against taxable income for tax deduction purposes, and buyers want as much as possible of the purchase price allocated to identifiable assets that can yield deductions.

A major problem with the evaluation of intellectual property that has no market-determined value exists with current accounting practices. Good will does not meet the above four tests and possibly should not be treated differently than it is now. However, the identifiable intangible asset (e.g., intellectual property such as a database that is subject to copyright) that has no market-determined value is the broadest gap requiring correction. "As with other types of intangible property, not all intellectual property has value. Its value is usually determined by the marketplace, either directly or indirectly, state Smith and Parr (1989, pp. 89–90).

The fact remains that when information is sold, the seller still has the information as well as the buyer. There is no reduction in capital assets from the seller's perspective, so traditional analysis of movable goods does not apply. There is no automatic protection available if the original parties resell the information to a third party in the absence of a contract or copyright. Cases in which the courts have grappled with such slippery concepts are inconclusive. The application of copyright-law protection for software has been established, but the specifics of such applications fill the courts today with litigants and inconsistent theories. The application of copyright protection to a database also has been established, but the actual methods used to protect such "compilations" of facts are subject to further analysis by the courts, especially since most databases are dynamic, with regular updates, and hard to define as a "work" at any particular point in time.

The scope of both software and database copyright protection will be profoundly effected by a landmark case decided by the Massachusetts federal District Court decision in June, 1990. *Lotus Development Corp. v. Paperback Software International and Stephenson Software Ltd.*, Civ. 88-1666K, includes a 113-page opinion by U.S. District Judge Robert E. Keeton, which not only decisively establishes liability for copying the nonliteral (noncode) elements of a computer program, it also serves as a treatise on software copyright protection. Judge Keeton formulated a three-part legal test for determining the scope of copyrightability of nonliteral elements of computer software. The test is by no means a "bright-line," but requires a formulation of the definition of the expression of idea in question to be made "along the scale from the most generalized conception to the most particularized." A determination must be made whether the identifiable elements of the expression are essential to every expression of the idea. And finally, a determination must be made whether such nonessential elements

of the expression are a qualitatively substantial part of the work in question. For the immediate future, questions of copyright will remain well within the discretion of individual courts which have been provided a methodology for analysis, but no clear standards.

Rate setting in the public and private utility world provides a model for society to deal with the valuation and distribution of public goods that do not automatically break down into unit values and quantifiable shipping charges. So long as it is clear that information in a government-owned database has value over and above its existence as public records, a policy and legal policy setting can be established to support broad GIS use. Although increasing recognition of the value of information will continue to generate many philosophical quandaries in such disciplines as law and economics (with their traditional and often inflexible rules and language), the immediate task is to evolve from the isolated GIS custodial agency of today into the community-wide information utility of the future.

Definitions of GIS "products and services" as intellectual property distinct from "records" must be formed, and access to such products and services tied to a fee. Maintaining the proper distinctions between the many constituent parts of "information," such as goods, records, data, programs, products, and services, is critical.

> Maintaining the proper distinctions between many constituent parts of "information," such as goods, records, data, programs, products, and services, is critical.

GOVERNMENT'S ROLE IN INFORMATION DISSEMINATION

Public access to government information is a fundamental right that operates to check and balance the actions of elected and appointed officials. Judges speaking for the United States Supreme Court and almost every superior state court in the United States have established a clear connection between the "right to know" and governmental accountability. When tax dollars are spent, or government takes action, citizens can review the information on which the decision was based. "The basic purpose of [the] Freedom of Information Act is to ensure an informed citizenry, vital to the functioning of a democratic society, needed to check against corruption and hold the governors accountable to the governed," the United States Supreme Court said in a 1978 case (*NLRB v. Robbins Tire & Rubber*).

The right to know the basis for the government's actions and decisions provides a deterrent to abuse of authority, but this general concept does not address myriad information access issues. The Office of Technology Assessment (OTA) reported in 1988 that the federal government spends, conservatively, $6 billion per year on information dissemination (not including the cost of collection, processing, or agency automation). Information is the lifeblood of state and federal government programs and activities, essential to accomplishment of agency missions as well as to informed public debate. Congress has enacted hundreds of specific laws that assign information dissemination and related functions to federal agencies.

Recognizing that the law has not kept up with technology, the OTA said congressional action is urgently needed to resolve federal information dissemination issues and set policy direction for future federal agency activities (OTA Report, 1988).

Increasing use of GIS at all levels of government makes the need to resolve information dissemination issues universal. Through information management technology, government has the opportunity not only to improve its productivity and cost-effectiveness, but also to empower individuals by increasing their access to information through the capabilities of GIS.

Open-Records Law and GIS Access

The information processing capabilities of GIS make it wise to anticipate an unprecedented number and variety of access requests soon to come from sister agencies, commercial vendors, and the general public as applications expand and more become aware of the resource. Most GISs will have the capacity and ability to produce a much more complex set of information products than was contemplated when the public records policies were set. Open-records laws deal quite adequately with a nominal number of requests made of a typical government agency, but have a different impact when applied to several hundred requests a month, including on-line access and complete data base updates.

The law does not change quickly by design, and it often fails to keep up with the dramatic changes brought by modern technology. This is evident in a review of open-records law and its impact on access to computer-managed information. The most significant effect on GIS access policy is embodied in the state and federal freedom of information statutes and case law.

Traditional rulings, statutory formulations, and current trends in this area are dominated by a struggle to balance privacy versus the right to know. The legal precedents direct custodial agencies to release or withhold records in consideration of this balance. The connection between accountability and the FOIA was more recently strongly confirmed in *U.S. Department of Justice v. Reporters Committee for Freedom of the Press* 489 US—, 103 L Ed 2d 774, 109 S Ct—(March 1989). ". . . FOIA's central purpose is to insure that the Federal Government's activities be opened to public scrutiny, not that information about private citizens that happens to be in the Federal Government's warehouse be so disclosed." The case recognizes distinctions between computerized data banks and the source records themselves, and discusses the possibility that departing from the original purposes of the FOIA could threaten to convert the federal government into a clearinghouse for personal information that had been collected about millions of persons under a variety of different situations. The limited decisions that have examined the universe of information encompassed by statutory definitions of "public records" do not pro-

Through information management technology, government has the opportunity not only to improve its productivity and cost-effectiveness, but also to empower individuals by increasing their access to information through the capabilities of GIS.

vide adequate guidance to assure a smooth transition to "information utilitarianism"—the greatest access to information for the greatest number.

Unfortunately, the impact of dated laws cannot be ignored. If distinctions between "records" and "products" are necessary, the burden of showing the proper distinction will fall on the custodial agency. In attempting to define GIS, D. J. Cowen (1987) writes, "The basic premise is that a true GIS can be distinguished from other systems through its capacity to conduct spatial searches and overlays that actually generate new information." The open-records law was not written to deal with the issues raised by this new tool. A sufficiently discriminating policy must be designed and implemented now before the wrong precedents are established either by courts or overly conservative managers.

Federal Information Dissemination

Technology has outpaced the existing statutory and institutional framework established by federal lawmakers largely during the preelectronic era. The Printing Act of 1895, Depository Library Act of 1962, and Freedom of Information Act (FOIA) of 1966 predate the era of electronic dissemination, and they have not been updated to explicitly reflect electronic formats. The Paperwork Reduction Act of 1980 was amended in 1986 to include information dissemination within its scope, but substantive statutory guidance on electronic information dissemination is minimal.

The status of computer software (including indexes, directories, and operating programs and codes) under the FOIA is uncertain, and few federal agencies mention software in their regulations. Agency practice is inconsistent, varying with the function of the software, its commercial potential, and general agency attitudes toward openness. No legal cases clearly address the issue of what classes of software constitute agency records. Some agencies have suggested that software is a *tool* used to manipulate information rather than a

record, while others relinquish software products when requests are perceived to be reasonable (OTA Report, 1988).

In *Yeager v. Drug Enforcement Agency* (1982), the Washington, D.C. Circuit Court examined whether an agency can refuse to deliver the code necessary to use a magnetic tape under exceptions to the FOIA. The court ruled that if the code is necessary to read magnetic tapes, the code would become more than trivial information, meeting the exception for intra-agency records. Using exceptions under the FOIA to refuse to release computer programs has not proven a successful strategy, and , if successful, would prove a two-edged sword. If the code is considered a record and meets an exception under the FOIA, the agency not only protects itself from unwanted releases of programs, it limits its options to release programs at all.

A more expansive view towards delivery of program code as records emerged in a 1982 decision by a Florida appellate court, where computer codes were compared to instructions accompanying a written document. In *Seigle v. Barry* (1982), the court said, "Information stored in a computer is analogous to information recorded in code. Where a public record is maintained in such a manner that it can only be interpreted by the use of a code, then the code book must be furnished to the applicant." As a result, if the public record requires a code book to be understood, then the code owner should be required to apply it to the database and extract the appropriate amount of information needed to produce the identifiable record that is the subject of the request.

The purposes of the FOIA are not further served by expanding the records availability requirement to extend to the delivery of intellectual property whether produced by the agency or purchased from a vendor. This exceeds any needs for accountability and effectively requires a subsidy of any commercial user that demands the software involved.

Limiting the format and scope of records requests in this way would protect public access and accountability but avoid subsidizing commercial information enterprises. It would require government agencies to draw a line between records as an identifiable segment of the raw database, and the *information delivery system*, which consists of the entire database and software. For these nonrecords (i.e., when the requestor wants the entire potential of the data base, but does not have a particular need for any identifiable segment), the agency should have the option either to exercise proprietary authority and charge user fees for defined products and services or to stay out of the services business and deny broad requests for the code or the entire database.

The Office of Technology Assessment (OTA) reported to Congress in 1988 that preexisting data can be demanded under the FOIA, but "further analysis" of data cannot. The distinction is blurred in the case of computer records.

Most federal agency regulations avoid these issues. Department of Defense regulations on fees and fee waivers adopted pursuant to the FOIA Reform Act (1986) specify that computer software is not a record ". . . unless created or used as primary sources of information about organizations, policies, functions, decisions, or procedures of a [Department of Defense] component." This definition expressly does not include the "underlying data." However, the release of information stored "within a computer for which there is no existing computer program or printout" is not required by the FOIA. The Department of Defense definition is too imprecise and limited to protect accountability, especially since much of the information sought under public information requests is printed only on demand.

Another issue is the extent to which fees may be charged for electronic information. "Given the trend toward cost recovery for federal agency information products, it seems likely that user fees will continue to help support federal on-line database delivery systems," the OTA said in its 1988 report.

One of the most interesting judicial ex-

aminations of fees vs. the right to know established by FOIA is the federal case *SDC Development Corporation v. Mathews* (1976). A private firm wanted to obtain the MEDLARS tapes (a database including health and medical research information and bibliography) for the nominal cost of search and copy directly from the National Library of Medicine rather than pay the National Technical Information Services (NTIS) $50,000 for the tape plus $50,000 per year for updates. The NTIS, part of the Department of Commerce, historically has served as the federal government's archive and clearinghouse for scientific and technical reports. NTIS maintains more than 1,000 databases and 2 million titles for sale, but SDC wanted to use the FOIA to establish a less-expensive route to MEDLARS. Based on a review of the National Library of Medicine Act and the Technical Information Act, the court found legislative intent for cost recovery and held that the database was not a public record. The Justice Department analysts have subsequently classified this as a FOIA displacement case. That is, it should be read as a precedent as to FOIA access generally, but the result depended on the specific cost recovery intent in the NTIS and NML's underlying legislation.

Dismukes v. Department of the Interior (1984), a more recent companion case, is very instructive. It cited the following language from *SDC Development Corporation v. Mathews:* "Congress specifically mandated the agency to prepare this system and hold it as stock in trade for sale to the public. As such the system constitutes a highly valuable commodity. Requiring the agency to make its delivery system available to the appellants at nominal charge would not enhance the information gathering and dissemination function of the agency, but rather would hamper it substantially. Contractual relationships with various organizations, designed to increase the agency's ability to acquire and catalog medical information, would be destroyed if the tapes could be obtained essentially for free. . . . The agency is seeking to protect not its information but rather its system for delivering that information. . . . MEDLARS information does not directly relate to the structure, operation, or decision-making functions of the agency, and where, as here, the materials are readily disseminated to the public by the agency, the danger of agency secrecy which Congress sought to alleviate is not a consideration."

The issues seem to vary depending on whether the object of a records request is the information delivery system, the database, or a computer program. Many agencies are guided by their view of programs as either tools or records. A fair distinction must be drawn between programs required to interpret records and programs that further analyze or manipulate data. The former need to be released, while the latter should remain subject to agency discretion. Some programs incorporate instructions that reveal agency decision-making techniques or information-gathering methods and may constitute records in their own right.

State and Local Information Dissemination

While the federal government long ago established its role as a major source and disseminator of information, this function traditionally has not been viewed as a primary responsibility of most state and local governments.

Compared to federal law, state laws are even more inconsistent and inappropriate in establishing public policy that promotes use of and access to automated geographic information. The variously labeled open-record laws, public record laws, and state freedom of information laws lie at the heart of the problem. State laws are being regularly amended to deal with particular issues raised by computers, but no model (outside of recent legislation passed in Alaska which has not been tested by time) exists that is adequate to handle the flexible power and commercial utility represented by GIS.

Almost all state open-records laws are based on the federal Freedom of Informa-

tion Act, but no two states have established exactly the same response to issues that arise from information management technologies. State laws generally govern the public access policies of local jurisdictions, but some local bodies govern the access policies of local government agencies within state-established guidelines.

However, in most states and local governments, requests for information are being granted and denied by agencies based on inconsistent theories and limited understanding of the capabilities of the technology, and without the technical vocabulary necessary to make sensible distinctions.

In 1987, the Office of the Massachusetts Secretary of State, Public Records Division, sponsored a national survey, conference, and report entitled "Issues Concerning Computerized Public Records." Several trends can be identified from the survey and review of applicable state statutes.

Problems. A majority of states reported they had or anticipated having problems with computerized public records. Only Illinois and Wisconsin said they were satisfied with their current statutory formulations for dealing with access to computerized records. States without problems had not experienced a proliferation of computer systems.

Litigation and confusion are building in the area of public records access because of computers. Broad definitions of "records" and statutes mandating limited time for response to requests are causing unanticipated effects.

Disclosure of Computerized Public Records. Public records stored in computerized form are subject to disclosure. An agency cannot rescind access that once was available because of the expense, form, or mixture with exempt information that comes with the computerization of paper records. Most states expressly (with general language) or through court interpretation include computerized information within

BOX 11-1 States Update Data Access Laws to Accommodate GIS

Alaska, Iowa, Kentucky, and Oregon are among the first states to pass laws governing public access to GIS data bases, allowing public agency data sales or fees for cost recovery. Similar legislation is being considered in at least a dozen other states.

The Alaskan law is the most comprehensive revision yet undertaken to incorporate electronic information into a state "open-records" statute and to allow cost recovery through sales of electronic services and products. The Alaska legislation "provides a model for open records legislation in other states and marks Alaska as a policy leader on the issue of public access to computer information," wrote *Government Technology* (September, 1990).

The law, effective in September 1990, allows local and state public agencies to recover a reasonable portion of costs associated with building and maintaining GIS and other public computer systems. Electronic services and products for which fees can be charged include the following:

- Manipulating data to meet a specific request
- Duplicating data in alternative formats
- Providing periodic updates of an electronic file
- Duplicating a GIS data base
- Generating maps or products from a GIS
- Providing on-line access
- Providing information that cannot be retrieved or generated by existing programs
- Providing software developed by a public agency
- Providing functional access or software capability

Fees may be reduced or waived for a public purpose, including public agency program support, nonprofit activities, journalism, and academic research. Fee reductions and waivers must be uniformly applied to persons in similar situations.

With the exception of requiring distribution of an electronic version of legislative voting records, providing electronic services and products is discretionary and cannot take priority over an agency's primary responsibilities.

Local and state agencies in Alaska will offer electronic services and products in contracts that are required to include provisions to protect the security and integrity of the information system, and to limit the liability of the agency.

The legislation establishes an electronic bulletin board within the state library system to catalogue public agency data bases. State and municipal agencies must notify the library of the format options and fees for electronic services and products.

In addressing privacy concerns, the Alaska law requires agencies to tell people what will be done with the personal information they collect from them, and establishes a procedure for challenging the accuracy or completeness of personal information in public files.

The law also authorizes state agencies and municipalities to copyright software developed by or for a public agency.

Legislation passed in Iowa effective in July 1989 authorizes bonding as a funding source to assist cities and counties in building geographic data bases. Local governments are authorized to issue "essential corporate purpose bonds," which do not require voter approval, to acquire, develop, and improve electronic geographic data base systems suitable for automated mapping and facilities management. The legislation allows local governments to charge access fees for these electronic data bases.

As a result of the legislation, a GIS file does not have to be disclosed under the open-records law except under terms and conditions acceptable to the local government; local governments must set their own reasonable rates and access procedures.

The city of Des Moines and Polk County, the state's largest metropolitan area, sought the legislation.

"We convinced the legislature that maintaining an information system is an essential corporate purpose. Local governments are a major repository of information, and making that information available to the citizenry is a central purpose of government," said Steve Gast, Polk County's interim director of planning. "GIS is an effective tool for small counties as well as large urban areas, but we didn't have any way of funding them. Bonding was seen as a good alternative."

Des Moines and Polk County now are pursuing a pilot GIS project that, if successful, will lead to issuance of $8 to $10 million in essential corporate purpose bonds, probably in 1991, Gast said. He added that more than a dozen other Iowa cities and counties are investigating such bonds to build a GIS.

The legislation will encourage development of more multiparticipant projects in Iowa, Gast maintained. "Without amendment of the open-records law, you don't get private-sector involvement." he said.

Kentucky amended its open-records law, effective in July 1990, to permit a public agency to charge a fee when providing GIS and other electronic information for a commercial purpose, and to exclude from disclosure most electronic information, if sought for a commercial purpose. When providing the information in such a case, public agencies can recover a fee that is based on the value of the intended commercial purpose and the acquisition, development, and maintenance costs of the system. Journalism is not considered a commercial purpose.

Electronic public records requested for other purposes are available on paper or in a flat file electronic ASCII format, at a fee not to exceed the actual cost of copying, excluding personnel time. For an electrical record in any format other than ASCII that is not for a commercial purpose, the fee can include computer and personnel time and printing costs.

"Kentucky's open-records law was so vague that you could infer that GIS files were just another record," said Bruce Seigle, a manager for the Metropolitan Sewer District, which runs a multiagency GIS for local governments in Louisville and Jefferson County, Kentucky's largest urban community. "We initiated this change to protect our GIS from being hit upon by developers and others through the open-records law. . . . I think [the law's] impact will be substantial," he said.

In Oregon, special legislation passed in 1989 authorizes the Metropolitan Service District in Portland to charge reasonable fees based on market prices for its GIS information, and exempts GIS databases from the state's open-records requirements. The law is applicable only to the Metropolitan Service District and to information in a GIS. Fees collected must be used to maintain the system or to provide services to cities and counties. Before the change, Oregon's open-records law prohibited charging more than the "actual costs incurred" in making copies available.

"We're interested in creating robust information systems with everyone sharing the cost," said Richard Bolen of the Metropolitan Service District. "Although the law authorizes fees to be set on a market basis, we're setting prices based on cost-recovery levels. We want to encourage profitmaking ventures to use our information. We see ourselves as enablers to entrepreneurs."

the definition of public records subject to disclosure.

Computerized records are available on magnetic tapes and disks. This traditional limit on choice of format is one area where courts are making quick fixes in the absence of legislation that allows custodians to avoid providing an undue commercial subsidy. The Ohio Supreme Court has changed the old rule to allow format decisions to remain in the hands of government computer custodians, to frustrate demands for digital assessor's files which were inspired by commercial rather than accountability needs.

Even when computerized records are kept in addition to paper records, the custodian must provide a digital version if requested. Agencies generally impose a fee to cover the costs of providing the requested format, but recovery of costs is limited to assure that the custodian does not abuse purse-string control to defeat the access request.

Custodial discretion is frowned upon. Definitions of "records" and "open records" are very broad. Formerly common release limitations based on the purpose of the requestor (i.e., commercial purpose) or the form in which the record is noted (i.e., electronically) were disappearing, although some states are reestablishing "commercial intent" tests and distinctions in their records laws to deal with computerized record access burdens. Arizona, originally, and, more recently, Kentucky and Oregon have used the commercial intent of requestors to distinguish appropriate treatment of a response by the custodial agency.

The time frame for meeting a request is specified to assure that delays are not used to defeat the public's right to know. The time frame for judicial review of any effort to restrict access is often given priority over other types of cases.

Legislators and courts typically are provided some relief from the impact of broad mandates to release information, but local governments and most property appraisal officers are not.

Software. Computer software programs are not available as an absolute right, and requests for computer programs are most often either expressly exempt or left to the custodian's discretion for release. The bases for the distinction are: 1) Software "programs" are not information and have not been available traditionally; 2) they represent an investment by government; 3) their disclosure would subsidize commercial enterprises; 4) their release does not further the principles of freedom of information; and 5) they often are subject to proprietary license restrictions.

However, one Florida court held that ". . . a public record is anything made or received in connection with the agency's business that is intended to communicate knowledge." (*Shevin v. Byron* 1980). This does not exclude software, and may implicitly include program software.

Fees. Fees for access to computerized records are charged in some states, but in others no cost is allowable. Some laws allow "appropriate" fees, while others allow consideration of "fair commercial value" in establishing fees. Most open-records laws limit the recovery of costs for providing "records" to the "actual" cost of reproduction, with variations on whether this includes staff time. In jurisdictions that allow a fee for access, a wide range of specific cost factors are included in laws, regulations, or policies, often with different combinations within agencies in the same jurisdiction. Factors on which fees are based include the costs of the printout, computer time, disk or tape, program creation, clerical time, and staff time to search for the record once a certain number of hours is exceeded.

Who Decides? Special commissions or express assignments to individual agencies to act as a clearinghouse for access issues are rare. Typically, the individual custodial agency deals with its own requests, and disagreements go to the Attorney General or directly to court, depending on the administrative procedures established by the custodial agency.

Trends. State open-records programs likely will be subject to substantive and frequent amendments brought on by rapid changes in the media for accessing, manipulating, and storing records, as well as by the increasing public awareness of the potential uses for government information.

STRATEGIES FOR CHANGE

Agencies planning for a GIS must take an active role in molding the information access policy and not wait for the legislative bodies to catch up with the technological progress. Without active guidance, they will not allow time to deal efficiently with the many systems that are currently reaching maturity. Literal application of the current statutory language in most states causes more potential restriction of information than it promotes in the GIS environment. In isolation from any policy formulated by the professionals in information management, the courts cannot be expected to find the best solutions based only on the current statutes and meager case-law precedents. Any case that comes before the court involving the sale of GIS products or the providing of GIS information should be based on a course of conduct established by the agency from the inception of the GIS program.

Rules that adequately protect the public's right to know in the world of paper records can prove counterproductive in the world of information management through computer technology. A map is now managed as a database and not as a picture.

The need to assure accountability for the decisions of government, especially in the area of expenditure of public funds, has overridden the potential problems that a broad access policy places on the custodial agencies. A balance was struck over the history of the evolving open-records

policy to allow the broadest possible access despite cost, specifying exceptions only in narrow rules protecting privacy for personal, national, and commercial security.

Given the economy of scale we face with the potential "market" for GIS access, however, the balance traditionally considered by the courts is shifting. The utility of these new tools for managing the government's information is such that outside requests could predictably reach a level that would severely limit any control over staff time and the agency's budget. The balance also is significantly affected by the ever-increasing costs to taxpayers to pay for increased access. Automated information management changes the impact of the old policy in ways that require immediate correction. This correction of the current trend with regard to GIS can be accomplished eventually through statutory changes, but now an interim strategy is needed, one that involves the creation of a defensible course of conduct under the existing open-records statutes and case law.

Policy Responses

Based on two basic and diverse policy approaches used in different jurisdictions, policy makers are trying to meet the requirements of applicable state open-records law. An agency dealing with open-records requests at its own expense (in terms of management of staff time, capital investment in the system, and especially operation and maintenance costs) has little incentive to operate as a utilitarian information processor.

In the absence of cost-recovery options, GIS managers may perceive it to be in their interests to refuse as many requests as possible by distinguishing the requested material from "records" or inventing some other basis to deny access. For example: The information is not final, the information requires proprietary software, the information is not available in the form requested, the request is not specific enough to identify the record, and so on.

Most GIS managers prefer to grant

> Agencies planning for a GIS must take an active role in molding the information access policy and not wait for the legislative bodies to catch up with the technological progress.

broader access to information, but they must be able to afford the exercise. The more successful the agency is in limiting access, the less expense that must be covered by increased appropriations or rate increases. Drawing distinctions between "records" and other information in order to limit access because of the agency burden (either in staff time or money) is exactly what the statutes and case law have resisted. However, the same types of distinctions are needed to facilitate access.

The context in which most GIS administrators are planning for record/product distinctions is one in which the broadest possible access to products and services is to be offered. Once a course of conduct to establish this primary difference is clear, the differentiations that promote broader access will merit more favorable judicial treatment than efforts to restrict it.

The challenge for many organizations is to establish practices limiting access within the confines of the current statutory framework. This may be possible if reasonable distinctions are made between information that constitutes a "record" and a product, publication, custom report, custom map, a service, and so on. Certain specific issues, particularly the characterization of a request for the entire data base, are subject to future legislation or court decisions; the custodial agency should form a strategy for dealing with such requests in advance (e.g., promoting new statutory language, interpreting the statutory records exception for "programs"). If the request meets certain tests, distinguishing the response as a product/service and not a record should be straightforward. Recommended tests include: Would the information provided exist but for the request? Does providing the information affect the accountability of government for its actions or expenditure of funds? Was the information used to formulate public policy or carry out the agency's duties?

A reasonable distinction need not—indeed, must not—erode the purposes of the open-records laws, particularly the accountability of government. A reasonable distinction will enable an agency to enhance access for the general public. Establishing an access policy based on these reasonable distinctions can act as a carrot to work with the stick represented by the open-records law.

Two Disparate Solutions

The most easily defended access policy at this time is to: 1) Grant all requests for traditional text material in printout form, 2) deliver a copy of the database in the form of a tape when asked, 3) absorb the real costs of the exercise, and 4) turn down all requests that don't fit that mold.

If the access request is analyzed exclusively under the open-records law, and an exception category fits well enough to refuse access, the exception will not allow for subsequent sale or distribution of the particular information or different information to satisfy a similar future request. If the access request is met merely because no open-records exception applies, then no subsequent control is retained for similar requests in the future, no matter how burdensome or expensive they might prove to be.

However, jurisdictions adopting this solution to treat all GIS access requests under the state open-records law are not serving the underlying utilitarian principles of the law, nor are they adequately planning for the increased demand on these systems that will come with time. In the big picture, they also fail to serve their taxpayers' financial interests by missing the opportunity to distribute some of the costs to specific users and thereby limit the impact on the average taxpayer.

A second solution is outlined by jurisdictions developing distinctions between certain products and services and traditional records. They are encouraging an environment where it will be an advantage to promote outside access by educating the public about the system's capabilities, marketing the existing applications to the commercial sector, looking for new applications and new products, and planning for staff and equipment to make broad access possible.

Given the state of the case law in most jurisdictions, an open-records challenge to any distinctions between records and other kinds of products and services is likely to have a chilling effect on a proposed strategy to allow for cost recovery. However, a cautiously constructed policy should be able to avoid premature challenges until a course of conduct can be established, regulations formulated, and eventual statutory changes enacted with a more discriminating breakdown of what constitutes a record and what does not. The answer to an access request should be either yes (e.g., "It is a public record and you may have it within three days") or yes, but (e.g., "It is a service/product we intend to provide, and when we have time to produce it, you may have it for fifty dollars"). Not turning down any reasonable request is critical to avoiding open-records challenges. This approach is most attractive and practical if the custodial agency decides to actively provide and sell access to the GIS from the beginning.

DEFINING SYSTEM ACCESS

The course of conduct that will support the necessary distinction between products/services and records depends to some degree on the careful definition of the forms of GIS access that will be allowed. Records will remain the universe of information "required to be kept" by a government agency, whereas information products will require distinguishing definitions. While the system configuration (inclusive of the quality and depth of the database) will place practical limits on the variety of information products, the limitation that must be anticipated in the long run is tied more to the policy adopted by the custodial agency.

Policy options include the degree of recovery for financing capital and operation and maintenance expenses; the relevance of the policy to potential users, the access method or product media to be used in response to a query, and the definition of use that distinguishes updating the database from one-time usage, among others.

Policy can be focused on the need for control, the need for funding, or a combination of these basic factors. Each policy approach involves a different level of specificity and marketing of defined GIS products and services. The following description of a system's functional access categories provides an example of the concepts that enter into defining products and services.

Functional Access

System access could be granted on the basis of licensing agreements (in the form of contracts, leases, subscription agreements, and conditional one-time sales), requiring clients to follow procedures designed to protect and strengthen system integrity and encourage wider participation and contributions of data to the system data base.

Both products and services could be made available by granting access to the system resources needed to produce them, either "hands-on" or indirectly through a GIS service center. System resources, including staff when necessary, as well as data, equipment, and software, could be available from a central facility; in distributed systems, functional access could be made available from multiple locations once system standards are firmly established. Clients also could be allowed to access GIS remotely using their own equipment and software, subject to demonstration of compatibility. Resources might be priced on the basis of time connected, thus establishing time as the single basis of cost-recovery calculation. Data volume is another parameter for setting costs. Clients could be offered "access packages," each sufficient to produce a variety of products or services within the full range of system capabilities. This approach would keep the cost structure integrated and manageable.

Three reasons commend this approach. Because the GIS resources can be applied in so many ways, any list of poten-

If GIS is conceived as an information utility, it is best organized in terms of successive degrees of access to the resources of the system, rather than in terms of more specific and isolated services and products.

tial products and services would be open-ended and complex to manage; whereas GIS access packages can be limited to a manageable number without inhibiting clients' use unnecessarily. An access orientation can integrate both products and services, simplifying management. Finally, an access orientation encourages clients to use their GIS resources freely and creatively, thereby expanding system applications and resultant benefits.

The proposed approach to system access is based on a number of assumptions, including: 1) The GIS custodial agency will not procure equipment for clients, though it may grant access to certain equipment on a temporary basis; 2) the agency will, in the general interest, retain firm control over its database's form, content, and integrity; 3) the agency will operate a central service bureau to provide routine short-term services and may contract for reproduction of standard hard-copy system products; and 4) the agency will recover from clients or allocate to participants all system costs on an annual basis.

If GIS is conceived as an information utility, it is best organized in terms of successive degrees of access to the resources of the system, rather than in terms of more specific and isolated services and products. It is reasonable to assume that most system clients will approach GIS with a functional objective, wanting to do, or have done, some task. Therefore, *functionality* is the primary adjective useful for defining system access offered at one of several functional levels, within which any process or product can be located or produced. Levels are cumulative. Basically, functionality is equal to software capability, so it is access to software that is at issue in this proposed access scheme. Functionality levels are defined for purposes of this example as follows:

- Plotting, printing, and reproduction of standard products only
- All of the above, with alphanumeric query and printing capability
- All of the above, with graphic query and plotting capability

- All of the above, with nongraphic data input and analysis capability
- All of the above with graphic data input and analysis capability

In addition, access could be granted on a case-by-case basis for requests to use compatible special applications software.

Data Base Access

Structuring data base access serves three purposes: 1) Limiting data base management effort and costs to the GIS; 2) meeting public and client interests and requests; and 3) enhancing data base security.

The data base security and integrity issues can be addressed in the read/write limitations of the functional access packages. These limitations can be defined for individual users by assigning "read only" or "read/write" access to the data base at the layer or feature level. Access limitations by area coverage and content are other options.

Each coverage set could be designed to be a "stand-alone" working subset of graphic and/or attribute data, containing a subset of base map data for orientation. The "offer of any coverage" option implies capabilities to extract, copy, and download the specified portions of the database on demand with minimal programmer or operator involvement. Since all federal and state agencies follow the rule that an agency is not required to "create a record" to respond to a request, the mere definition of programming as equivalent to creation can limit most of the GIS access that is desirable to the community. Many agencies have formally defined all programming as creation of a record, but current Justice Department recommendations include the distinction between mere operation of the computer, which should be part of a required reasonable search, and programming that would constitute the creation of a record to meet a request. No clear guidelines exist, however, and federal and state agency practice remains inconsistent, even within jurisdictions operating under the same legal precedents and statutes.

Data base access would deal differently with requests for a product versus a service. A digital version of an identified portion of the database could be sold as a product. However, in order to avoid proliferation of varying copies of the data base, no digital reproduction of the entire database would be authorized unless updating is required and restrictions are placed on reproduction and use by third parties. This can be accomplished through a subscription agreement that deals with access to the database as a service, rather than the sale of the database as a product.

The equivalent of copyright protection can be established through contract law by inclusion of restrictions on third-party use or commercial use by the buyer in an access agreement. The extent of formal protection that can be obtained under the Copyright Act remains unresolved. Databases that constitute a compilation of facts can be copyrighted, but the registration process is cumbersome and not suited to the dynamic nature of a database. A new registration would be required frequently under current procedures in order to protect regular database updates.

Support Access

GIS could support clients both with staff operators and programmers, and with client staff training. Under normal circumstances, the GIS service center generally would provide staff resources, but in specific cases, the center might be used directly by trained client staff. The GIS center could provide operator training and, where appropriate, programming and data base management training as well. Training would augment that given by vendors of the client system, and focus on GIS procedures and system standards.

Mode of Funding Access

To encourage wide participation, a GIS agency could offer a range of funding mechanisms for the various access options, each connoting a different level of mutual commitment and granting the client a different level of participation in GIS governance.

Casual Funding. The GIS could offer casual services of short duration on a basis of standard hourly or daily fees varying by access and support level. As indicated previously, standard hard-copy products could be offered for fees derived from aggregate system usage. Casual-services clients could have nonvoting membership in a GIS users' group.

Subscription Funding. GIS center resources needed for more substantial but time-limited projects could be offered under renewable variable-length subscriptions. Subscriptions agreements would commit the GIS center to provide, and the client to fund, specified levels and schedules of access during the subscription period. This option is aimed at clients with application projects not large or long enough to justify equipment and software purchase.

Similarly, clients could subscribe for connection time with the GIS from client-owned facilities remote from the GIS service center. Such remote subscription requires installation of communications links. Subscription clients could have a voting membership in a GIS users' group.

Lease Funding. For major clients with data inputs, applications, or a scope of operations that will affect system capacity to a major degree, but that do not wish to become GIS sponsors, the GIS should offer full-time access from client-owned resources under long-term leasing arrangements. GIS lease clients could hold positions on a GIS technical advisory committee.

Sponsorship Funding. For clients with high levels of data input, applications, or operations who desire a primary role in GIS management and are willing to budget dollars on a continuing, long-term basis to GIS support (including a guarantee of cost-recovery shortfalls), the GIS center could offer funding access through sponsorship. Sponsorship would carry all the privileges and responsibilities of the founding agencies, including a voting position on the GIS policy committee or appropriate participation in other institutional arrangements for managing the GIS.

Grants. At management's discretion, subject to policy committee approval, the GIS center would grant access to clients, for specified purposes and time periods, without cost recovery or with reduced standard fees. Grants could be made for the purposes of expanding system capabilities and extending system use.

USER FEES

User fees are the fastest-growing source of local government revenues. User-fee revenues have tripled since the mid-1970s, from $30 billion in 1976 to $98 billion in 1987. As a portion of the total revenues of local governments, including state and federal payments, they rose from 17 to 21 percent. Almost three-fourths of all local jurisdictions have user fees in some form, and they are being applied to a rapidly expanding number of public services and facilities (Lemov 1989). Policy decisions about the application of user fees will have a significant impact on the formation of a legal setting that promotes the broadest use and custodial control of the GIS resource.

Government-owned geographic information systems sometimes are financed by bond issues, which traditionally have been used to finance public facilities construction projects. Statutes establishing the rules for administering these bond issues promote the lease or sale of excess capacity of the public facility to offset the interest payments or principal of the debt, when practical. For example, leasing parking spaces to private individuals at a garage built for government employees with bond money is not only allowed but promoted in most jurisdictions. Since a GIS will have continuing operation and maintenance expenses as well as an extensive front-end

capital cost, with or without financing charges, its financing may be more analogous to highway construction and maintenance projects. GIS differs in either case in the lack of specific precedents. Traditional roles for user fees deserve more study as analogies for their applicability to GIS access.

The applicability of user fees requires that lines be drawn between *necessary* public services financed by general revenues and *available* public services that are provided on a fee-for-service basis. For example, garbage collection remains "free" in some jurisdictions, while leaf collection and police escorts for funerals are provided for a service fee. Supplementing the costs of general garbage collection with leaf collection fees from those who require this additional service gives the community an overall benefit. A similar synergy can be built into the GIS access fee structure by careful planning to match costs to particular products and services offered.

The overall user-fee plan should have express exceptions to deal with known inequities and the flexibility to allow for social considerations. The attitude toward user fees is dependent on the attitude of the community. The simpler the fee system, the greater the inequity (and the more difficult it is to apply in support of social policy), but the easier the administration. Fee structures for GIS access also must find the appropriate balance.

The establishment of fees for GIS access will vary with the historical course of conduct reflected in the legal setting in the jurisdiction, the system configuration, and the database design and quality. The anticipation of funding from "outside users" of a GIS through some sort of user-fee program without a clear path for system administrators to follow in creating information products, or in implementing a marketing plan, has been a major roadblock to the progress or survival of several early GIS projects.

A proposal gaining substantial support today is that GIS should pay for itself at least to the extent it is used to supply

User fees are the fastest-growing source of local government revenues. The applicability of user fees requires that lines be drawn between necessary public services financed by general revenues and available public services that are provided on a fee-for-service basis.

nonessential public services that benefit individuals or corporations, as opposed to the public as a whole.

COLLATERAL ISSUES

Even if the access policy for a particular GIS tracks the example outlined in this section, the policy will work only with the cooperation of the user community, and with the resolution of certain issues specific to the jurisdiction covered by the GIS:

- Projection of demand
- Acceptance of clients' applications software
- Data base protection
- Development and acceptance of input data standards
- Quality control
- Promotion of cooperative system participation
- Scheduling implementation of an access program
- Access/Participation license language
- GIS software availability beyond the GIS service center facility in stand-alone setups
- Integration of client-owned equipment with GIS system

In addition to the practical issues of standards, compatibility, the form of supporting agreements, and so on, many policy decisions will be necessary to provide a framework for system access to GIS products and services. For example, promotion and management of public and private access to GIS will involve a major marketing initiative. Development of a sophisticated access program is merited only by a diverse and extensive set of clients. However, government is not overly experienced in this kind of "private-sector" marketing activity.

Extensive marketing and sale of information products will bring several other legal issues to light as government is thus placed in the role of a private corporation.

Proprietary Authority

In order to sell GIS products and services, the custodial agency must find either express or implied proprietary authority to define and provide GIS access for a price. In governmental law, *authority* means legal power—the right and power of public officers to command, to act, or to require obedience to orders lawfully issued. The most obvious authority granted to a GIS will be incorporated into the charters, enabling legislation, and into state constitutional provisions applicable to the participating sponsoring agencies. These powers are express, implied, governmental, proprietary, or some combination. It is not critical to identify express language in a particular statute that allows the sale of products and services before considering such activity. However, it is critical to identify any express prohibitions that would limit the necessary authority, and to construct from express language and implied powers an argument that such authority does exist.

A distinction between *governmental* and *proprietary* powers is also important to the issue of adequate authority. The duties imposed on the executive agency or the municipal corporation as an agent of the state are said to be governmental. The other kind of authority exercised by agencies and municipalities is proprietary, which includes powers not conferred primarily from responsibilities connected with the governance of the jurisdiction-at-large, but for the private advantage of the community. The provision of police protection is governmental, while the operation of a railroad has been ruled proprietary. Proprietary powers do not merit the same strict treatment because they are conferred not for the purpose of government, but for the private advantage of the state or town and its inhabitants. A majority of courts have held that in the exercise of proprietary powers, the agency or municipality is governed by the same rules that govern private individuals or corporations, and it may conduct its business in the manner that promises the greatest benefit to the community.

A majority of states follow the rule of *ejusdem generis*. Under this legal maxim, the courts will in most cases require a clear, if not express, manifestation of any powers that would not traditionally be read into these broad grants of power. This rule directs courts to read participants' general powers as covering only areas of the same kind, class, or nature as the agencies' express powers. In a majority of states, case law supports the assumption that municipal corporations possess only those powers that are expressly given by the constitution and statutes, or are necessarily implied by those grants of authority. Several cases state that ". . . if there is any reasonable and substantial doubt as to the existence of a particular power, such power will be deemed not to exist." (McQuinlan 3rd ed.). The rule of *ejusdem generis* is subject to limitations in a corollary *implied powers* rule, which states that the general powers should not be construed in a way that destroys the purpose for which the underlying grant of power is made. If it can be shown that the control of commercial access to the GIS is critical in order for the system to be used for its primary purposes of existence, the argument for implied purse-string power over access may be supported.

Case law is sparse relative to establishing limits on proprietary authority, but the existence of proprietary authority is often recognized in the context of sovereign immunity discussions. Case law has supported the sale of toasters by a public electric utility, and even the operation of the local railroad by a municipality, but the scope of the proprietary control over products and services that will allow the GIS to evolve into an information utility is a dramatic step in this direction, reaching beyond existing precedents.

The future commercial applications for government-supported GIS data bases demand a new role from government that is suggested but not entirely covered by the court-tested examples of government assumption of implied proprietary authority. It should not be difficult in most jurisdictions to establish that proprietary authority to engage in the sale of GIS products and services exists under the general governmental authority derived from the custodial agencies' charter or home-rule provisions.

Privacy

Limitations to protect information that fits under open-records act privacy exceptions also must be established. The traditional balance between providing unlimited right to information and withholding release involves the constitutional right to privacy. Some courts have found a compromise that is not a good precedent for GIS technology. When an academic researcher wants information about test scores and income levels, courts have required the custodial agency to perform a statistical analysis of the "private" data, and provide this analysis in response to an open-records request for otherwise private information. This so-called "redaction" requirement places a burden on the agency to produce information that did not exist, and was not necessary for agency purposes, to meet an outside request. This kind of result seemed a reasonable compromise when the issue was considered only in light of issues of privacy versus open access to government information. It has unintended results in the context of a GIS utility, however.

Both federal and state laws are being changed dramatically to take into account malfeasant use of a computer. The awareness of this possibility has come to the forefront through the invasion of public and private data bases by "hackers." These acts have covered everything from changing grades in a school's computers to accessing bank and credit card accounts.

Security issues involving the transmission or access of data are being addressed on both technical and legal fronts. Specially designed operating systems, fault-tolerant computers, and new methods for verification of the identity of the person seeking access all provide protection from hackers, criminals, duplicitous insiders, and acts of God. In the legal setting, revised criminal statutes and interpretations of the exceptions to the open-records law provide the tools for protection of privacy.

It was much simpler in the preelectronic age to review a specific request and determine whether a privacy issue was present. When offering the potential resources of a GIS data base for commercial application, the issue becomes much broader. The custodial agency must anticipate what could be done through different forms of access, and any privacy invasion that might result.

The same factors that make GIS an appropriate foundation for the information utility of the future make it vulnerable to invasions of privacy. These problems have not surfaced for GIS thus far because the nature of the typical data base is not significantly "personal," and the profit motive is thin (unlike the personal gain incentive for moving funds by unauthorized manipulation of bank records).

Liability for GIS Products/Services

Since the GIS participants in a given jurisdiction will operate under proprietary authority when marketing and selling products and services, a court would likely consider claims of negligence and product liability claims. Appropriate disclaimer language can avert most potential products liability claims from being filed and provides a good defense against most such claims should they be filed. Before government agencies can be brought to court for negligence or products liability claims, an investigation of sovereign immunity, governmental immunity, legislative limited waivers of sovereign immunity, application of the waivers to a particular agency, and whether immunity should apply at all to the particular activity must occur.

The great shield against federal, state, and local government liability is the ancient doctrine of *sovereign immunity*, often paraphrased as "the king can do no wrong." In practice, this means a government cannot be sued without its consent (*Lalehite v. United States* 1953). Mapping functions traditionally have been official

government undertakings. Governments make maps and have not consented to be sued for errors on them, at least not in specific terms. Sovereign immunity is waived by legislative bodies. Both the federal government and most states have some version of a Tort Claims Act that establishes limited waivers of immunity for tort claims against the government (29 USC 1346, et seq.).

One line of demarcation is the potential for liability if the undertaking is *discretionary*. Two principles are involved: The government's free right to engage or not engage in discretionary functions; and the government's liability when, by its conduct, it has created a justified reliance on its adequate performance. As to the first principle, the government not only has the discretion of whether or not to engage in proprietary activities like marketing and selling GIS products, but also the discretion to determine the extent to which it will do so. As to the second principle, there can be no initial justified reliance upon or expectation of any particular degree of performance, and this element will not be automatically imputed. For example, justified reliance could be found as a result of particular actions, statements, knowledge of the product limitations and the intended use, and the particular representations that were made that caused damage.

Once the decision is made to engage in a discretionary activity or function, the sovereign may be held liable in the same manner as private parties for the negligence of its employees' and agents' actions in the performance of these activities. Negligence may take the form of incorrect information entered into the GIS database during data conversion that later produces an incorrect line or label on the map. If that map is sold under proprietary authority and leads to damages resulting from construction in the wrong place, *and* the actions of the participants' agents in marketing or explaining the use of the map caused the inferred party to justifiably rely on the information, liability is likely. It would be inappropriate to characterize

Marketing activity that can be characterized as proprietary requires the investigation to ensure protection from liability that would normally be taken in the private sector.

actions based on implied proprietary powers as other than discretionary activity. Because the proposed marketing activity is necessarily characterized as proprietary, it is also necessary to investigate steps to ensure protection from liability that would normally be taken in the private sector.

By issuing maps and selling other information products and services, the GIS custodial agency assumes a duty to invest sufficient resources necessary to achieve what the government and the courts will agree to be proper care. The duty of care depends upon the facts of the particular case. When the user is induced to rely upon the map, in whole or in part, for a forseeable decision and subsequent action, then the government may be held liable. Failure to provide an adequate warning, which creates an unreasonably dangerous condition, also can result in liability.

The absence of a contract often has resulted in a finding that no duty exists in cases of unintentional dissemination of incorrect information that is not defamatory. However, a duty of care has been found without privity of contract in unusual situations leading to personal injury. Dissemination of inaccurate aeronautical charts whose subsequent use resulted in airplane accidents has been treated as strict liability situations (*Jeppesen v. Brocklesby* 1986). These implications can be summarized as follows:

- Liability is based on the user's justified reliance on the map or other information product quality for the user's purposes. The map quality and the quality of the mapmaking may be factors considered by a court. Disclaimers do not automatically prevent liability, but they show that the agency has met its duty of care by specifying where the user's reliance on the information is not justified.
- Failure to warn the user that a defect is possible can create liability when the user is induced to place inappropriate reliance on the map.
- The objective of a map disclaimer is to alert users that errors are possible and

that they need to consider the degree of reliance appropriate for their purposes.
- The disclaimer should indicate when the map or other information was compiled, the source material and its nature, and direct the user to that material.
- The disclaimer's location, style, and size should make it noticeable to the user.

If there is a contractual relationship between the parties, the extent of the government agency's liability for providing inaccurate data should be defined by the express and implied agreements between the parties. This will apply if subscription agreements are used for on-line access or in the case of data base update agreements. Proper disclaimers should be included in the contract, as well as a requirement for disclaimers on products produced under the contract. Such agreements also should specify that the GIS access being provided under a data base subscription agreement is merely a service and not "goods."

Uniform Commercial Code. Applicability of the Uniform Commercial Code (UCC) to the provision of information is still unclear. As a general rule, Article 2 of the UCC applies only to the sale of goods and not to the provision of services. However, it is possible to characterize some information products as goods and not services. Courts have found electricity transmitted over wires or natural gas pumped through pipes to be "movable goods" subject to the UCC. Electronically transmitted data may also be subject to similar treatment by analogy. The remedies available under the UCC are more extensive than those available through the common law of negligence, and disclaimers should include language to exclude the implied warranties of merchantability and fitness for a particular use. These disclaimers, under the UCC, must be conspicuous and mention the term *merchantability*.

Without the insulation of sovereign immunity, the custodial agency must look at each product or service and develop procedures to protect it from undue liability in the same manner as a private corporation. This requires an extensive analysis, but generally "products" will involve implied warranties for product liability established under Article 2 of the UCC, while "services" will involve proof of negligence, duty of care, and proximate cause under tort law. Liability for errors and omissions for services also could involve contract law in situations where system access to applications is allowed, subject to contract or subscription agreements. Warranty disclaimers and purchase of liability insurance are related policy considerations.

User Fees, Antitrust Law, and Undue Competition

When drafting contracts, subscription agreements, or leases to provide access, protection of proprietary system value must balance restrictions establishing restraints on trade and unfair competition. In *City of Lafayette v. Louisiana Power & Light* (1978), the United States Supreme Court ruled that there is a presumption against implied exclusions from coverage under antitrust laws for the actions of a municipal utility, and the intent is not to limit protection only from abuses of private power. In other words, antitrust law can be applied to government agencies and public utilities. The doctrine of "state action" does protect government from some antitrust actions. State action will only provide exemptions from antitrust pursuant to a legislated state policy to displace competition with a monopoly service. The GIS custodial agency needs to establish a clear course of conduct that does not exceed the legal limits of establishing a monopoly, and this need is increased by the typical lack of express authority to sell information products.

Private industry can set prices for goods and services and adjust them up and down to fit the market, subject to limitations established by unfair trade practices law. Setting fees for government information services is not so simple when giving due consideration to accountability requirements.

Fee setting for different forms of functional access to a GIS should be documented in such a manner that satisfies requirements for "public notice and affirmative consideration" (standards for regulation promulgation) in the absence of express statutory provisions for setting rates. Selling GIS products and services is the first step on the road to the eventual establishment of a new form of public utility—one that provides information products instead of gas or electricity. It also may raise undue competition claims from private-sector map makers and other value-added GIS users who would do better economically if the information utility does not evolve.

In the absence of a legislative policy establishing such an information utility, state and local governments are subjected to the same principles prohibiting unfair trade practices as a private corporation. This primarily involves restrictions on price setting with the intent to destroy competition from established dealers of the same commodities. The revenues and expenditures for most jurisdictions should be balanced. Documentation of an "annual equivalent cost" for GIS, and connecting the established charges for products/services to the actual system costs for the appropriate types of functional access, should satisfy the custodial agency's need to defend the price for access. Documentation of a one-to-one ratio of revenues to expenditures also will offset any private-sector claim of inappropriate competition from government.

Most private-sector enterprises should support making GIS available to them for value-added use under a products and services policy. There is little precedent for use in the analysis of government as competitor under antitrust law, and the case law in the area deserves careful analysis before making final policy decisions concerning pricing.

Equal Treatment

The constitutional guarantee of equal protection under the law, or equal treatment, will come into play if and when GIS participants set different prices for the same products to implement policy. The law for local government generally provides that "equal protection of the laws is violated by municipal exercise of power which arbitrarily, unreasonably, or invidiously discriminates. Stated affirmatively, local governments can classify, categorize, and even discriminate so long as the differentiation of treatment is reasonable." (Antieau). Even though the GIS may be treated like a private corporation for most purposes when acting under proprietary authority, a government agency may not always exercise the same freedom afforded the private sector.

One area restricting government action that does not similarly affect private-sector corporations is the constitutional requirement for equal protection. Government may discriminate, but it may not invidiously discriminate. If price-breaks or special services are granted or made available to one group, the custodial agency must be prepared to grant the same treatment to other groups that reasonably fit the same classification. It is much easier to establish a special classification, intentionally or unintentionally, than it is to do away with a special classification once established. Procedures for special exceptions need to be established with full consideration of the limits equal protection imposes.

Equal protection does not inhibit manipulating user fees to allow for good public policy results. For example, all statutory program support, nonprofit use, journalistic use, and purely academic research could be provided on a least-cost basis, as long as the required staff time and computer access is available for the request. All commercial requests for discretionary services and products could receive secondary consideration after such public service use of the system is provided.

Copyright, Contracts, and Control of Proprietary Interest

The traditional view of disseminating government information left little room for use of copyright protection by local and state governments, but such protection is more frequent today. Copyright has an important role in the exercise of proprietary rights for any government-owned GIS.

Copyright. Copyright is a federal grant of monopoly that is intended to maximize the availability of creative works to the public. It also is designed to afford copyright creators a fair return for their creative works and to give copyright users (e.g., the publisher of a work copyrighted by the writer) a fair income as well. Generally, a copyright holder is entitled to the exclusive use of all material copyrighted. This includes permission for the owner to reproduce copyrighted works in any manner; prepare derivative works; lend, lease, rent, or transfer ownership; and perform or display the works publicly. A number of exceptions apply, including "fair use" and "government publications."

The monopoly is granted so that distribution of ideas and information will not be restricted or hidden away to preserve the creator's proprietary rights in intellectual property. Public records principles have traditionally preempted restricting dissemination of a "public creator's" product, so copyright has not typically been an issue for government agencies.

However, the use of copyright protection can strengthen the distinctions that are critical to marketing GIS services and products within the public records limitations. Use of copyright is a valid part of the overall course of conduct, since it will show consistency in the anticipated exercise of powers that require the GIS participants to assume the rights and liabilities of a private corporation.

It may be convenient to limit private-sector competition for sale of the products and services available from the GIS by using copyright protection, and it may well be critical to prevent the proliferation of numerous versions of the "official" data base. The federal Copyright Act of 1976 expressly prohibits federal government publications from being copyrighted, and only allows the federal government to hold copyrights transferred to it by assignment (Public Law 94-553). The express prohibitions against use of copyright in the federal legislation technically apply only to the federal government, and not to other government agencies.

The use of copyright by local or state government is so unusual that no cases examining the interaction between copyright and public records law have been identified. Initially, the idea of state or municipal government's use of copyright protection seems contrary to the public records principles. However, it is clear that in the exercise of proprietary powers, the agency is governed by the same rules that govern private individuals or corporations. This would imply, at least in the exercise of proprietary powers, that state and municipal government agencies can use copyright protection in the same manner as the private sector.

Without a clear public policy supporting use of copyright protection by government, the federal legislation's bias against government use of the protection has provided a strong precedent against use of copyright by any government agency. The evolution of information economics has provided these new opportunities to expand the use of copyright to implement public policy, and the first precedents are now visible. In 1984, the Minnesota Data Privacy Act was amended to provide that ". . . nothing in this chapter or any other statute shall . . . prevent a state agency, statewide system, or political subdivision from acquiring a copyright or patent for a computer software program or political subdivision from acquiring a copyright or patent for a computer software program or components of a program created by that government agency. In the event that a government agency does acquire a patent or copyright, the data shall be treated as trade secret information [nonpublic]. . . ." (see 3A, MSA 13.03).

GIS custodial agencies across the country must establish new precedents for proprietary action in order to gain control over the voluminous nonrecord access requests these systems will generate. The GIS custodial agency's proprietary powers should include the same tools available to private corporations, particularly copyright protection. Use of copyright and patent protection in the computer industry has followed growth of "value."

In the 1960s, value in the computer industry was associated with hardware. In the 1970s and 1980s, this focus has shifted to software. Almost all of the "computer law" today is focused on the issues of defining the applicability and limits of software copyrights. As software copyright issues are resolved, a new shift to the more generic issue of control over distribution and use of information in automated form is evident, and the focus is shifting to protection of the proprietary rights in a data base.

The most important use of copyright at this time is to protect the "raw" data base from unlimited access, especially in the case of requests from an organization that did not share in the capital expense of establishing the system. The data base represents much of the capital investment for any GIS and the ability to manipulate it comprises much of the commercial potential of the system. If on-line subscriptions to the data base are offered, a subscriber could create a derivative data base by downloading the information available on-line and thus remove the need for a subscription to the original data base. The subscriber could then select, store, and rearrange the information in the data base, repackage it, and compete for potential GIS customers.

In the private sector, commercially valuable data bases have been copyrighted, but this has not provided sufficient protection in the face of improved technology available to end users that allows downloading and storage rather than requiring continuing on-line access. A more common practice in the private sector is to limit "use" and redistribution rights through subscription agreements.

The Copyright Act of 1976, as amended, includes among the list of works protected by copyright law "compilations," or works "formed by the collec-

Almost all of the "computer law" today is focused on the issues of defining the applicability and limits of software copyrights. As software copyright issues are resolved, a new shift to the more generic issue of control over distribution and use of information in automated form is evident, and the focus is shifting to protection of the proprietary rights in a data base.

tion and assembling of preexisting materials or of data that are selected, coordinated, or arranged in such a way that the resulting work as a whole constitutes an original work of authorship." (17 USC 101). The legislative history of the Act makes it clear that the protection extends to "compilations of data," including "computer databases." Several cases have found protection for on-line data bases under copyright law (*West Publishing Co. v. Mead Data Central* 1985).

Computer programs, maps, and photographs technically can be copyrighted once they have been distinguished from public records. However, whether the data base itself can be excluded from public records access is still in question, and the practical effect from use of copyright for a raw data base is particularly subject to question. Works made up of common information with no original authorship cannot be copyrighted. Copyright protection does not extend to any "idea, procedure, process, system, method of operation, concept, principle, or discovery" contained in a work (17 USC 102b). Also, since the GIS data base is not static but is subject to continuous updates, a single copyright would have limited practical value. It would not attach to a "changed" work, so a new copyright technically would be necessary after each update.

When the printing press was first introduced, intellectuals of the time were fearful of the ultimate results of putting reading material "into the hands of the great mob" (Boorstin 1978). Some have similar fears about broad access to GIS, but these are unlikely to be realized. Achieving information utilitarianism will be a long process, made difficult by the volume and potential creative rearrangement of information that continues to outgrow our ability to use it.

An organization's institutional and policy position must achieve the balance between cost recovery, control, public access, and an appropriate private-sector role for information dissemination. Government will remain a collector of information, controlling the updates to the most extensive and useful data bases, and thus is the initial "owner" of the value of a significant portion of source information.

For two kinds of financial reasons, GIS custodians in public organizations are choosing to pursue an active policy of providing and selling information. The first reason is to obtain reasonable purse-string control over access to limit frivolous requests and to avoid the unpredictable expense of meeting the potential demand. The second reason is to establish a method for offsetting the costs of establishing, operating, and maintaining the system.

Few GIS custodians are beyond the point of initially deciding to pursue a policy for providing and selling certain products. For the most part, the complex results of deciding to sell a particular product for a particular price have not been faced. However, certain choices will support the chances of surviving a legal challenge while other decisions might invite one.

These questions become sore points when the public-agency GIS manager researches the open-records law and other statutes for guidance, and thinks about how the agency should be organized to deal with access requests. The structure today restricts government from inappropriately limiting access to certain information. In the case of establishing new tools for managing information, the same rules serve to hinder government's ability to grant broad access to information and information processing abilities.

As embodied in the Alaska law (Box 11-1), one solution is to formalize the distinction between "records" and "products." This solution upholds the underlying principles of the Freedom of Information Act, to keep government accountable for its actions and expenditures while simultaneously promoting utilitarian access.

This approach authorizes proprietary government action to make information *products* available for a price calculated to produce a balance between revenues and expenditures. Information that is used to form public policy is treated as a *record* and made available for a nominal cost.

REFERENCES

Antieau, *Municipal Corporations Law*, sec. 5.20.

Boorstin D. J. (1978). *The Republic of Technology—Reflections on Our Future Community*. New York: Harper & Row.

City of Lafayette v. Louisiana Power & Light, 98 S.Ct. 1123 (1978).

Cowen D. J. (1987). GIS vs. CADD vs. DBMS: What are the differences? Proceedings of GIS/LIS '87 Conferences, sponsored by ASPRS/ACSM, URISA, AAG and AAG., San Francisco, pp. 46–56.

Dismukes v. Department of the Interior, 603 F. Supp. 760 (D.D.C. 1984).

FOIA Reform Act of 1986, P.L. 99-570, 100 Stat. 3207-49.

OTA Report (1988). Informing the nation. Federal information dissemination in an electronic age. Washington DC: Office of Technology Assessment.

Jeppesen v. Brocklesby, 106 S.Ct. 882 (1986).

Lalehite v. United States, 346 U.S. 15, 73 S.Ct. 956, 97 L.Ed. 1427 (1953).

Lemov P. (1989). User fees, Once the answer to city budget prayers, may have reached their peak. *GOVERNING* (March).

McQuillin Mun Corp. sec. 10.19, 3rd Ed.

NLRB v. Robbins Tire & Rubber Co., 437 U.S. 214, 242 (1978).

SDS Development Corporation v. Mathews, 542 F.2d at 1116 (9th Cir. 1976).

Seigle v. Barry, 422 So. 2d 63 (Fla. 4 D.C.A. 1982).

Shevin v. Byron, Harless, 379 So. 2d 633 (Fla. 1980).

Smith G., Parr R. (1989). *Valuation of Intellectual Property and Intangible Assets*. New York: John Wiley & Sons, pp. 6–7, 89–90.

West Publishing Co. v. Mead Data Central, 616 F.Supp. 1571 (D. Minn. 1985).

Yeager v. Drug Enforcement Agency, 678 F.2d 315 (D.C. Cir. 1982).

SUGGESTED READINGS

Administrative Conference of the United States. Recommendation 88-10, "Federal Agency Use of Computers in Acquiring and Releasing Information," 1C.F.R. §305.88-10 (1989).

American Bar Association. Resolution No. 102 (adopted Feb. 12–13, 1990) (published in 15 *Admin. L. News 1* (Spring 1990)).

Benton Foundation (1990). Electronic public information and the public's right to know. Report of Conference.

Bureau of National Affairs (1990). Federal information in the electronic age: Policy issues for the 1990s.

Reporters Committee for Freedom of the Press (1990). Access to electronic records: A guide to reporting on state and local governments in the computer age.

U.S. Congress, Office of Technology Assessment (1988, October). *Informing the Nation: Federal Information Dissemination in an Electronic Age*.

Chapter **12**

Horizons

Geographic information management systems provide the tools for government agencies to become more efficient and to improve the quality and timeliness of services to the public. Private companies understand the productivity gains in efficiency that the technology provides and the competitive advantages that can result.

What does the future hold for geographic information management systems?

GIS technology will continue to evolve toward full integration in the workplace and broader usefulness in a wide range of disciplines and endeavors. Practitioners and decision makers in many organizations understand the value of managing and processing information from a geographic perspective, as well as the limitations of traditional information systems to provide this perspective. Geographic information management systems provide the tools for government agencies to become more efficient and to improve the quality and timeliness of services to the public. Private companies understand the productivity gains in efficiency that the technology provides and the competitive advantages that can result.

PERSPECTIVE ON THE STATUS OF THE TECHNOLOGY

To understand technological trends and assess future prospects, it is helpful to view GIS technology in the context of its historical development. To summarize briefly, four phases characterize the evolution of data processing since computers

became commercially available in the late 1940s (Fig. 12-1).

The first phase was characterized by large mainframes that used vacuum tubes and complex wiring. These early computers were single-task devices, capable of performing one operation, usually a complex mathematical calculation. Stored programs or programming languages had not yet been invented and, therefore, "programming" the computer required setting actual switches and circuits to control a particular operation. These computers were large, expensive, dependent on strict environment controls, and required a staff of skilled operators to keep them running.

The development of transistors in the early 1950s was a milestone in computer architecture. During the late 1950s, general-purpose computing and the perfecting of programming languages gave rise to the second major data processing phase. This phase was characterized by batch-program execution, wherein users submitted "jobs" to a mainframe computer that were executed in a timesharing manner. In the 1960s and early 1970s, technicians punched programs and data onto cards, the card "deck" was given to operators, and the users waited to see if the program ran successfully or "bombed" during operation. During this second phase, programming languages were invented, making it

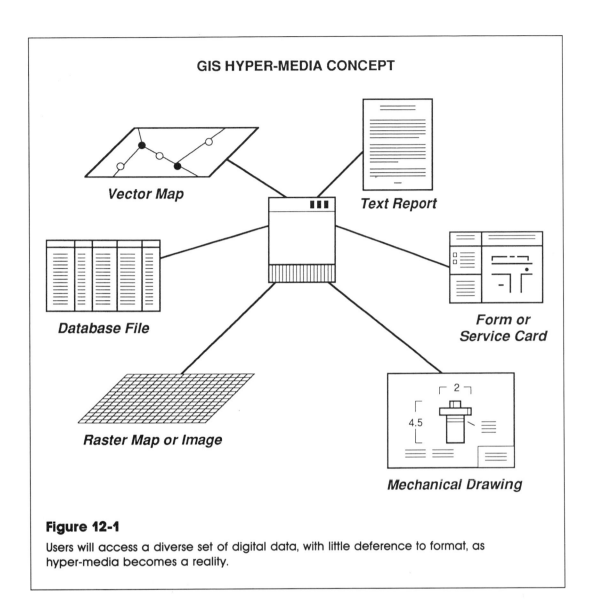

GIS HYPER-MEDIA CONCEPT

Vector Map

Text Report

Database File

Form or
Service Card

Raster Map or Image

Mechanical Drawing

Figure 12-1

Users will access a diverse set of digital data, with little deference to format, as
hyper-media becomes a reality.

possible for a larger user group to have access to computer resources.

Development of the integrated chip and other technological advances in computer architecture and communications helped bring about the third phase of computing, characterized by interactive computing and single-vendor networks. From its beginnings in the early 1970s, interactive computing (using mainframe computers, minicomputers, microcomputers, and intelligent workstations) has become the norm in geographic information processing. Tremendous progress in networking computer devices for the physical dis-

tribution of processing power and data has allowed users to share resources regardless of location.

Sophisticated networks now connect devices that understand specific operating systems and communication protocols, and because of this, single-vendor networks are quite popular. It is still more difficult and expensive to create flexible networks of hardware components from many different vendors.

Although networking is yet in its early stages and data processing generally remains in the third phase, impressive strides are being made to bring computing into its fourth phase. Flexible and transparent networking of all types of computer devices will become standard. Advances in operating systems, transmission speeds, security and integrity of networks, and above all, the adoption of standards in software and data communications, will make it possible to link computer devices from multiple vendors easily.

Like any application of computer technology, geographic information management systems are advancing through a cycle of development that will lead to its general acceptance and wide use in operational settings. Although geographic information management systems borrow concepts and techniques from many established disciplines, the technology itself is still immature. The status of GIS is comparable to a child's passage into adolescence: Fundamental discipline has been established, but more development lies ahead before maturity.

One illustration of the relative immaturity of GIS is the confusion and imprecision that exists today in the terminology used to describe various components and concepts of the technology. What is the difference between a *GIS* and an *AM/FM system*? What is a layer? Users, researchers, and commercial developers alike struggle with an imprecise lexicon that inhibits the development and adoption of the technology. Over the past several years, researchers have become aware of the need to establish an overall "model" on which to base the development and define the use of GIS. Impressive efforts are underway toward this end.

The National Research Council of the National Academy of Sciences sponsored research in the late 1970s and 1980s on the topic of *multipurpose cadastre*. The panel assigned to this research produced two documents that described many fundamental concepts important in the development of GIS (National Research Council 1980, 1983).

The National Committee for Cartographic Data Standards (NCDCDS), sponsored by the American Congress on Surveying and Mapping (ACSM), was formed in 1982. This group has coordinated its activities with the Federal Interagency Coordinating Committee on Digital Cartography (FICCDC), created by the Office of Management and Budget and the Digital Cartographic Data Standards Task Force (DCDSTF), established by the United States Geological Survey. Several years of research by these groups have culminated in proposed standards for the structure, content, terminology, and quality of digital cartographic data, establishing a foundation for the adoption of GIS standards (*The American Cartographer*, 1988; Rossmeissl 1989).

The future seems bright for continued research and refinement of GIS concepts and standards. In 1988, the National Science Foundation awarded a three-year grant to establish the National Center for Geographic Information and Analysis (NCGIA). The NCGIA is a consortium of research groups at the University of California–Santa Barbara, State University of New York–Buffalo, and the University of Maine–Orono. The NCGIA has outlined a series of research tasks that are designed to evaluate the social, legal, and institutional impacts of the spreading use of GIS technology. Other professional organizations, such as the Urban and Regional Information System Association (URISA), AM/FM International, the American Society of Photogrammetry and Remote Sensing (ASPRS), American Congress on Surveying and Mapping, the International Cartographic Association (ICA), and others, continue to examine issues and provide recommendations that contribute to the understanding and effective application of GIS technology.

The 1990s will bring greater acceptance and incorporation of geographic information technology and management techniques in many organizations' overall information management strategies. The obstacles to full integration of geographic data resources are both organizational and technological. Advances on both fronts are resulting in less redundancy in effort, greater consistency, and decision making that is based more on facts and systematic analysis than on intuitive reasoning or "seat-of-the-pants" estimates.

Users, researchers, and commercial developers alike struggle with an imprecise lexicon that inhibits the development and adoption of the technology.

A discussion of technological trends can give the impression that advances occur in a passive environment and that users are bystanders who accept what developers provide. Technological developments, however, do not occur in a vacuum. The user community can have a strong voice in the direction of commercial system development, but active and potential users must voice their needs to the data processing industry.

SOFTWARE AND DATA STRUCTURE TRENDS

Graphic Data Structures

Over the past decade, significant advances have occurred in the development of graphic data structures. Modern graphic data structures allow depiction of the spatial relationships of map features as opposed to their simple graphic depiction. The term *topologic* has come to describe geographic data structures in which the inherent spatial connectivity of features is implicitly stored and maintained. Such a structure supports efficient spatial analysis in contrast to basic graphic or mapping operations.

While topologically structured data bases have allowed more sophisticated spatial analysis than is possible with simple graphic structures, their complexity also has inhibited their use. The complexity of the file structure introduces additional "overhead" in data entry, update, and analysis that often results in slow response time and works against use of the systems in an interactive environment. Response-time problems that now exist will be reduced and perhaps eliminated during the 1990s through the increasing availability of low-cost computing power and further refinement of the data structures themselves.

A problem facing GIS developers is how best to logically store a geographic data base to provide for maximum efficiencies in queries and retrievals. Since geo-

graphic features are identified by a two- and sometimes three-dimensional position, the issue of logical data storage is more complex than with tabular data bases.

Various software vendors have developed schemes to partition geographic data bases into fixed geographic areas. These techniques address some of the problems inherent in processing geographic data but are not completely adequate in themselves. Most systems employ one of several techniques to logically divide the data bases for indexing and query purposes in a way that is obvious to users. As an example, some software uses a quadtree approach that successively subdivides rectangles to logically partition and index geographic features as a function of data density and to provide efficiencies in spatial analysis (Westwood and Brinkman 1988; Vanzella and Cabay 1988). More research and development is needed to fully optimize geographic data structures and increase the efficiency of spatial queries on map features and associated attribute data. Major strides in this area likely will be made over the next decade.

Few commercial geographic information systems have the capability to efficiently store and process geographic data in three dimensions. Applications that involve terrain analysis, subsurface modeling, underground utilities, multistory architectural plans, inside-plant management, and other analyses involving three-dimensional space will demand more sophisticated solutions from software developers. GIS technology will borrow more techniques from computer-aided design and three-dimensional modeling systems and apply them to geographic problems.

Nongraphic Database Management and Manipulation

Since the early 1980s, the data processing community as a whole has embraced the relational model, and relational techniques will continue to increase in popularity in geographic information management. GIS

The integration of different types of geographic data and documents will provide users with a flexible environment for complex decision making. Over the next decade, the traditional perception of GIS as consisting of a vector map and associated nongraphic data will be replaced with a more comprehensive view of *hypermedia*.

software developers have succeeded in using existing relational data base management packages (frequently developed by third-party vendors) for managing nongraphic data associated with map features. For much of their history, relational packages have been hailed for their flexibility but criticized for the inefficiency that results in slower response time when compared to more highly structured data models.

Recent advances in relational data base software have greatly reduced these inefficiencies, and the architecture of computer processors has become more suitable for relational data structures. The near future will see data base management systems that combine the flexibility of relational models with an ability to perform interactive transaction processing on large data bases. Developers of geographic information management packages will rely to a greater degree on third-party relational data base packages. Many vendors will offer the option of integrating or linking their software to a variety of these existing relational packages.

Advances in high-level-query and programming languages have accompanied the increases in popularity of relational systems. Data query languages use English-like commands to perform complex retrievals from the data base. A recently accepted standard is structured query language (SQL). Initially developed by IBM, SQL is used by many GIS software developers.

High-level programming languages have been adopted by the data processing community as a means to more quickly develop and modify applications. Fourth-generation languages, usually associated with a particular data base management package, have greatly increased the efficiency of programmers in application development work. In the near future, GIS will commonly provide integrated high-level programming languages. In the more distant future, fifth-generation languages using less structured syntax, sometimes referred to as *natural languages,* will be a favorable addition to geographic information technology.

Hypermedia

The integration of different types of geographic data and documents will provide users with a flexible environment for complex decision making. Over the next decade, the traditional perception of GIS as consisting of a vector map and associated nongraphic data will be replaced with a more comprehensive view of *hypermedia.* In a geographic data context, systems will have the ability to reference and retrieve different forms of data, including traditional vector maps, tabular databases, freeform text, raster maps, engineering drawings, standard forms, photographs, or audio records, among others (Kindleberger 1989).

One example of the hypermedia concept, under development now, is the linkage of property maps to a database of scanned photographs of the property to aid in appraisal. The map provides a digital index to site photographs. This concept can be extended to many data elements, such as road log or utility photographs, building permits, architectural plans, or text description. Advances in computer processors, optical disk storage, scanning technology, and software will continue to enrich hypermedia approaches.

Another trend is toward the integration of raster and vector data for geographic analysis and display. For some GIS applications, raster/vector overlay capabilities can be a powerful tool. For example, scanned orthophotographs or satellite imagery in raster form can serve as base maps for certain data best represented in vector form. Raster/vector integration routines are just beginning to be offered by commercial vendors (Wallace and Clark 1988; Spencer and Menard 1989).

Expert Systems, Artificial Intelligence, and Object-Oriented Approach

Much press attention has been given to the applications of expert systems and artificial intelligence in many information management environments. Rule-based logic, characteristic of expert systems, is being applied in the development of geographic data management applications. An example may be a polygon overlay analysis to assess suitability for a type of development (e.g., sanitary landfill suitability). Suitability categories are generated based on user-defined combinations of data in the overlay analysis. An operator "programs" the logic for this analysis and the GIS performs the work and generates a predefined product (Maggio 1988).

The potential role of artificial intelligence in geographic data management systems is somewhat uncertain. Artificial intelligence systems, capable of simulating inductive reasoning based on a pool of knowledge, likely will have application in defined areas. One of the most promising is data conversion and the scanning of maps that would reduce the now manually-intensive quality control and editing process (Dunaway 1989).

As a database architecture, the concept of "objects" is not completely defined and its application in GIS is just beginning. One of the fundamentals of object orientation is that data and instructions associated with objects are maintained together as the object's definition. The object, therefore, has "intelligence" and "knows" how to act in various applications, freeing a programmer from the need to write detailed code that acts on a separate database. A second common theme of object-oriented databases is *inheritance of attributes*. This means that objects are arranged hierarchically into classes and subclasses; subclasses inherit the rules associated with their parent classes. In GIS, objects may be map features along with their nongraphic attributes (Egenhofer and Frank 1989).

Some GIS software vendors are developing systems using an object-oriented approach. This concept promises to provide feature independence so that strict requirements to structure data into layers will be eliminated. Potentially, object-oriented approaches applied to GIS will make application development much more flexible and will provide better ways to manage data accuracy and quality characteristics of geographic data.

COMPUTER HARDWARE TRENDS

The Desktop Revolution

Chapters 7 and 9 discuss the concepts of centralized and distributed systems and the growth in popularity of inexpensive but powerful personal workstations. The tremendous explosion in the use of microcomputers throughout the 1980s and, more recently, intelligent 32-bit workstations, will continue to affect geographic data management systems (Van Demark 1988; Croswell and Clark, 1988).

Microcomputers that rely on 16-bit processors and operating systems such as DOS have become common platforms for many mapping and database management operations for those who require inexpensive, single-user systems. Many users are becoming more sophisticated, however, demanding capabilities that exceed the limits of microcomputers. Although microcomputer capabilities have been enhanced by the addition of special processing boards and greater main memory and mass storage, microcomputers generally are not able to handle complex spatial analysis requirements or large volumes of data. While DOS-based and comparable microcomputers will continue to have a strong niche in the low-end, single-user market, the popularity of 32-bit workstations will increase rapidly.

These workstations, employing UNIX or other operating systems optimized for 32-bit processors, deliver great processing speed and can support sophisticated software, high-resolution graphics, main memory, and mass storage. Price decreases over the next few years will make these systems much more attractive to organizations that formerly would have purchased a microcomputer system.

Advances in Computer Processor Performance

The 1970s and 1980s have been characterized by an astounding climb in the speed and performance of processing units in comparison to their cost. During the last ten years, the performance/cost ratio of computer processing units has approximately doubled each three years; this trend likely will continue if not accelerate in the foreseeable future (Kucharvy 1988).

As discussed in Chapter 7, parallel processing and RISC are two approaches that will fuel the increased processing power. While they are not used extensively today in GIS, these established technologies can play a more improtant role both at the intelligent workstation level and in larger, multiuser processing units (Datapro, 1988).

Over the next decades, some of the following important areas of current research will yield extraordinary advances in computer processor performance.

Neural networks, which simulate information processing of the human brain, extend the concept of parallel computing to the design of processing units consisting of tens, hundreds, or thousands of individual interconnected processors.

Current research in *superconductivity* promises to reduce the inefficiencies of computer circuitry based on traditional conductive materials. The electrical resistance and resulting generation of heat in computer circuitry is a major barrier to increasing the speed of data flow and further miniaturization of computer components. New materials being developed that exhibit high conductivity (low electrical resistance) at normal temperatures will increase computer performance.

Another technological advance that may be applied to computer architecture is *optical circuitry*, which relies on impulses of light through optical fibers in place of electrical conductors. The feasibility of optical circuitry in commercial computer processors is still unproven, but it has the potential to increase processing performance at levels comparable to superconductivity.

Mass Storage Technology

Considerable advances in mass storage technology were accomplished in the 1980s. Users' demands for large data bases were met by mass storage devices of increasingly higher capacity and higher data-access rates. Until recently, it was common for a minicomputer-based GIS to use 1 gigabyte of on-line mass storage. Users in many organizations require 2 to 5 gigabytes, and the trend suggests tenfold increases in demand during the course of the next decade.

Many important improvements are being made to traditional magnetic disk storage media as well as optical disk drives. Since 1980, storage densities of magnetic drives have nearly tripled, and the devices have become faster. Intelligent disk-control devices have increased the efficiency of disk access, and advances in the drive mechanisms have improved disk-access speed.

The challenge of the future for very high mass storage capacities will be met by optical disk storage units. Optical storage technology, still in its infancy, has started to gain a niche in certain information management fields. Optical drives now exhibit storage densities up to twenty times that of magnetic media and allow several gigabytes to be stored on a single disk. Storage densities of optical media are likely to increase substantially. WORM (write once, read many) drives, which do not allow erasing and rewriting of data, are used for records that do not require frequent modification. Optical disk systems are now used in GIS for storing and retrieving scanned drawings, raster images, and miscellaneous forms.

Erasable optical drives have only recently appeared on the market. This technology is still largely in the development phase, and it will be the mid-1990s before use of erasable optical technology becomes common. Currently, the lack of optical disk format standards and low disk-access rates inhibit their general acceptance. (Healy and Desa 1989). The trend toward hybrid mass storage environments using magnetic, WORM, and erasable optical drives will grow in the 1990s.

Graphic Display and Hard-Copy Output

For those who remember the crude graphic display devices based on low-resolution vector storage tube technology in the mid-1970s, it is astounding how far graphic display technology has progressed in fifteen years. Today, 19-inch raster-scan monitors with resolutions exceeding 1,200 × 1,000 pixels are common. Larger and higher-resolution monitors for specialized uses are available. As impressive as this sounds, demands for larger and higher-resolution display systems will continue. The popularity of video projection systems has illustrated the power of large display systems.

Large-size, high-resolution interactive display devices could substitute for hard-copy maps in many cases. As an example, a *softplot* device using laser-driven liquid crystal techniques produces a high resolution, D-size display. Additional improvements in resolution and display speed are required before devices of this type become cost-effective.

Advances are also ahead for hard-copy graphic output devices. Electrostatic plotting technology will grow in popularity, particularly for the production of large-format plots. Today, large-format electrostatic plotters are available for about one-third of their 1980 cost. This cost drop, along with an increase in resolution, make them important competitors with pen plotters. Moderate cost decreases for electrostatic plotters can be expected through the 1990s. Laser printers are also likely to gain popularity in large-format graphic output over the next several years.

In addition, technological advancements likely will be made in local controllers used to rasterize plot files and drive electrostatic and laser plotters. Color electrostatic plotting will move toward single-pass color plotting not available from many vendors today.

The demand is high for color output from GIS. Color technology now available, however, is limited by relatively low resolution, slow production speed, a small range of colors and shades, and comparatively high cost. Improvements in color production processes initially focused on small-format (A- and B-size) plots. Color hard-copy production devices currently use impact printing, thermal transfer, elec-

trostatic, and ink jet printing techniques. Of these, it is likely that electrostatic and thermal transfer plotters will gain in popularity as resolution increases to over 400 dots per inch and cost continues to decrease.

Color production techniques that are emerging include color laser plotting and microencapsulation. Although it will take several years for color laser plotters to be perfected and to become cost-competitive, they promise high resolution and rapid production rates. Microencapsulation uses a special medium with embedded capsules of light-sensitive dyes. An optical head exposes the capsules, creating a color image. This technique potentially provides quick and inexpensive color plots and a greatly expanded range of colors. Other color hard-copy production techniques are in experimentation and will compete with established technology over the next five years.

COMMUNICATIONS AND NETWORKING

It is difficult to read any computer trade journal today that doesn't include an article on some aspect of the "revolution" in data communications and networking. *Networking* has become a buzz word for the 1990s. As one manufacturer puts it, "the network is the computer." GIS has been influenced greatly by breakthroughs in networking, and future advances in data communications will continue to shape the field.

Continual technological developments over the next decades will result in many advances in computer communications and networks, including these:

- Higher-speed networks capable of being extended over larger geographic areas
- Integration of data communications to support a wide range of user needs
- Greater distribution of processing power and data
- Improved software for managing complex networks

For those who remember the crude graphic display devices on low-resolution vector storage tube technology in the mid-1970s, it is astounding how far graphic display technology has progressed in fifteen years.

- Flexibility in networking of multivendor hardware
- Operating system and data communication standards

At present, it is common to make a clear distinction between local area networks (LANs) and wide area networks (WANs). In the future, this distinction will become less important as high-speed data networks using a variety of communication media become feasible over long distances. LANs currently used in GIS are dominated by the Ethernet standard, a coaxial cable usually serving a single building or multiple buildings in close proximity. The 10-megabit-per-second speed limitation, although adequate for most users today, will be too limiting for future networks, which will support transmission speeds of over 100 megabits per second. The hypermedia concept discussed above will require speeds in this range.

These greater speeds and distances will be made possible by new communication media, most notably fiber optic cable, and improvements in existing techniques for remote data communications, such as satellite, microwave, and radio transmission.

Integrated communication networks in the 1990s will be capable of transmitting data between computer systems, along with voice, video, and other special transmissions. Current efforts by private companies to create fiber optic "backbones" within metropolitan areas and on a national scale eventually will provide high-capacity data "highways" for networks of the future. Software improvements will allow a smooth interface for transmissions using a variety of communications media, and this will blur the distinction between LANs and WANs. The net effect of these advances in GIS and data processing in general will be greater access to information in a transparent, open environment.

Chapter 9 discussed the trend toward distributed systems, in which processing power and data are dispersed throughout a network. Fueled by advances in data communication media, software, and the gradual adoption of standards, the popularity of distributed systems will continue to increase. As noted previously, users will have a wide variety of options in the way a network is configured—not simply a choice between a centralized or distributed approach. Hybrid systems that combine centralized and distributed characteristics will be created and customized to the users' particular needs (Dangermond 1988). As higher network speeds become available in the mid-1990s and more sophisticated network management software is developed, distributed data configurations will become more common.

The greatest impetus to the development of flexible networks will come from the full adoption, by computer vendors and users, of software and data communication standards. Flexible networking wherein users can access processing units, databases, and peripheral devices in a multivendor environment is constrained now by the lack of common operating systems and communication interfacing procedures (Tom 1988).

True portability of application software can occur through the use of a standard operating system by different vendors. Ideally, this would allow GIS software to operate without modification on processing units developed by different vendors. Computer firms are investing significantly in the development of operating system standards, while at the same time seeking to protect their bases of users dependent on proprietary operating systems. Currently, the UNIX operating system holds perhaps the greatest promise for an acceptable standard. Over the next four or five years, it is likely that UNIX, with major enhancements of the versions available today, will become an accepted standard in general purpose computing. Most GIS software will probably use this operating system. This does not mean that proprietary operating systems supported by major hardware vendors such as IBM, Digital Equipment, and others will cease to be important. The trend toward operating system standards will, however, give users a choice when software portability is a key concern.

A major push is underway on an international basis to develop data communication standards that will make it possible to link a wide variety of computer devices. This thrust toward communication standards, initiated by government and private user groups, is receiving active attention by the computer vendors.

The open system interconnect (OSI) model, as discussed in Chapter 9, is the best illustration of a system development of communication standards. The OSI model being defined and promoted by the International Standards Organization (ISO) establishes a structure for communication protocols. This model addresses low-level communications (physical and electrical interfaces, basic network protocols) as well as higher-level communications (network routing, error correction, communication between application software).

In the United States, the National Institute of Standards and Technology (NIST) is leading the way toward promulgation and adoption of computing standards. The NIST is developing the Government Open System Interconnect Profile (GOSIP), which incorporates a wide range of open system standards. GOSIP will be used to guide future procurements of computer hardware and software and therefore will be a strong inducement for the development of compliant products by vendors.

Although the proprietary interests of large computer vendors have inhibited the adoption of industrywide standards, user interest in open communications is strong, and vendors will respond with increasingly flexible solutions for complex networking.

SYSTEM PLANNING IN A DYNAMIC ENVIRONMENT

How can one plan a computer system now that will take advantage of the new developments in hardware and software right around the corner? Will a new computer system become obsolete a year or two after it is installed? Is the best approach to wait until that "new and improved model" is released? Although these concerns are valid, it is not wise to delay GIS development because technological advancements are on the horizon. This philosophy only delays the benefits an organization can realize from a system. With proper planning, a system can be designed to take advantage of future developments and respond to changes in the technology and the demands of users.

New advances in technology can be incorporated with a coordinated implementation plan that looks to the future and assumes that the system will be built through a series of incremental upgrades. Smaller, inexpensive processing units and more flexible networks provide users with a greater ability to grow as demands dictate. Users must adopt a realistic attitude about the notion of obsolescence. A computer device is not obsolete because a more powerful unit enters the market; it is obsolete when it no longer meets the demands of the users.

Several basic guidelines can help in the design and implementation of system configurations that meet the challenges of the existing dynamic computer industry:

- Put sufficient efforts and resources into system planning to fully evaluate and anticipate the short- and long-term needs and applications of the users
- Plan for an incremental hardware and software procurement. Add hardware and software to the system as needed and avoid the tendency to acquire large amounts of hardware and software before the data base reaches a sufficient maturity
- When selecting hardware and software vendors, consider their support for standards, the ability to easily upgrade their systems, and their capability and commitment to research and development
- Incorporate GIS planning with other information system developments in the

Users must adopt a realistic attitude about the notion of obsolescence. A computer device is not obsolete because a more powerful unit enters the market; it is obsolete when it no longer meets the demands of the users.

organization and use existing or planned networking facilities when appropriate

The adage "To fail to plan is to plan to fail" holds true with GIS and relevant technologies. A key element of the planning process provides a migration path for the incorporation of emerging technologies as they prove reliable and pertinient to a user's requirements. Technology provides much promise for the management of geographic information.

REFERENCES

Croswell P., Ahner A. (1990). Computing standards and GIS: A tutorial. *URISA Proceedings*, Vol II. Urban and Regional Information Systems Association, pp 88–105.

Croswell P., Clark S. (1988). Trends in geographic information system hardware. *Photogrammetric Engineering and Remote Sensing 54:* 1571–1576.

Dangermond J. (1988). A technical architecture for GIS. *Proceedings of GIS/LIS '88 Conference*, Vol. 2. Congress on Surveying and Mapping. San Antonio: The American Society of Photogrammetry and Remote Sensing/American Congress on Surveying and Mapping, pp. 561–570.

Datapro Research Corporation (1988). RISC technology: An overview. *Datapro Reports on Minicomputers*, Section M07-010, Delran, NJ.

Dunaway D. K. (1989). Knowledge-based automation for map conversion. *AM/FM International Conference, XII Proceedings*. AM/FM International Inc., pp. 144–151.

Egenhofer M. J., Frank A. U. (1989). Object-oriented modeling in GIS: Inheritance and propagation. *Auto-Carto 9 Proceedings*. Baltimore.

Healy R. G., Desa G. B. (1989). Transputer-based parallel processing for GIS analysis: Problems and potentialities. *Auto-Carto 9 Proceedings*. Baltimore.

Kindleberger C. P. (1989). Hypermedia systems: Implication for state, provincial, and local government. *URISA Proceedings*, Vol. III. Urban and Regional Information Systems Association, pp. 143–152.

Kucharvy T. (1988 September/October). Wrestling with the differences between PCs and workstations. *Computer Graphics Review.*

Maggio R. (1988). The role of the geographic information system in expert systems. *Proceedings of GIS '87 Conference*, Vol. 2. San Francisco: The American Society of Photogrammetry and Remote Sensing/American Congress on Surveying and Mapping, pp. 685–692.

National Research Council (1980). Need for a Multipurpose Cadastre. Washington, DC: National Academy Press, p. 112.

National Research Council (1983). Procedures and Standards for a Multipurpose Cadastre. Washington, DC: National Academy Press.

Rossmeissl H. J. (1989). The spatial data transfer standard: A progress report. *GIS/LIS '89 Proceedings*, Vol. 2. Orlando: ACSM, ASPRS, AAG, URISA, and AM/FM International, pp. 699–706.

Spencer R. C., Menard R. D. (1989). Integrating raster/vector technology in a geographic information system. *GIS/LIS '89 Proceedings*, Vol. 1. Orlando: ACSM, ASPRS, AAG, URISA, and AM/FM International, pp. 1–8.

The American Cartographer (1988, January). The proposed standard for digital cartographic data.

Tom H. (1988). Standards: A cardinal direction for geographic information systems. *URISA Proceedings*, Vol. II. Urban and Regional Information System Association, pp. 142–151.

Van Demark P. (1988). Desktop mapping: A glimpse into the future. *URISA News 94:* 20–21.

Vanzella L., Cabay S. (1988). Hybrid spatial data structures. *Proceedings of GIS/LIS '88 Conference*, Vol. 1. San Antonio: American Society of Photogrammetry and Remote Sensing/American Congress on Surveying and Mapping, pp. 360–371.

Wallace T., Clark S. (1988). Raster and vector data integration: Past techniques, current capabilities, and future trends. *Proceedings of GIS/LIS '88 Conference*, Vol. 1. San Antonio: American Society of Photogrammetry and Remote Sensing/American Congress on Surveying and Mapping, pp. 418–426.

Westwood K., Brinkman J. (1988). Toward a successful integration of relational and quadtree structures in a geographic information system. *URISA Proceedings*, Vol. III. Los Angles: Urban and Regional Information System Association, pp. 181–189.

SUGGESTED READINGS

Hazelton N. W. J., Leahy F. J., Williamson I. P. On the design of temporally-referenced, 3-D geographical information systems: Development of four-dimensional GIS, *GIS/LIS '90 Proceedings*, Vol. 1. Anaheim: American Congress on Surveying and Mapping/American Society for Photogrammetry and Remote Sensing, pp. 357–371.

Parsaye K. et al. (1989) *Intelligent Databases: Object Oriented, Deductive Hypermedia Technologies*, John Wiley & Sons, Inc., New York.

Stock R., Robertson, B. The next wave. *Computer Graphics World* 13:82–88.

Worboys M. F., Hearnshaw, H. M., Maguire, D. J. Object-oriented data modeling for spatial databases. *International Journal of Geographic Information Systems* 4:369–383.

Glossary

Accuracy—In the context of map accuracy, the closeness of results of observations, computations, or estimates of graphic map features to their true value or position. *Relative accuracy* is a measure of the accuracy of individual features on a map when compared to other features on the same map. *Absolute accuracy* is a measure of the location of features on a map compared to their true position on the face of the earth. *Mapping accuracy* standards generally are stated as an acceptable error that must be achieved and the proportion of measured features that must meet the criteria. In the case of some plotting and display devices, accuracy refers to tolerance in the display of graphic features relative to the original coordinate file.

Accuracy of geographic information database also takes into consideration the accuracy of content (e.g., errors of commission and omission of features), the correctness in identification of features, the currency or temporal characteristics of the data, and the topologic integrity of graphic and nongraphic information.

Address Matching—Relating street addresses to point locations or areas such as census blocks, tracts, administrative units, or the location of buildings where permits are pending or the location of emergency response incidents.

Analog—A medium or mode in which data is represented by continuously variable quantities such as amplitude, frequency, shape, or position. Hard copy or screen displays of maps, drawings, records, and photographs are analog images. In analog phone communications, voices are transmitted as an electrical signal that continuously varies in frequency and amplitude.

Analytical Triangulation—A process used in photogrammetric mapping that establishes the mathematical relationships among control points and adjusts them as a network of location values. Analytical triangulation involves a computer-aided extension of the control points in order to mathematically densify the control network and to provide a positionally accurate structure for the photographs.

Annotation—The alphanumeric text or labels plotted graphically on a map, such as street names, place names, identification numbers, and dimensions.

Application Software—Computer programs that enable uses (e.g., mapping, analysis) to be made of computerized information.

Area—A bounded, continuous two-dimensional object that may or may not include its boundary.

Arithmetic and Logic Unit (ALU)—A central part of the computer within the Central Processing Unit (CPU) that performs numerical operations fundamental to the execution of programs and other functions.

Artificial Intelligence—The capability of a computer to simulate inductive reasoning based on a set of logical rules.

ASCII (usually pronounced *as-key*)—An acronym for American Standard Code for Information Interchange, an 8-bit code established by the American National Standards Institute which assigns 8-bit codes to all standard keyboard characters to provide compatibility for data communications.

Assembler—A "low-level" symbolic programming language in which individual statements represent discrete machine operations.

Asynchronous Communications—The transmission of data via serial lines without a specific timing pattern.

Attribute—A type of nongraphic data that describes the entities represented by graphic elements. The term *attribute* frequently is used to cover all types of nongraphic, usually alphanumeric, data that are linked to a map element. For example, a map depicting parcels, each identified by a parcel number, may be linked to an attribute data file containing information about ownership, land use, and appraised value. In the context of engineered facilities, attribute data describe the facilities themselves, recording the characteristics of individual devices such as the type, size, material, and manufacturer.

Automated Mapping (AM) System—A computer system used to draw and produce maps. AM systems are designed for efficient graphic data processing and display; they have little or no geographic analysis capability and limited ability to store and manipulate any data other than graphic images.

Automated Mapping/Facilities Management (AM/FM) System—A term used predominantly by individuals and organizations involved in the infrastructure management disciplines (e.g., utilities and public works) to describe a computer system that processes graphic and nongraphic data for a variety of purposes, such as managing geographically distributed facilities, overlaying combinations of features and recording resultant conditions, analyzing flows or other characteristics of networks, and defining districts to satisfy specified criteria.

Auxiliary Storage Device—A device used to maintain large volumes of digital data and software that is directly accessible through high-speed connections to a processing unit. Sometimes called mass storage devices, they include disk and tape drives.

Bandwidth—A term used to express data transmission speed. A "bandwidth of 10 megabits" means the line can transfer data at the rate of 10 million bits per second.

Batch Processing—A mode of processing in which the user is not in direct communication with the processing unit while a program is being executed. In *batch* or *offline* processing, the user submits a *job* to the processing unit for execution when time is available. The job is placed in a *batch queue*, a temporary storage area. The job contains information necessary to execute the program, the destination of products to be generated, and identification of the user.

Baud—One *signal unit* per second, a measure of analog data transmission speed that describes the modulation rate of a wave, or the average frequency of the signal. If an analog signal is viewed as an electromagnetic wave, one complete wavelength or cycle is equivalent to a signal unit. Baud often has been used synonymously with *bits per second*. The baud rate may equal bits per second with some transmission techniques, but special modulation techniques frequently deliver a bits-per-second rate higher than the baud rate.

Binary—Having two possible states or values (e.g., "on" and "off"). Computer instructions and data are represented as sequences of binary digits.

Binary Synchronous Control (BSC)—A protocol used for synchronous communications, often referred to as *bisynch* or *bisynchronous protocol*. This protocol, originally developed by IBM, was adopted by many mainframe and minicomputer vendors in the 1970s and early 1980s. Bisynch is classified as a *character-oriented* protocol because a data transmission consists of frames of characters (8-bit strings), each of which represents a control signal or a portion of the message. Significant disadvantages of this protocol for interactive processing in complex networks led to the development of *bit-oriented* protocols in the late 1970s.

Bit—Abbreviation for *bi*nary digi*t*. The basic element of a binary code, a bit is a unit of computer information represented by an electrical pulse (binary value 1) or no pulse (binary value 0) or a similar code equivalent to the result of a choice between two alternatives (e.g., "yes" or "no").

Broadband Network—A communications facility that operates with uniform efficiency over a wide band of frequencies with the capability to transmit multiple channels simultaneously over the same physical connector.

Buses—Parallel data paths that carry data at high speed between certain components of a computer system. The information-carrying capacity of a bus depends on its *width*, or the number of bits that can be transferred simultaneously. Mainframe buses usually are 24, 32, 48, or 64 bits wide, while minicomputer and microcomputer buses typically are 8, 16, or 32 bits wide.

Byte—A group of adjacent binary digits that a computer stores and retrieves as a unit.

C Language—A third-generation *source code* programming language often used to develop application software for computers using the UNIX operating system.

Cache Memory—Temporary storage for frequently accessed data and programs. When a program is executed by the CPU, the cache is first examined for the required data or instructions; auxiliary storage devices are accessed only if the information is not found in cache.

CAD—Acronym for Computer-*A*ided *D*rafting; software that supports civil, mechanical, and other engineering design ac-

tivities, including interactive graphics display, engineering calculation and analysis, and limited attribute processing.

Also an acronym for *computer-aided dispatch*, which describes a number of automated tools used by organizations to identify the physical location of a query for service or emergency assistance and assist in identifying the most appropriate vehicle and route of response.

CAD/CAM—Acronym for Computer-*A*ided *D*esign/Computer-*A*ided Manufacturing. CAD/CAM software differs from GIS in that it can only create displays but cannot analyze or process base data.

Cadastral Features—Map elements created from legal descriptions that depict boundaries of property ownership and other rights to land.

CADD—Acronym for Computer-*A*ided *D*rafting and *D*esign; software that supports civil, mechanical, and other engineering design activities, including interactive graphics display, engineering calculation and analysis, and limited attribute processing. In addition, CADD systems enable some modeling of graphic relationships and analysis of logical relationships.

CAM—Acronym for Computer-*A*ssisted *M*anufacturing and Computer-*A*ssisted *M*apping.

Cartesian Coordinate System—A coordinate system in which the location of a point on a plane is expressed by two coordinates (x, y) that measure the point's distance from two intersecting, often perpendicular, straight-line axes along a line parallel to the other axis, or three (x, y, z) coordinates that locate a point in space by its distance from three fixed planes that intersect one another at right angles.

Cartography—The science or art of making maps.

Centralized Processing—Centralization of a computer system's "intelligence" and central execution of all instructions and functions of a system.

COBOL—Acronym for *C*ommon *B*usiness-*O*riented *L*anguage; a *source code* computer programming language.

Command—A user's directive to a computer system to initiate an action or operation.

Compiler—A computer program that translates a *source code* language into *machine code*, which is the binary representation of the program, for execution by a computer.

Computer-Assisted Mass Appraisal (CAMA)—A computer-aided analysis of data describing property characteristics that is used in establishing property values for tax assessment.

Computer Output Microfilm—Systems which store computer-generated microfilm images of maps and other documents. Images can be stored on aperture cards and used with a microfilm printer to generate hard copies quickly.

Coordinate Geometry (COGO)—A program used to generate digital map features from geometric descriptions. Mathematical algorithms compute coordinates from geometric descriptions such as bearings and distances; the coordinates are stored and used to generate graphic map displays. COGO has become the common shorthand name for the technique.

Coordinate System—A framework used to define the position of a point, line, curve, or plane and derivative map features within a two- or three-dimensional space.

Core Memory—A computer's main memory unit that provides high-speed access to data and program statements during program execution. In the microcomputer industry, *random access memory* (RAM) is the term used for main memory or core memory.

CPU—Acronym for *Central Processing Unit*, which directs and supervises all of a computer's functions.

Cursor—A hand-held device consisting of a glass or plastic lens with a cross-hair target and multiple buttons. A cursor is used to *digitize* a hard-copy map mounted to the surface of a tablet by recording the position of objects on the map. Conductors in the tablet receive electrical signals emitted by the cursor and convert these to relative tablet positions.

Data Base—A large collection of data in a computer, organized so that it can be expanded, updated, and retrieved rapidly for various uses.

Data Communications—The movement of data between devices in a computer network.

Data Dictionary—A database that serves as a catalogue containing information about map features or attributes. The catalogue may define data file and element names, sources, accuracy, date of entry or update, and other characteristics of the data and its sources.

Database—A specific grouping of data within the structure of a data base management system that has been defined by a particular use, user, system, or program.

Decentralized Processing—Distribution of computer processing tasks among separate processors.

Device Driver—A program that provides an interface between the operating system and application software to support output on a specific peripheral device. The device driver translates commands issued through application software into instructions that a device, such as a plotter, can interpret to perform a certain function.

Device Server—A hardware component used to connect multiple nonintelligent devices, such as nonintelligent terminals, query stations, plotters, and printers, to a local area network. A device server contains computer processors dedicated to supervising communications between the devices and the network.

Diapositive—A positive photographic transparency (the mirror image of a photograph negative).

Digital—The discrete numerical representation of information, contrasted with analog representation; a method of defining numbers or values in increments.

Digital Capture Device—A device that converts an analog image, such as a hard-copy map or aerial photograph, to digital form by optically reading the tone or color variation of the analog image and converting it to a raster form suitable for display and analysis.

Digital Elevation Model (DEM)—The digital cartographic representation of the surface of the earth or a subsurface feature through a series of three-dimensional coordinate values.

Digital Line Graph (DLG)—A standard file structure for cartographic digital data established and used by the United States Geological Survey that contains point coordinates describing planimetric and contour data such as boundaries, drainage lines, transportation routes, and other features.

Digital Service Unit—A peripheral device that enables data transmission between computer devices using an outside communication carrier. These units connect devices across digital lines that support both voice and data transmission.

Digital Terrain Model (DTM)—Most often used synonymously with DEM.

Digitizer—A peripheral device used to convert graphic information from analog to digital form for use in a computer; sometimes called a *digitizing tablet* or *digitizing table*.

Digitizing—A technique used to convert graphic information from analog to digital form, in which maps are retraced by a person using a digitizing tablet and cursor. The map to be digitized is mounted on a digitizing tablet. The scale, angle of rotation, and area coverage are registered with the computer system through a standard procedure; each map feature is then digitized by pointing the cursor and pushing appropriate buttons.

Digitizing Tablet—A peripheral device, also called a digitizing table, used to convert graphic information from analog to digital form for use in a computer. Positional points are selected by a cursor on a hard-copy map mounted on the surface of the digitizing tablet and transmitted by electrical signals from cursor to tablet. Conductors in the tablet receive the signals and convert these to relative tablet positions.

DIME File—Acronym for *Dual Independent Map Encoding File*, a geographic file based on line segments produced by the United States Bureau of the Census for each Standard Metropolitan Statistical Area in the United States. Also called the *Geographic Base File/Dual Independent Map Encoding (GBF/DIME)* when street segments are assigned geographic coordinates.

Disk Drive—A peripheral mass storage device using magnetic or optical technology that enables direct and quick storage and retrieval of data without loading data.

Distributed System—A system in which processing tasks and/or data are distributed among separate computers on a network.

DOS—An acronym for *Disk Operating System*, a microcomputer operating system developed by IBM and Microsoft Corporation that has become a de facto standard in the microcomputer industry.

Dumb—A descriptor associated with the lack of intelligence of a database element or user workstation; annotation incorporated into a map or engineering drawing that exists only as a graphical element and is not linked to another graphical element or a corresponding attribute database. Workstations that have little or no local processing power and rely on the processing power of another device e.g., mainframe or minicomputer are also considered unintelligent or dumb.

EBCDIC—An acronym for *Extended Binary Coded Decimal Interchange*. A standard which assigns 8-bit codes to standard keyboard characters.

Edgematching—The process of eliminating locational and content discrepancies in the representation of features at the edges of adjacent map sheets to create a continuous map.

Electronic Distance Measuring Device (EDM)—A surveying instrument using laser and computer technology to record, digitally, distance and angles; replaced traditional optical surveying transits.

Electrostatic Plotter—A peripheral device that produces hard-copy graphic images in raster format.

Ethernet—A network protocol defining the physical and data link layers in the Open Systems Interconnect (OSI) model, a framework for developing standard interfaces and communication protocols. The Ethernet standard uses a bus topology to transfer data at a maximum speed of 10 million bits per second.

Executable Code—The binary representation of a computer program that can be executed by the computer processor; also called *machine code*.

Expert System—A computer system using rule-based logic to simulate human intelligence.

Federal Geodetic Control Committee (FGCC)—A committee established by the federal government that has developed and published geodetic control standards and specifications for the United States describing the methods, equipment, and procedures used to achieve specified accuracy levels.

Federal Interagency Coordinating Committee on Digital Cartography (FIC-CDC)—A group created by the Federal Office of Management and Budget to promote the exchange of information and ideas on technology and methods for collecting and using digital cartographic data.

Fence—A border that defines an area on a computer monitor or digitizer in which all enclosed features may be used for graphic operations.

Fiber Optics Transmission—The technique of transmitting data through an enclosed bundle of thin, transparent fibers of glass or plastic as pulses of light generated by lasers.

Flat File—A two-dimensional data structure commonly used as an interchange format for the loading and exchange of digital data; also called a *sequential file*.

FORTRAN—A *source code* programming language.

Fourth-Generation Language (4GL)—A high-level computer programming language used to develop applications. It is based on an English-like, nonstructured language that supports the development of programs with relatively few statements pertinent to third-generation languages. A 4GL must be translated into *source code* and then into *machine code* for execution by a computer.

GBF/DIME File—Acronym for *Geographic Base File/Dual Independent Map Encoding File*, a geographic file based on line segments produced by the United States Bureau of the Census for each Standard Metropolitan Statistical Area in the United States.

Geocode—A spatial index code (e.g., addresses, parcel numbers) identifying unique points, lines, or areas that is stored in both graphic and nongraphic data.

Geodesy—A branch of applied mathematics that determines by observation and measurement the shape and size of the earth or a large part of its surface, and the exact location of points on its surface.

Geodetic Control—A network of surveyed and monumented points on the earth's surface whose location is established in accordance with national accuracy standards.

Geoprocessing and Network Analysis System—A computer system incorporating geographic identifiers, or geocodes, used primarily for geographic analysis. The most common application is the assignment of planning, census, or statistical area identifiers to data records containing addresses, using a geographic reference file such as a DIME file.

GIS—Acronym for *Geographic Information System*, a computer system that stores and links nongraphic attributes or geographically referenced data with graphic map features to allow a wide range of information processing and display operations, as well as map production, analysis, and modeling.

Global Positioning System (GPS)—A constellation of nine satellites originally developed by the United States Department of Defense as a navigational aid that recently has become available for geodetic control surveying. The satellites transmit signals that can be decoded by specially designed receivers to determine positions precisely (within centimeters). These measurements can be made at a fraction of the cost of traditional first-order surveying techniques. Eventually, the GPS will be replaced by an eighteen-satellite constellation that will allow 24-hour, all-weather operational capacity in both navigation and relative positioning.

Graphic—In the GIS context, a term that

refers to digital descriptions of map features, logical geographic relationships among features, drawings, and images. Graphic features or geometric forms are described by seven elements: points, lines, areas, grid cells, pixels, symbols, and annotation. The GIS uses graphic data to generate a map or cartographic "picture" on a display device (computer screen), paper, or other media.

Graphic Terminal—A workstation capable of projecting and manipulating graphics with little or no local processing capability.

Grid Cell—An area enclosed within a network of uniformly spaced horizontal and perpendicular lines. In raster processing, data is represented spatially on a matrix of grid cells, also called *pixels*, which are assigned values for image characteristics or attributes.

Hardware—Any tangible or physical device used as part of a computer system, including the processing units that execute programs, auxiliary storage units that maintain data and programs, and peripheral devices such as terminals, plotters, and printers.

Hierarchical Database—A database structure in which data is stored based on *parent–child* or *one-to-many* relationships with explicit pointers that define the relationship between segments of a record.

High-Level Language—A computer programming language used to develop applications, sometimes used synonymously as a fourth-generation language.

High-Performance Workstation—A computer with a 32-bit processor, large main memory, and extensive graphics capability.

Hydrography—The description and study of water bodies such as seas, rivers, and lakes.

Hypermedia—In a geographic data context, an emerging capability to reference, retrieve, and integrate different forms of data, including vector maps, tabular databases, freeform text, raster maps, engineering drawings, standard forms, photographs and other images, or audio records.

Information Resource Management (IRM)—A discipline dealing with the creation, production, collection, management, distribution, and retrieval of information.

Ink Jet Plotter—A peripheral device used to make hard-copy plots in which colored inks are forced through small jets onto the print medium.

Intelligent Workstation—A 32-bit microcomputer, high-performance computer, or assemblage of devices with a substantial amount of processing power and often mass storage capability that works alone or interactively with a computer to perform specific tasks.

Interactive Processing—A processing mode in which a user issues commands directly to the processing unit for immediate execution, receiving output as soon as the operation is complete.

International Standards Organization (ISO)—An organization of government and industry representatives from many countries that developed the *open system interconnect (OSI)* model, a structure for communication protocols.

Key Entry—A technique commonly used to enter nongraphic data into a computer.

Land Information System (LIS)—A database describing physical or legal characteristics of land areas, sometimes called a *land records system (LRS)*. The most common forms of land information systems describe property ownership, land value, tax assessment, and property boundaries.

Landsat—A satellite program initiated by the National Aeronautics and Space Administration (NASA) in the early 1970s; the term also refers to the individual satellites that are part of the program. The early Landsat platforms produced images that aided regional analysis, although the sensors lacked the resolution required for most detailed mapping or analysis. The increasing spatial and spectral (light band) resolution of satellite sensors in the late 1970s and 1980s extended the usefulness of the images. In particular, the Thematic Mapper sensor of Landsat 4 and 5 increased ground resolution from the 79 meters of previous Landsat sensors to 28 meters. Since 1986, Landsat has been owned and operated by Earth Observation Satellite Company (EOSAT), a private firm. Landsat 6 is to be launched in 1992.

Laser Plotter—A peripheral device that makes raster graphic plots of computer data by exposing photosensitive material to a thin laser beam of light.

Laser Printer—A peripheral device used to produce high-volume, high-quality hard-copy text reports and graphics from computer data. A laser is used to apply charges to a rotating drum. The drum is exposed to a dry toner, which causes toner particles to adhere to charged portions of the drum. Paper is pressed against the drum, and the toner fuses to the paper. A page can be produced at each rotation of the drum.

Latitude/Longitude Coordinate System—A global coordinate system in which locations are expressed by geographical coordinates (the geodetic latitude and geodetic longitude) that depict angular measurements relative to the earth's ellipsoid.

Layers—A term commonly used to describe the graphic component of a GIS database. Each layer contains a set of homogeneous map features registered positionally to other database layers through a common coordinate system. Data is separated into layers based on logical relationships and the graphic portrayal of sets of features. The term *layer* is used in a cognitive sense to describe a grouping of map features which are organized and stored by the GIS software in many, and typically proprietary, ways.

Line Printer—A peripheral device that prints text or images line-by-line and provides either fully formed characters or dot-matrix characters. Line printers with fully formed characters arrange raised characters on a chain or band that rotates horizontally (parallel to the print line orientation) at very high speeds. Hammers strike the characters over a ribbon, and the image is formed on the paper being fed through the printer on a sprocket mechanism. Dot-matrix printers have print heads that consist of multiple wires or pins that form a character with a pattern of dots as the head strikes the paper.

Line—A one-dimensional object. A line segment is a direct line between two points. Special forms of lines include *string*, a series of line segments; *arc*, a locus of points forming a curve defined by a mathematical function; and *chain*, a directed sequence of nonintersecting line segments or arcs with *nodes* at each end.

Local Area Network (LAN)—A system for connecting computer hardware and software to provide shared access and data movement among multiple devices within close proximity, normally not exceeding a radius of several miles.

Macro Programming—Creating a sequence of software commands that, in combination, perform a complex operation. The sequence of commands is stored in a file (the macro program) which is executed by a single user command.

Main Memory—A computer's main or core memory unit that provides high-speed access to data and program statements during program execution. In the microcomputer industry, *random access memory* (RAM) is the term used for main memory or core memory.

Map Projection—A mathematical model used to transform positions on the surface of the earth, which is curved, onto a flat map surface.

Map Scale—The ratio of units of linear measurement on the map to units of measurement on the earth. Scale often is stated as a representation fraction such as 1:2,000, where one part or unit of measurement on the map is equal to 2,000 parts or identical units of measurement on the earth. Scale also can be stated in specific measurement units such as 1″ = 100′, where 1 inch on the map is equivalent to 100 feet on the ground.

Mass Storage Device—A computer peripheral that increases storage capacity, sometimes called *auxiliary storage*, such as a disk drive or tape drive. Sometimes called *auxiliary storage device*.

Megabyte (MB)—A unit measure of computer memory or storage equivalent to 1,048,576 bytes.

Menu—A display of command options from which a user selects by pointing with a mouse, digitizing cursor, arrow keys, or other mechanism without having to type the command as text. Menus may be displayed on a computer screen where they may be prompted by movement of the cursor or mounted on a digitizing table and prompted by a digitizing puck or pen.

MFLOPS—An acronym for *Million Floating Point Operations Per Second*, a measure used to indicate power or speed of a computer processing unit.

MIPS—An acronym for *Million Instructions Per Second*, a measure used to indicate power or speed of a computer processing unit.

Modem—An abbreviation for *modulator–demodulator*, a peripheral device that translates digital computer data to analog form for transmission over a telecommunications network, usually an analog telephone line. A modem at the end of the telecommunications line restores the information to digital form.

Mouse—A palm-sized device connected to a workstation that is used to control a cursor or pointer on the screen of a graphics monitor. A small, free-rotating ball on the mouse's underside or a light-sensing device tracks its movement over a surface. As the mouse is dragged over a surface, the screen pointer moves in relation to the movement of the mouse. Using buttons on the mouse, an operator can invoke commands from a screen menu and enter and edit graphic features.

Multiplexor—A device that splits a high-speed communication line into multiple channels supporting multiple devices and recombines multiple channels into a single high-speed channel. Multiplexors maximize the efficiency of a communication line by allowing multiple users to communicate on one high-speed link.

Multipurpose Cadastre—An integrated land information system containing legal (e.g., property ownership or cadastre), physical (e.g., topography, man-made features), and cultural (e.g., land use, demographics) information in a common and accurate reference framework. The reference framework typically is established with rigorous geodetic and survey control standards, such as the state plane and latitude/longitude coordinate systems.

National Center for Geographic Information and Analysis (NCGIA)—A university research consortium formed to evaluate the social, legal, and institutional impacts of the spreading use of GIS technology. The NCGIA, established in 1988 with a grant from the National Science Foundation, is a consortium of research groups at the University of California–Santa Barbara, State University of New York–Buffalo, and the University of Maine–Orono.

National Committee for Digital Cartographic Data Standards (NCDCDS)—A group sponsored by the federal government and composed of government, university, and industry representatives that proposed standards for the structure, content, terminology, and quality of digital cartographic data.

National Digital Cartographic Data Base (NDCDB)—A collection of digital data maintained and distributed by the USGS and organized by scale-specific, product-specific cartographic units. The NDCDB contains information on transportation, hydrography, boundaries, Public Land Survey System (PLSS), topography, land use and land cover, and geographic names. Two major types of digital data are produced: elevation data, distributed as digital elevation models (DEMs), and planimetric and contour data, distributed as digital line graphs (DLGs).

National Geodetic Reference System (NGRS)—A network of monumented geodetic control points in the United States maintained by the National Geodetic Survey.

Natural Language—A computer command and programming language, sometimes called fifth-generation language, that is less structured and closer to conversational English and will accept commands from users in forms that require less stringent rules for command syntax and specific command vocabularies.

Network—A total computer system that integrates the functions of hardware and software, and the support provided by data communications. Also, a data structure used to model relationships of interconnected linear features, such as roads and utility transmission systems.

Network Analysis System—A computer system used primarily for geographic or statistical analysis of relationships and flow that are dependent on the connectivity of segments and nodes in a linear system, such as the circuits of a utility network.

Network Data Model—A data storage structure in which records of the same type are grouped in conceptual files, in owner-coupled sets reflecting one-to-many relationships, or in many-to-many relationships.

Network Protocol—A fixed set of rules specifying the format of data exchange. A particular protocol defines how the individual bits are to be arranged in transmitting a message so that it can be received and interpreted correctly by another node on the network.

Network Topology—The arrangement or layout of the devices and communication lines in a computer network.

Neural Networks—A computer network wherein individual processors are linked similarly to neurons in the human brain; processing power and memory are completely distributed but efficiently connected. An extension of the concept of parallel processing, neural networks may have tens, hundreds, or thousands of individual, interconnected processors.

Node—A zero-dimensional object, and a special type of point, that is a topological junction or end point and that may specify a geometric location.

Nonalgorithmic Digital Records Systems—Systems designed to store, retrieve, and manipulate raster databases.

Nongraphic Data—Digital representations of the characteristics, qualities, or relationships of map features and geographic locations, usually stored in conventional alphanumeric formats. The term *nongraphic* is often used to differentiate those data that do not describe the graphic images of the map features. Nongraphic data are often called *textual data* or *attributes*. They are related to geographic locations or graphic elements and linked to them in the GIS through common identifiers or other mechanisms. Often nongraphic data are managed separately from the graphic data due to their different characteristics or their maintenance and use in other systems. Sometimes referenced as tabular data.

Operating System—The software that interacts directly with the processing unit to control all basic functions of a system, such as allocating memory and disk space, loading programs for execution, and controlling input/output operations, user access, and security.

Optical Disk Technology—The capability to store vast amounts of data on optical disk drives. Also called optical storage technology, optical drives now exhibit storage densities up to twenty times that of magnetic media and allow several gigabytes to be stored on a single disk. Most optical drives use a technique in which a laser burns small impressions on a disk surface. The drive reads data by sensing variations in the light reflected from this surface. However, with the laser technique, data can be written to a particular portion of the disk only one time, and no data can be erased; this is called *write once–read many (WORM)*.

Optical Fiber—An enclosed bundle of thin, transparent fibers of glass or plastic through which data is transmitted as pulses of light generated by lasers.

Optical Scanner—A peripheral device that digitally encodes information in a raster form from hard-copy maps and documents by optically scanning an image consisting of line work, text, and symbols. The scanner senses variations in reflected light from the surface of the document.

OSI Model—An acronym for *Open System Interconnect Model*, a structure for standard interfaces and communication protocols developed and promoted by the International Standards Organization (ISO). The model addresses low-level communications (physical and electrical interfaces, basic network protocols) as well as higher-level communications (network routing, error correction, communication between application software programs).

Parallel Communications—A broad classification of digital communications in which a sequence of bits is transmitted simultaneously via cables consisting of multiple wires, each carrying one data bit at a time. Parallel lines transmit high volumes of data very quickly, and are used in the internal circuitry of computer processors and to connect devices requiring very high transmission speeds (e.g., mass storage devices). Parallel lines have major distance restrictions and are used most often to connect components and devices in close proximity, usually in the same room.

Parallel Processing—A computer architecture in which more than one processor is embedded within a single computer to execute tasks more quickly. Computers using parallel-processor architecture execute multiple programs or tasks of the same program in a parallel fashion, as opposed to the sequential execution of instructions that occurs in single-processor systems.

Parcel Identification Numbers (PIN)—A numbering scheme for identifying parcels in a computer system. Parcel identification schemes range from a simple sequential number to a geocode that also defines location by incorporating x and y coordinates for the parcel.

Parcel—A map feature depicting land ownership and rights. Parcel boundaries usually are described in narrative form on a deed as metes and bounds or bearings and distances. In much of the United States, the descriptions may refer to the Public Land Survey System (PLSS) township, range, section, and aliquot parts (half or quarter sections).

Pen Plotter—A peripheral device commonly used to generate large-format, hard-copy maps or drawings. Devices such as pen plotters draw line elements in vector form, while most other devices produce hard-copy graphics using a raster display or plotting technology.

Peripheral Device—Any hardware component that is physically and logically separate from the central processing unit. Peripheral devices commonly used in geographic information systems include digitizing stations, graphic query stations, alphanumeric terminals, plotters, and other devices used for entry, analysis, and generation of products from a tabular or graphic database.

Photogrammetric Digitizing—A technique used to compile new maps from aerial photographs. Similar to manual digitizing, in this approach, the table digitizer is replaced by a photogrammetric instrument such as an analytical stereoplotter. Photogrammetric digitizing most often is used to record very precise and accurate digital planimetric features and elevation data from stereophotography.

Photogrammetric Mapping—The process of compiling maps by measurement of controlled aerial photos.

Photogrammetry—The science of making reliable measurements by the use of photographs (usually aerial) in surveying and map making.

Pixel—Abbreviation for *picture element*, the smallest indivisible element that makes up an image. In raster processing, data is represented spatially on a matrix of grid cells, called pixels, which are assigned values for image characteristics or attributes.

Platform—a term used generically to denote the host device for GIS software, e.g., a workstation platform, frequently associated with a specific vendor or vendor and device model, like an IBM RS6000 workstation platform.

Planimetric Map—A map that shows the horizontal position on the earth's surface of natural and cultural physical entities such as roads, buildings, and water bodies that are visible and identifiable on aerial photographs and can be compiled into map features through photogrammetric or surveying procedures.

Point—A zero-dimensional object that specifies a map location through a set of coordinates.

Polygon—A closed, two-dimensional figure with three or more sides and intersections; an enclosed geographic area such as a land parcel or political jurisdiction.

Polygon Overlay—Software that generates derivative polygons based on the union, intersection, or subtraction of polygons in the original layers.

Primitives—Device commands to generate a hard-copy product. Primitives go to a device driver, which translates them into instructions to pen plotters, electrostatic plotters, and other graphic output devices. The device driver interprets the specific plot generation commands that describe the appearance and format of the plot, including positioning of the pen on the plot media, line weights and line types, shading and colors, annotation, and scale.

Processing Unit—A hardware device that executes programs and controls functions of a system.

Protocol—A set of rules that defines a function. In synchronous communication, blocks of information are structured to adhere to a particular protocol understood by devices on the network. A network protocol is a fixed set of rules specifying the format of data exchange. A protocol defines how the individual bits are to be arranged in transmitting a message so that it can be received and interpreted correctly by another device or node on the network.

Public Land Survey System (PLSS)—A reference scheme for recording property ownership by township, range, section, and aliquot parts (half or quarter sections) in the United States. The PLSS was laid out during the settlement of the country, dividing land areas into townships of thirty-six 1-square-mile sections.

Random Access Memory (RAM)—A term used in the microcomputer industry for the main memory unit which provides high-speed access to data and program statements during program execution.

Raster—A format for storing, processing, and displaying graphic data in which graphic images are stored as values for uniform grid cells or pixels.

Read Only Memory (ROM)—Memory that is protected from modification and that can only be viewed or reproduced.

Real Time—Computations or data communications that occur instantaneously or nearly so with the execution of a command or transaction.

Reduced Instruction Set Computer (RISC)—A computer with an operating system that executes complex commands very rapidly in one processing cycle, in contrast to conventional processing unit architectures that require many machine cycles to complete complex instructions. Although they offer increased performance for many applications, RISC-based systems require large main memories and special compilers to translate program source code into machine language.

Refresh Graphics—A technique used in raster-scan graphic display monitors. Using an approach similar to color television sets, raster-scan devices create an image that is redrawn or *refreshed* as a series of horizontal scan lines, at a time interval fast enough that no flicker is perceived by an observer. The refresh is required because of the diminishing sensitivity of the video screen phosphors to the initial electrical stimulation. Common refresh rates used in monitors today are 30 and 60 hertz (cycles per second). An *interlaced* monitor refreshes every other scan line in a refresh period, totally refreshing the image every two cycles. Noninterlaced monitors refresh the entire image in each cycle.

Relational Database—A database structure commonly used in GIS in which data is stored based on two-dimensional tables where multiple relationships between data elements can be defined and established in an ad-hoc manner. This structure allows great flexibility in the range of queries that can be handled and is particularly useful in situations where nonstandard inquiries are common.

Remote Communications—Communications among devices that are physically remote from the central computer site and rely on commercial or private communication facilities, such as a telephone system, fiber optic channels, microwave transmission, or satellites for the transmission of data.

Remote Sensing—The analysis and interpretation of images gathered through techniques that do not require direct contact with the subject. A discipline that evolved from photogrammetry, remote sensing of the earth's resources uses aerial or space photographs, electronic scanners, and other devices to collect data about the earth's surface or subsurface.

Resolution—A measure of the accuracy or detail of a graphic display, expressed as dots per inch, pixels per line, lines per millimeter, etc.

RS-232—A cabling configuration for asynchronous transmission common in minicomputer or microcomputer systems, in which peripheral devices such as terminals, printers, or plotters are connected with individual lines to the host computer.

Rubber Sheeting—A process that geometrically adjusts map features to "force" a digital map to fit a designated base. It is called rubber sheeting because the map is mathematically stretched to fit the base, given a set of known coordinate values whose locations are defined on both the base map and the map to be rubber sheeted. The process uses mathematical operations to minimize distortion.

Scanning—A technique for capturing and converting analog data to digital form. Scanners use an optical laser or other electronic device to "scan" an existing map and convert its images to digital format. Most scanners produce digital data in raster format, recording a value of dark (e.g., representing a line or symbol) or light (no line or symbol) for each grid cell or pixel of the scan.

Schema—The characteristics of computer files that contain nongraphic attributes. The schema holds information such as the name of data elements, the size of element fields in bytes or columns, the data element format (e.g., alpha, integer, binary), and other data required by the software to process attribute data.

Screen Copy Device—A device used to generate a small-format hard copy of a GIS display screen.

Serial Communications—A broad classification of digital communications in which data is transmitted one bit at a time, either asynchronously or synchronously. Asynchronous communication lines transmit data without a specific timing. In synchronous communications, blocks or packets of bits containing data are sent according to an established timing sequence. Serial lines are commonly used to connect peripheral devices in computer networks.

Source Code—A computer program written in a third- or fourth-generation programming language, such as FORTRAN, COBOL, BASIC, or C, that uses English-based commands and structured syntax to direct a computer to perform a certain task. A source code program must be translated by an interpreter or compiler to *machine code*, the binary representation of the program, that can be executed by the computer processor.

Standard Interchange Format (SIF)—A graphic data exchange format developed by Intergraph Corporation as an intermediate structure for the exchange of data from systems having different proprietary structures and formats.

State Plane Coordinate (SPC) System—A system of x,y coordinates for each state maintained by the National Geodetic Survey and commonly used in GIS technology in the United States. One of two map projections—Transverse Mercator (for states with a north–south orientation) or Lambert Conformal (for states with an east–west orientation)—is usually used to define one or more specific zones for each state. The geographic extent of the zones is limited to a distance of 158 miles (east and west in the Transverse Mercator or north and south in the Lambert Conformal) to limit the amount of distortion that accumulates due to the curvature of the earth.

Stereoplotter—A device used in photogrammetric mapping to digitize the horizontal position and elevation of selected points and features visible in aerial photographs.

Structured Query Language (SQL)—A data query language developed by IBM for use with several of its mainframe database management packages. Many developers of GIS software have adopted SQL.

Surveying—A branch of applied mathematics that provides techniques for determining the area of any portion of the earth's surface, the lengths and directions of the bounding lines, the contour of the surface, and accurate delineation of the whole on paper.

Symbols—Graphic elements that represent features as points on a map. A GIS is equipped to display a wide variety of symbols for points, lines, areas, and facilities.

Synchronous Communications—The transmission of data via serial or parallel lines according to an established timing sequence. Synchronous communication is used in many mainframe computer systems and in computer configurations using high-speed local area networks (LANs). Synchronous communication is used, for example, in IBM configurations using Systems Network Architecture (SNA) and in local area networks employing Ethernet standards.

Synchronous Data Link Control (SDLC)—A bit-oriented communication protocol developed by IBM as part of its Systems Network Architecture concept. Many major computer vendors support the SDLC protocol.

Systems Network Architecture (SNA)—A proprietary data communications architecture developed by IBM that has become predominant for data communications between IBM mainframe computers and locally or remotely attached microcomputers and data terminals. The SNA design is based on a tree structure; ultimate control of the network is assigned to one or more communication processors that support all network nodes.

Tape Drive—A peripheral mass storage device used to archive data on magnetic tape for later access by system users. Tape drives, commonly available in a reel-to-reel or cartridge format, typically are used to store data or software that is not needed frequently, perform routine backups of data and software on disk drives, and transport data or software between different computer systems.

Thematic Mapping—Portrayal of some geographic variable or "theme," such as land use, geology, or population distribution, by shading or symbolizing areas or map features to represent the value or class of that theme.

Thermal Transfer Plotter—A peripheral device that makes graphic plots of computer data using a heating element to melt dots of waxy ink from a film substrate to a specially treated paper. Most thermal transfers require three or four passes to plot a color image; one primary color or black is plotted on each pass.

TIGER System—Acronym for *Topological Integrated Geographic Encoding and Referencing System*, a major new digital database developed by the Bureau of the Census and the USGS in preparation for the 1990 Decennial Census of the United States. TIGER is the first comprehensive digital street map of the United States, containing digital data at a scale of 1 : 100,000 for every street and road in the nation, the range of address numbers located along each section of every street in the 345 largest urban areas, all the railroads and names of their operating companies, all significant hydrographic features and their associated names, and other information. TIGER covers the fifty states, the District of Columbia, Puerto Rico, the Virgin Islands of the United States, Guam, American Samoa, and the Northern Marina Islands.

Token Ring—A ring configuration network topology developed by IBM to link personal computers and other devices on a local area network. In ring topology, devices are oriented along a closed path and data is transmitted in one direction around the ring; each device on the ring has some intelligence and can receive and send messages through a mechanism and topology in which a supervisory frame or token is passed from station to station in sequential order. Stations wishing to gain access to the network must wait for the token to arrive before transmitting data.

Topographic Features—Map features that define elevation information, which often is called third-dimension or *z* values. Topographic data usually are compiled from aerial photographs at relatively large scales and high levels of positional accuracy. The features recorded typically include contour lines, point elevations, and elevation values.

Topographic Mapping—Delineating on maps the position and elevation of natural and man-made features of a place or region.

Topography—The configuration of a surface, including its relief and the position of its natural and man-made features.

Topologic—A term that describes a geographic data structure in which the inherent spatial connectivity and adjacency relationships of features are implicitly stored and maintained.

Topology—A branch of geometrical mathematics dealing with two types of objects—points (called *nodes*) and lines (called *edges*)—and one type of basic relation between them (called *incidence*). Topology is used to record and manipulate the logical relationships of map features and geographic information in a GIS.

Transverse Mercator Projection—A map projection that is used to establish zones in the State Plane Coordinate System.

Turnkey System—As a marketing concept, turnkey represents an approach wherein a system with components from various vendors is assembled by an integrator who guarantees it will work regardless of the component sources. Turnkey also is used to mean a system that can be immediately operational.

Universal Transverse Mercator (UTM)—A map projection and plane coordinate system based on sixty north–south trending zones, each 16 degrees of longitude wide, that circle the globe. The UTM grid is used commonly in GIS technology in the United States.

UNIX—A portable operating system developed by scientists at the University of California at Berkeley and at AT & T Bell Laboratories. Originally developed for scientific applications on large processors, UNIX is becoming increasingly popular on multipurpose computers. Many computer system manufacturers have adopted one of a variety of UNIX versions and sometimes offer UNIX as an alternative to their proprietary operating system.

URISA—An acronym for *U*rban and *Re*gional *I*nformation *S*ystems *A*ssociation, an international, professional organization for GIS users.

Vector—A format for processing and displaying graphic data. Vector data are represented by strings of coordinates representing the true position of features represented by points, areas, and their boundaries. Vector data define polygons, objects, and other complex entities that can be manipulated or displayed on the basis of attributes.

Vertical Survey Control Monuments—Monuments or benchmarks that provide the control structure from which elevations can be measured, used in compilation of topographic data. When aerial photography is the source for compilation, these monuments or other vertical control points are marked for clear visibility on the photographs. Specific densities of vertical control points must be available to achieve particular levels of accuracy.

Viewshed—The area that is visible from a specific location, e.g., a ridge top or building location, considering obstructions to sight caused by terrain and other physical features.

Virtual Memory—The capability of a computer system's main memory to access and efficiently use data and programs stored separately on disk drives. Virtual storage techniques allow a processor to handle large programs and data sets from multiple users. Programs are segmented into "pages" that can be accessed by main memory and retrieved from auxiliary storage in separate pieces.

Wide Area Network (WAN)—A general term referring to a computer network where devices are connected via remote communication links (e.g., analog and digital phone lines, microwave). The term is often applied to describe networks that extend throughout a defined geographical region, such as a metropolitan area or service district.

Word—A discrete group of adjacent bits or bytes.

Workstation—An imprecise term whose

meaning varies with the context in which it is used. Generally, a device or assemblage of devices that works interactively with a computer to perform specific tasks. All forms of stations and terminals are called workstations, including digitizing and query stations, alphanumeric terminals, microcomputers, and high-performance workstations. The term is frequently used to describe a high-performance device with a 32-bit processor, large main memory, and high-resolution graphics capability.

WORM Drive—An optical mass storage device that utilizes a laser to record and read data. WORM, an acronym for *write once–read many*, means data can be written to a particular portion of the disk only one time, and no data can be erased. WORM technology, which recently became cost-effective, is used to archive large volumes of data that must be stored and retrieved but rarely changed.

x **and** *y* **Coordinates**—The elements of a Cartesian coordinate system that specify the location of a point on a plane. Most maps are based on a geographic reference grid that determines the horizontal (x, y) coordinates of all features included on the map. The grid is defined by an origin point, and coordinates are defined as increments from this origin along the x and y axes. The reference grid may be an arbitrary local coordinate system, or it may be based on a global reference grid (e.g., latitude/longitude) that accurately references map features to absolute positions on the earth's surface.

z Value—The third element of a Cartesian coordinate system that locates a point in space (typically elevation).

Index